Redeeming REDD

It is now well accepted that deforestation is a key source of greenhouse gas emissions and of climate change, with forests representing major sinks for carbon. As a result, public and private initiatives for reducing emissions from deforestation and forest degradation (REDD) have been widely endorsed by policy makers. A key issue is the feasibility of carbon trading or other incentives to encourage landowners and indigenous people, particularly in developing tropical countries, to conserve forests, rather than to cut them down for agricultural or other development purposes.

This book presents a major critique of the aims and policies of REDD as currently structured, particularly in terms of their social feasibility. It is shown how the claims to be able to reduce greenhouse gas emissions as well as enhance people's livelihoods and biodiversity conservation are unrealistic. There is a naïve assumption that technical or economic fixes are sufficient for success. However, the social and governance aspects of REDD, and its enhanced version known as REDD+, are shown to be implausible. Instead, to enhance REDD's prospects, the author provides a roadmap for developing a new social contract that puts people first.

Michael I. Brown is the founder and President of Satya Development International LLC (www.satyadi.com), a consultancy based in Washington DC. He has over 30 years of experience in Africa and other regions working with non-governmental organizations and for-profit groups primarily on USAID-funded projects across diverse development sectors and in conservation.

"Michael Brown recognizes that there can be no environmentalism without representation – especially local representation. If there is salvation for REDD, this book may be looked upon as the benediction that brought it around. By presenting the case for social protections in accessible language, Brown gives REDD a redeeming chance."

Jesse Ribot, Professor of Geography, University of Illinois, USA

"Recent years have seen the emergence of REDD and its variants which have provided great hope in avoiding deforestation, coupled with combined livelihood and biodiversity benefits. It has become a pervasive and persuasive paradigm, beguiling policy-makers, academics and practitioners alike. Despite the international attention on REDD, it seems to have been accepted with very little critical analysis. In this book, Michael Brown challenges the very essence of REDD and REDD+, particularly from the social and institutional perspectives. His analysis is detailed, revealing, and timely; asking many questions of the REDD mechanism that has thus far been largely, and perhaps conveniently, ignored. This book should guide the real policies behind REDD and how local communities and indigenous people engage in its possible implementation. It is a welcome counter to the largely positive and unquestioning way in which REDD has entered our collective consciousness."

Terry Sunderland, Principal Scientist, Forests and Livelihoods Programme,
Center for International Forestry Research, Bogor, Indonesia

Redeeming REDD

Policies, incentives, and social feasibility for avoided deforestation

Michael I. Brown

Routledge
Taylor & Francis Group
LONDON AND NEW YORK

earthscan
from Routledge

First published 2013
by Routledge
2 Park Square, Milton Park, Abingdon, Oxon OX14 4RN

Simultaneously published in the USA and Canada
by Routledge
711 Third Avenue, New York, NY 10017

Routledge is an imprint of the Taylor & Francis Group, an informa business

British Library Cataloguing in Publication Data
A catalogue record for this book is available from the British Library

Library of Congress Cataloging-in-Publication Data
Brown, Michael I.
Redeeming REDD : policies, incentives, and social feasibility in
avoided deforestation / Michael I. Brown.
    pages cm.
  Includes bibliographical references and index.
  1. Carbon sequestration. 2. Deforestation – Control. 3. Forest
  conservation. 4. Forest protection. I. Title. II. Title: Redeeming
  reducing emissions from deforestation and forest degradation.
  SD387.C37B76 2013
  333.75′16–dc23   2012049172

ISBN13: 978-0-415-51787-4 (hbk)
ISBN13: 978-0-415-51786-7 (pbk)
ISBN13: 978-0-203-12365-2 (ebk)

Typeset in Baskerville
by HWA Text and Data Management, London

Printed and bound in Great Britain by
TJ International Ltd, Padstow, Cornwall

Contents

Illustrations

Figures

Tables

Boxes

Acknowledgements

A number of reviewers need to be thanked for their incisive comments that helped in turning an unwieldy manuscript into something that hopefully is more presentable.

Sylvia Tognetti provided early encouragement for the original concept and helped in initially framing the issues despite holding many different opinions. Janis Alcorn provided an early review of an overly ambitious and unwieldy manuscript that helped me feel there may be enough value in this to keep on to the finish. Alain Karsenty provided literature and information on technical aspects to REDD that was mind broadening. Neal Hockley provided very useful exchange on CBNRM and REDD in the Malagasy context that spawned broader thinking of issues. James Acworth exchanged ideas on conservation, NRM, and REDD over a long period of time. Jonathan Adams provided insightful comments towards the end of the process that helped immeasurably in tying up overly loose conceptual ends, while highlighting many troubling editorial details. Erica Rosenberg's watchful eye helped in reminding me of substantive and organizational issues that I had presumed had been handled, that still in fact required attention. Hank Cauley provided very helpful critique of why impact investing may or may not make sense in REDD, and why a new social contract and personalized stories in REDD are important. T. Paul Cox helped considerably with the final editing.

Tim Hardwick of Earthscan had the willingness to take this project on, along with patience in providing the time to get the work done, while Ashley Irons has helped with many nuts and bolts issues. What was finally produced is a book that significantly differs from what was originally envisioned in 2010.

Clearly, all faults in the current text are exclusively mine. For those intrepid readers of the entire book, apologies for any apparent "over-underscoring" of social feasibility and carbon finance across chapters.

My loving family – Mona, Pearla, and Arthur – deserves great thanks for their patience with me through this process as well. Their humor and encouragement as they often found me sitting behind a desk muttering to myself helped me make it through what seemed to them, if not to me, an interminable process.

Finally thanks to my mother and father, Pearl and Arthur, who while no longer here, encouraged me to write a book about Africa as I understood it. While perhaps broader and different than that, their prodding provided me with the spark I needed to tackle a big topic.

Michael I. Brown
Washington, DC and Westport, CT

Acronyms and abbreviations

3Es	effective, efficient, and equitable
AA	analytical assessment
AB 32	California's Global Warming Solutions Act, 2006
ACC	anthropogenic climate change
AD	avoided deforestation
ADP	Ad Hoc Working Group on the Durban Platform for Enhanced Action
AFOLU	agriculture, forestry, and other land uses
AGW	anthropogenic global warming
AI	appreciative inquiry
AIDESEP	Asociación Interétnica de Desarrollo de la Selva Peruana (Interethnic Peruvian Jungle Development Association)
AWG-LCA	Ad Hoc Working Group – Long-term Cooperative Action
BAP	Bali Action Plan
BAU	business as usual
BCI	Bonobo Conservation International
BDS	benefit distribution systems
BINGO	big international non-governmental organization
BRIC	Brazil, Russia, India, and China
BSP	Biodiversity Support Program (World Wildlife Fund)
CAR	Central African Republic
CARPE	Central Africa Regional Program for the Environment
CAZ	Ankeniheny-Zahamena Corridor
CBA	cost–benefit analysis
CBD	Convention on Biological Diversity
CBFM	community-based forest management
CBNRM	community-based natural resource management
CBO	community-based organization
CCBA	Climate, Community and Biodiversity Alliance
CCC	Catastrophic Climate Change
CCPF	Central Cardamom Protected Forest
CCX	Chicago Climate Exchange
CDM	Clean Development Mechanism
CDO	collateralized debt obligation

CER	certified emission reduction
CFM	community forest management
CGIAR	Consultative Group on International Agricultural Research
CI	Conservation International
CIFOR	Centre for International Forestry Research
CMP	Conservation Measures Partnership
CO_2	carbon dioxide
COAIT	Community Options Analysis and Investment Toolkit
COBAs	community based associations (in Madagascar)
COONAPIP	National Coordinating Body of Indigenous Peoples
COP	Conference of the Parties
CSO	civil society organization
CSR	corporate social responsibility
DOE	Designated Operational Entity
DRC	Democratic Republic of Congo
ELC	Economic Land Concession
ENGO	environmental non-governmental organization
ER	emission reduction
ERA	Ecosystem Restoration Associates
ETS	Emissions Trading System
FAC	forest-adding country
FAO	Food and Agriculture Organization of the United Nations
FCPF	Forest Carbon Partnership Facility (World Bank)
FIP	Forest Investment Program
FLEGT	Forest Law Enforcement, Governance and Trade Action Plan
FPIC	free, prior, and informed consent
FSC	Forest Stewardship Council
GCC	global climate change
GCF	Governors' Climate and Forests Task Force
GDP	gross domestic product
GFC	Governors' Forest Conference
GHG	greenhouse gas
GIS	geographic information system
ICDP	integrated conservation and development project
IEA	International Energy Agency
IEC	information, education, communication
IISD	International Institute for Sustainable Development
ILO	International Labour Organization
IPCC	Intergovernmental Panel on Climate Change
IPs	indigenous peoples
IRM	Innovative Resources Management
IRS	Internal Revenue Service
IUFRO	International Union of Forest Research Organizations
IWGIA	International Work Group for Indigenous Affairs
JFM	joint forest management

JI	Joint Implementation
KFCP	Kalimantan Forest Carbon Partnership
KP	Kyoto Protocol
LLC	limited liability company
M&E	monitoring and evaluation
MDG	Millennium Development Goal
MEA	Millennium Ecosystem Assessment
MIT	Massachusetts Institute of Technology
MRV	measurement, reporting, verification
MT	metric ton
NASA	National Aeronautics and Space Administration
NGO	non-governmental organization
NRM	natural resource(s) management
NTFP	non-timber forest product
ODA	official development assistance
OPIC	Overseas Private Investment Corporation
PA	protected area
PDD	project development document
PES	payment for ecosystem services
PFM	participatory forest management
PIN	project identification note
PNG	Papua New Guinea
PRA	participatory rural appraisal
PTMAs	policies, tools, methods, and approaches
R-Package	Readiness Package
R-PIN	Readiness Plan Idea Note
R-PP	Readiness Preparation Proposal
R-Ps	Readiness Plans
RED	reducing emissions from deforestation
REDD	reduced emissions from deforestation and forest degradation
REDD+	reducing emissions from deforestation and forest degradation in developing countries (the addition of a "+" refers to conservation, sustainable management of forests, and enhancement of forest carbon stocks)
REL	reference emission level
RGC	Royal Government of Cambodia
ROI	return on investment
SBIA	social and biodiversity impact assessment
SBSTA	Subsidiary Body for Scientific and Technological Advice
SCS	Scientific Certification Systems
SDI	Satya Development International
SESA	Strategic Environmental and Social Assessment
SFM	sustainable forest management
SIA	social impact assessment
SLM	sustainable land management

SRC	social responsibility contract
TA	technical assistance
TI	Transparency International
TIK	traditional and indigenous knowledge
TMAs	tools, methods, approaches
TNC	The Nature Conservancy
TUK	tradition and indigenous knowledge
UN	United Nations
UNDP	United Nations Development Programme
UNDRIP	United Nations Declaration on the Rights of Indigenous Peoples
UNEP	United Nations Environment Programme
UNFCCC	United Nations Framework Convention on Climate Change
UN-REDD	United Nations Collaborative Programme on Reducing Emissions from Deforestation and Forest Degradation in Developing Countries
USAID	United States Agency for International Development
USG	United States Government
VCM	voluntary carbon market
VCS	Verified Carbon Standard
WCS	Wildlife Conservation Society
WWF	World Wildlife Fund
ZOPP	Zielorientierte Projektplanung

Our comforting conviction that the world makes sense rests on a secure foundation: our almost unlimited ability to ignore our ignorance.

Daniel Kahneman, *Thinking, Fast and Slow*, 2011

In terms of the lessons learned, what we've seen is that especially in Indonesia, REDD is not addressing the real problem, yet. One project here, one project there, one project somewhere else and with lessons learned from the project sites… But nothing has really changed.

Bustar Maita. REDD-Monitor, Interview with Bustar Maitar and Yuyun Indradi, Greenpeace: "REDD is not answering the real problems of deforestation, yet", 2012

[E]very year there is hope that REDD will be sorted and every year people come home disappointed.

Senator Christine Milne, Proof Committee Hansard, Government of Australia, 21 May 2012

We lie to ourselves the better to lie to others.

Robert Trivers, *The Folly of Fools: The Logic of Deceit and Self-deception in Human Life*, 2012

Beware of listening to this imposter; you are undone if you once forget the fruits of the earth belong to us all, and the earth itself to nobody.

Jean-Jacques Rousseau, *Discourse on Inequality*, 2004

Introduction

This book is written for those with an open mind on how best to address deforestation's contributions to global climate change (GCC). The topic is important, because deforestation is estimated to generate around 20 percent of anthropogenic greenhouse gas (GHG) emissions globally. These emissions in turn contribute to GCC and, with it, a progressive rise in temperature and climate anomalies. In an attempt to shock the world into action, the World Bank recently commissioned a report (Potsdam Institute for Climate Impact Research and Climate Analytics 2012) on the implications of what the world would be like if it warmed by 4 degrees Celsius (°C), which is what scientists are nearly unanimously predicting by the end of the twenty-first century. The prospects are not pretty.

It is known that the last eleven years (2001–2011) were "among the top warmest years on record" (World Meteorological Society 2012). 2012, for example, brought extreme precipitation in West Africa, rain deficits in eastern Africa, drought in Yunnan and Sichuan provinces in China, massive wildfires in Spain, episodes of extreme cold, snow events, precipitation and devastating floods in eastern Europe, increased tropical cyclone activity globally, drought across two thirds of the US, and the well publicized Hurricane Sandy in the Caribbean and northeastern US (World Meteorological Society 2012). Equally troubling, the ice melt rate in Greenland has grown from about 55 billion tons a year in the 1990s to almost 290 billion tons a year recently (Shepherd et al. 2012), contributing to worsening flooding scenarios and a 20 percent rise in sea levels overall. Ironically and symbolically, the horrific typhoon Bopha that left over 500 dead, hundreds missing, and over 250,000 homeless in the Philippines, occurred as United Nations (UN) convened climate change negotiators in Doha were struggling to reach any agreement on financial commitments and next steps to deal with global climate change. Combined, these events simply foreshadow what will be experienced in a world that is 4 degrees warmer.

The major initiative for approaching the deforestation component of GCC internationally is known as REDD – reduced emissions from deforestation and forest degradation. While REDD does not touch on the biggest contributing factor impacting climate change overall – fossil fuel extraction and its burning – for many, REDD nonetheless represents a potentially viable, incentive based path to reducing the deforestation and forest degradation component to GHG emissions and thereby contributing to GCC mitigation. For others, REDD is a proposed solution that either will fail, or is irrelevant in the face of the need to dramatically *reduce* GHG emissions, as opposed to simply offset them.

For simplicity's sake, I use the term REDD throughout the book as a cover for all things RED/REDD/REDD+ (see List of acronyms). Only in instances when REDD+ is clearly at issue do I use that term. With no formalized international mechanism for REDD+ as of 2013, use of the term REDD appears reasonable.

My position is that regardless of one's position on the urgency to mitigate climate change, REDD as currently approached will not work. At the end of the day, sustained buy-in at local levels in tropical forested countries is needed for REDD to succeed. Yet the combined policies and incentives, along with the tools, methods, and approaches (TMAs) now employed remain thoroughly unconvincing at this scale.

The main reason is that REDD is replicating a well rehearsed, top-down agenda that while technocratically glossy, lacks a credible "social contract" that most Pygmies, Papua New Guineans, or indigenous Amazonian communities can agree to. This top-down agenda is being negotiated *on* forest communities by nation states participating in the United Nations Framework Convention on Climate Change (UNFCCC) process at international and national levels. It also is offered through the many so-called "voluntary" initiatives related to the Voluntary Carbon Market (VCM) that bring official development assistance (ODA) for establishing "enabling conditions" in specific countries signatory to the UNFCCC process, together with private sector funding for site specific actions. While independent of the UNFCCC, these initiatives are often predicated upon the UNFCCC framework. In both cases, the views of forest communities are officially represented and negotiated by their governments, or may be validated and verified by industry leading standard setters whether or not communities understand or have actually consented in an "informed" and "free" manner to the terms that are presented to them. These consent designations have in many cases been obtained independent of serious local input.

The specifics of the technocratic gloss forwarded by REDD thought-leaders at UNFCCC and VCM levels are therefore not the sole issue. Deeper flaws in current approaches to REDD involve inadequate decision-making and governance arrangements, which result from questionable assumptions and weak analysis of deforestation drivers and solutions. The global community is largely delaying broad REDD implementation until after 2020, though a deal is presumably to be struck by 2015. The so-called Ad Hoc Working Group on the Durban Platform for Enhanced Action (ADP) is deliberating on the specifics of this new incoming climate agreement for the UNFCCC that, as of December 2012, may or may not formally include REDD.

Paradoxically, with so much in flux, this could actually be the right time to introduce the notion of *new social contracts* into international UNFCCC negotiations and national level scenarios where greater room is offered for creativity. These social contracts would inject a transformational break from the premises and assumptions now on offer under REDD, which appear increasingly unrealistic with each passing day. For this, however, vision, leadership, and innovation in addressing REDD is required that has been lacking to date.

The contention is regularly made by REDD proponents that free, prior, and informed consent (FPIC), a pivotal but non-binding principle in UNFCCC and VCM processes, has been systematically obtained from local communities in REDD scenarios for both national policy setting and local projects. I believe that focused analysis shows that, in most cases, the contention is dubious. When either elites or ill-capacitated communities "consent" to issues they do not fully understand, or do so for lack of better alternatives, consent is largely semantical or spurious.

Firsthand accounts from Papua New Guinea, the two Congos, Mali, Uganda, and Cameroon, among other developing countries, provide a basis for why I firmly believe that more is being made of the "genuineness" of any FPIC obtained, and the safeguarding of community interests under REDD, than is warranted. It may not all be smoke and mirrors, but too much is being predicated on the recycling of questionable logic and unverified assumptions about what has and has not worked in terms of past policies and practices in biodiversity conservation and development. These in turn have influenced standards of excellence promoted under REDD.

Arguably, the leading industry standards setters have set the bar *so low* when it comes to defining what participation and consent mean, for example, that so long as chiefs and community members sign off for whatever reasons on a proposed initiative, participation and FPIC boxes will be checked. So too, implementing partners may interpret the FPIC process as "sufficient," theories of change framing monitoring and evaluation components will be established, and donor and investor funding will begin flowing. Accountability as to the meaning of these key terms, however, remains vague. Unfortunately, this approach contributes little to establishing REDD feasibility or long-term success.

Meanwhile, the so-called "Readiness Phase" of activities that has been facilitated by the World Bank's Forest Carbon Partnership Facility (FCPF), with support from the United Nations Collaborative Programme on Reducing Emissions from Deforestation and Forest Degradation in Developing Countries (UN-REDD) has done little to resolve some of the more fundamental issues most likely to constrain REDD's success.[1] Instead of challenging and verifying key premises and assumptions, they are largely accepted and serve as the foundation for moving forward. While the Readiness Phase focuses on "building the legal, institutional and technical capacity countries will need to implement and monitor REDD as well as holding consultations with affected stakeholders, including forest-dwelling communities" (Bank Information Center 2009), there is nary a sign that a transformational (or inspirational) approach to REDD is in the offing at any level. For example, Panama's Indigenous Peoples Coordinating Body (COONAPIP's) withdrawal from the UN-REDD over how UN officials and the Panamanian government "are dividing indigenous communities with money from the Programme to force supposed consultations" in REDD planning (quoted in

Lang 2013) indicates significant perceptual gaps remain among stakeholders as to participation and FPIC.

That said, there is hope for REDD *if* a significant shift in strategy and approach could be made. REDD can still be redeemable. This book explores what the main foci in a shift in approach should comprise.

REDD's most recent incarnation currently in negotiation is known as REDD+. Yet there remains more than a fair degree of ambiguity as to what REDD+ comprises. For example, while it is suggested that REDD+ is meant to be "a national level mechanism" (UN-REDD 2012d) that is being negotiated under an international framework through the UN through its UNFCCC, any number of individual, VCM project developers use the term REDD+ liberally to refer to their activities at sub-national, project levels. At the same time, others use the short-hand REDD and REDD+ interchangeably.

Despite its voluminous literature and support, and given that "shortage of clear information about VCM, REDD+ and forest carbon finance is a challenge in most countries" (Go-REDD+ 2012), confusion over REDD is both semantical and substantive in nature. Since REDD/REDD+ remains a work in progress yet to be fully formalized, I opt in the pages that follow for referring to REDD, unless greater specificity is warranted.

After some seven years in the making, REDD remains a highly contentious activity. This is because everyone and no one "owns" REDD; while the UNFCCC still remains the predominant facilitator internationally and nationally, a host of other UN agencies, international panels, multilateral development banks, bilateral donors, non-governmental organizations (NGOs), private sector groups, along with civil society organizations (CSOs) are all interested stakeholders. Some go so far as to suggest that the UNFCCC has been so ineffectual that it is high time that voluntary initiatives take the lead, as "the UNFCCC may not be the best option or the best use of time and energy" (Niles 2011).

It is fair to say that the stakes for thought leadership, and control over the REDD agenda still remains in play. The UN system and multilateral development banks, big international NGOs (hereafter BINGOs)[2] along with many private sector participants shaping the VCM, and civil society groups and other critical stakeholders opposed to the current process, are all vying and active in attempting to shape REDD's future. Given the palette of perspectives that is brought to REDD, the topic provides fertile grounds for passionate debate, along with great potential for local, site-level conflict.

Substantively, REDD's complexity is related to the array of disciplines and issues that are being tackled. The expanded REDD+ includes "policy approaches and positive incentives on issues relating to reducing emissions from deforestation and forest degradation in developing countries; and the role of conservation, sustainable management of forests and enhancement of forest carbon stocks in developing countries" (UNFCCC 2010).

The topic directly implicates various disciplines ranging from climate science, finance, and conservation biology, to political science, development economics, sociology, anthropology, and conflict management. In the future, REDD might well receive welcome contributions from other fields such as moral philosophy, and psychology as well, as both values and the reasoning behind stakeholder negotiating positions is worthy of attention too.

While the locus of concern is in developing countries, and mainly tropical forested ones at that, so-called Northern country stakeholders including governments, private sector interests, and big NGOs have held considerable leverage to date over the evolution of REDD. In this regard, while the scope of REDD is extraordinarily broad, and the stakes at international and local levels significant, REDD has largely evolved through traditional channels for official development assistance (ODA) programming, which is both inevitable and explanatory for where REDD has gotten stuck, and why.

The problems are not simply due to lack of progress in the faltering compliance market where international negotiations have been most focused, or in the rapidly evolving voluntary market where much activity and optimism is now placed. Fundamental issues involving participation and decision making remain doggedly resistant to the type of transformational change required for REDD to succeed.

This book makes the case for why the current approach cannot work in either the compliance or voluntary contexts, why a new approach is needed, and what its components must be. While the compliance and voluntary contexts for REDD differ in terms of scale, the key challenges as regards enabling conditions and sustainability arguably converge.

Just as biodiversity conservation is ultimately about people adopting salutary conservation practices, REDD is similar. Both require policies, incentives, and effective mechanisms to work. In REDD, neither the compliance mechanism nor the VCM approach addresses "people issues"– livelihoods and incentives, governance arrangements, and a bevy of constraints (resource tenure, administrative and grand corruption, elite benefit capture, etc.) – at all convincingly. Whether it is irrational exuberance that risk-taking entrepreneurs can relate to, or simply a leap of faith, the foundation underpinning REDD remains visibly shaky, hence the uncertainty and resistance to REDD on the part of many.

I have written the book with multiple target audiences in mind. REDD policy makers at international levels, practitioners in the field, community groups and activists, potential financiers or investors in REDD, students and professors in academia, and the interested and informed lay public may each find interest in what follows. As a result, *Redeeming REDD* may appear to be part (a) primer on REDD ,(b) exposé of its flaws, (c) a guide as to how the social dimensions of REDD may be better approached at a practice level, and (d) assessment of policy and finance issues. Ideally, readers will find the work in the aggregate useful, leading to a rethinking of premises, assumptions, and possible solutions.

For those looking for a rehashing of all things related to UNFCCC, UN-REDD, and FCPF activities there will be disappointment. While these programs must be considered in any book on REDD, abundant literature already covers the major REDD programming initiatives.

I have tried to write a book that addresses issues that others have not focused on, to fill in gaps I believe are important, and to be constructive about moving forward. The current focus on finance mechanisms and technical measurement and reporting issues for REDD has, I believe, ultimately undermined REDD's viability by diverting attention away from social and governance issues that are clearly pivotal. This is true for both evolving compliance and voluntary market settings, be it at national and subregional levels, or individual projects. So too, while policy issues such as tenure reform along with social safeguards are clearly of great relevance and have received the lion's share of attention in terms of social issues, clearly other facets of the social agenda are important too. These include governance and decision making, the quality and extent of stakeholder participation required, and the ability of people to participate credibly in complex, high stakes fora where few successful precedents are available in most countries to serve as signposts along the way.

My interest is in identifying what I believe is required for avoided deforestation (AD) and its flip side of the coin, carbon sequestration, to work in practice. My focus is on what I personally believe are the deeper issues that will make or break REDD, or any initiative attempting to tackle the broader deforestation side of GCC issues. These are first and foremost social, institutional, and political in nature. My main argument is that if, and only if, community level stakeholders to REDD become integrally involved in all phases of REDD programming – participating effectively in establishing international governance regimes; participating effectively in national program strategies to verifiably influence "drivers of deforestation and forest degradation, land tenure issues, forest governance issues, gender considerations and safeguards" (UNFCCC 2011a) – can REDD be achieved. If tackled well, I believe that REDD could be redeemable. This remains a big "if".

While acknowledging that issues pertaining to technical feasibility in establishing reference baselines and verifying emissions reductions and the like are obviously important under either carbon market scenarios or other forest sector initiatives, my personal opinion is that REDD is faltering most dramatically on the social dimensions that have historically troubled both conservation and development programming.

I am less concerned with preserving REDD in the form that we now know it, though I do not anticipate that decision makers and thought-leaders will agree or react positively to such a suggestion. Nonetheless, the book is about why change is needed, and offers suggestions for what may be done differently. If stakeholders can come to agree on the urgency, and agree in particular that establishing a new social contract that enables both participation in planning along with risk to be equitably shared among key stakeholders, REDD stands a chance of succeeding.

I have had to draw on my own field experience often, along with the literature on lessons learned from development and biodiversity conservation, to illustrate what is wrong or missing. This experience began on a year-long trip across Africa in 1973, and has continued on to the present day professionally in Africa, and the other continents where REDD is being implemented. For lack of a better term, this book is therefore somewhat of a hybrid. Narrative styles are mixed. More "academic" text is juxtaposed with personalized presentation. My hope is that this does not prove too disconcerting.

I believe that a major problem with REDD has been its virtual impenetrability due to the technical jargon and abundance of literature that has, in the main, been primarily accessible to REDD policy makers and other technocrats with vested interest in the topic. Yet the underlying issues of tropical deforestation and forest degradation *should* be important to tens (if not hundreds) of millions of people. In the event REDD were to work, it would indirectly become important to billions more globally. These are big round numbers, but the reality is that the policies and underlying issues are of broad relevance.

As poor people stand to be most impacted by REDD policies, putting a human face on what is happening on the ground seems to me justifiable. Too much of REDD has been, and remains, abstract.

While the narrative on REDD needs to break out of the near impenetrable technospeak that camouflages important issues to people, it is also impossible to break away from what REDD is – a highly technical topic that has been dominated far too long by technocrats and administrators. *Redeeming REDD* is my own attempt to bridge between highly diverse communities of potential readers, with the full knowledge that this approach probably will not satisfy any one group.

A tremendous literature to build upon

Writing a book about REDD in 2013 is daunting, if for no other reason than REDD literature has exploded into a booming industry that is challenging to keep abreast of, either in its support or opposition.

So too, numerous comprehensive, "everything you want to know" about REDD materials already exist online, so any book on REDD must be assiduous to not duplicate ground that others have already covered adequately. In this regard, the work of Angelsen (2008), Angelsen et al. (2009), Meridian House (2011), and Gregersen et al. (2011), all nominally works that can reasonably be argued fall on the "constructive proponent" side, are noteworthy. On the highly critical side, Lohmann (2006, 2008, 2010, 2011a) provides a series of critiques of the REDD project as it has been, and is evolving, while The Munden Project (2011) is frequently cited as an inside-the-industry contribution for why carbon offset markets under REDD are unworkable from the start. Llanos and Feather (2011) meanwhile provide an excellent grassroots, case study perspective on all that is wrong with REDD. To get a sense of the breadth of REDD projects globally, CIFOR's (2012) interactive map is very useful.

My own motivation for writing this book is to fill several significant gaps or inconsistencies in thinking and programming in REDD that appear so glaring that it is hard to imagine why others have not already hammered these points home. So too, I believe that the hubris regarding perceived achievements from biodiversity conservation practice, despite the challenges from Chapin (2004), Brockington et al. (2008) and numerous others in recent years, bodes ill for REDD as its current influence on REDD's evolution is clear. The fact remains that indigenous peoples have never "been given the chance to design and run their own projects" (Chapin 2004). Verifiable results and impacts from conservation remain largely elusive. Even from inside the industry, it has been confirmed that progress has been painfully slow, yet clearly needed (Kareiva et al. 2012). Yet methodologically, conservation best practice has significant influence on REDD.

Any comprehensive appreciation of why and where REDD is likely to go wrong, and what to possibly do about it, must therefore begin with success and failure in the biodiversity conservation and broader development arenas. The disconnect between the rhetoric imploring strong monitoring and evaluation (M&E) and capturing lessons learned, with the reality that in biodiversity conservation anyway, M&E and lessons learned has been far from what it should be (O'Neill and Muir 2010; USAID 2012), must somehow come to inform and guide REDD planners. How this is being operationalized remains unclear.

To succeed, REDD must begin by challenging orthodoxies and conventional wisdom, leading to a more structured, if not committed public debate. This is different than the enabling of diversity of opinion simply as a means to confirm that stakeholders are allowed to speak, with top-down, business-as-usual decision-making then proceeding.

While being very sympathetic to the reasoning of REDD skeptics, I try to carve out what could be termed an "agnostic middle ground" where the imperative is to identify what to do to enable REDD, or avoided deforestation under another label, to work. The book may best fit readers who in fact wish REDD well, of which I consider myself to be one, yet remain highly skeptical of whether REDD is worth trying given the prevailing approach.

I believe that while vested stakeholder interests may constrain the type of evolution I suggest is needed, given the urgency of climate change issues, a shift in strategy and approach is not impossible. Two preconditions for this are for greater objectivity in problem analysis coupled to commitment to broader stakeholder negotiation to assume a role. Figuring out how to do this at international and national levels will be key.

Here, local groups in developing countries, governments in developing countries, and highly principled "impact investors" with the means of capitalizing avoided deforestation initiatives for the *right reasons*, not simply high return on investment (ROI), will all have a role to play. If leadership is to come primarily from the stalled UNFCCC process, or from BINGOs, transformational change is highly unlikely. Meanwhile, imploring and relying upon policy changes on moral grounds, however right and worthy, may well not be sufficient either.

For many REDD critics who blame the capitalist system for GCC and the current approach to REDD, any suggestion that so-called "impact investors" may be pivotal to avoiding deforestation seems, on the surface, anathema if not contradictory in terms. Yet, must it always and forevermore be the case that the private sector and AD are categorically, mutually exclusive? My preference, given the stakes, would be to let negotiations around a level table determine if this is true or false, as by 2020, it has been suggested that between US$400 billion and US$1 trillion dollars will be invested through impact investing.[3] Moreover, scaling entire sectors by enhancing collaboration among policy makers, local entrepreneurs, philanthropists, and commercial funders is seen as key.[4] If responsibly negotiated, could REDD not both qualify as a "sector" of opportunity while benefitting from this?

Should it turn out that such highly principled investors potentially exist in REDD, they will need technical help from intergovernmental agencies and BINGOs in more of a technical assistance role, versus as primary thought-leaders and implementing agencies in the process. The past twenty years' experience with big biodiversity conservation managed projects has not demonstrated consistent success in the face of declining biodiversity and forest habitat metrics to justify continuing down the current pathway. Nor have governments or donors demonstrated great effectiveness in the forest sector either. Figuring out if and how new configurations of stakeholder interests can be aligned which offer a break from unproductive current arrangements is clearly an issue that while not on the agenda *now* at major REDD international fora, needs to be.

My hope is that this book will spur debate and a rethinking of how best to proceed from here. To get there, the book presents background and scientific arguments for REDD.

I support the goal of AD but believe that the strategies employed are inappropriate and unfeasible. The particular failing is that the front line communities who will make or break sustainable forest use that REDD is predicated upon, remain marginalized players in setting policy, identifying practical approaches, and receiving commensurate benefits given the risks they bear. This is as true for broader policy setting for addressing deforestation drivers that traditionally have been the concern of governments and international organizations, as it is for any project-based solutions at finer spatial scales. The book suggests why the current technocratic fixation on measurement, reporting, and verification (MRV) tools and finance through a carbon market mechanism provides unrealistic incentives to keep communities constructively involved, and explains from a stakeholder analytical and institutional perspective why this has come to be.

My own framework for a theory of change is provided, and the need for addressing the human side of REDD based on nearly forty years of field experience from around the world is used to substantiate arguments as needed. While grounds for pessimism and naysaying arguably outweigh those for optimism, I believe that if stakeholders could look open-eyed at the issues, the basis for approaching a new social contract to frame REDD could be established. With this, grounds for optimism would become more justified.

1 Grounds for pessimism and optimism

This chapter reviews the basis for both pessimism and optimism in REDD. It considers a range of stakeholder opinions as well as my personal perspective. The premises underpinning the book are presented, along with sources of influence on the arguments.

Josephstaal and REDD

In 1992 I led the social sciences team for Papua New Guinea's (PNG's) Conservation Needs Assessment. I was under contract to the World Wildlife Fund's (WWF's) Biodiversity Support Program (BSP), working in collaboration with the United Nations Development Programme (UNDP) and the Government of Papua New Guinea.

We had been allotted three weeks to determine what the main social issues and "needs" related to biodiversity conservation in that country of 850 distinct languages, correlated to a similar number of distinct clan groupings, might be. It was assumed that this input would help in promoting more effective conservation strategies and practices in a country of extraordinarily high biodiversity values.

Our task was quite implausible – three weeks to pin down how and where people fit into biodiversity conservation in PNG. The challenge was of course exciting, and at the end of the day, represented an opportunity to make a practical contribution all the same. But one thorny operational question remained: given the extreme time and logistical constraints in the face of a daunting mission, what on earth could the social scientists' strategy be to escape superficiality when speaking to the social needs in conservation?

The cobbled, shoot-off-the-hip strategy developed by our two-person expatriate team, assisted by a small group of NGO and journalist facilitators native to PNG, was to identify types of situations that were representative of the major socioeconomic challenges that clans appeared to be facing at the time. One challenge involved the evolution of relations between landowning clans and logging companies. Our sense was that customary clans, and their "members", were being systematically duped into signing all nature of bogus agreements. This represented a challenge to conservation. The tentative "conservation need" was to determine if capacity building should be undertaken to help clanspeople avoid being duped, and if so, what?

As 97 percent of land in PNG is legally owned by clans under constitutional writ, clans in PNG maintain leverage over resources and their management that many tenure-insecure Africans or Asians would certainly envy. But this leverage is actually nominal, insofar as clanspeople are often unable to defend their rights, or capitalize on them proactively. And as corruption is rife across public administration in PNG (Transparency International (TI) notes it being 154th out of 183 countries in 2011),[1] this further complicates matters.

To move forward, we theorized that logging companies were out doing deals in remote parts of the island that negatively impacted people and biodiversity. The one requirement companies had to legally meet was to obtain signatures from clan representatives in which clans signed off on their rights to forests in exchange for what euphemistically was known at the time as "spinoffs" – e.g. small-scale community development projects of one sort or another. We felt this was one dynamic worthy of investigation, as anecdotal evidence had it that the phenomenon was widespread. We felt that the apparent proliferation of the spinoff phenomenon could capture an array of the social challenges that biodiversity conservation would be facing in PNG in the years to come (Brown and Holzknecht 1993).

In many respects, understanding how clans reacted to spinoffs that were offered by logging companies in the 1990s provides insights into how today's communities in PNG and beyond may react in REDD in 2013. This is because the context facing PNG clans in 1992 is not so different from the one now confronting clans in PNG. Moreover, it is not so different than the situation for hundreds of millions of forest peoples across developing countries today in areas where early phase REDD activities (REDD Readiness) have been introduced. Everywhere, expectation levels have been raised through awareness raising or simple misinformation about REDD processes.

With all due respect to indigenous peoples, and traditional knowledge systems aside, many poor forest dwellers are sick and tired of living in poverty, and potentially see their access to forested resources as expendable if that leads to poverty alleviation, as observations from some early phase REDD outreach programs suggest. Some community members may be willing to sell their house if it means escape from poverty; others will remain true to the logic and customary values that have sustained their forebears for generations, in which carbon transacting plays no role. Across the board generalization is impossible.

The point is this: the situation confronting rural forest communities faced with REDD prospects combines desperation, dynamism, and risk. Getting a grip on this dynamism is fundamental to any success that REDD may experience in years to come. This dynamic is not solely based on economic and technical considerations. It involves an array of social, cultural, and political criteria. Reducing it to profit maximization is a mistake.

The hypothesis employed to evaluate the significance of "spinoffs" on conservation in PNG was simple: *Papua New Guinean clans in remote areas that had signed or were about to sign agreements with logging companies, by and large had no idea what they were actually and legally agreeing to in these agreements.* In acting out of desperation, clans were getting hoodwinked. Corrupt government officials were facilitating contacts between First World capitalists and still-stone-age Melanesians from a technology and literacy standpoint.[2] While 56 percent of the PNG population in late 2012 was literate – dismally ranking 148th out of 182 countries surveyed[3] – in 1992 the literacy rate was considerably lower, opening the door to widespread abuse.

As conservation biology has established, there is an extremely high correlation between intact primary forest and biodiversity values as measured by species richness, endemicity, etc. Take away the forest, and biodiversity in all respects suffers. It seemed to us that if the trends about the rapidly evolving interface between loggers and forest dwelling clans in PNG could be ascertained, perhaps we could speak to the social aspects of conservation needs more coherently.

My own hypothesis concerning REDD with remote forest peoples in PNG *and* elsewhere in 2012, two decades later, is equally simple: *local forest peoples with customary rights to forest resources who are being offered REDD incentives in 2013 are for all intents and purposes unaware of the full cost, benefit, and risk implications of REDD as being offered. This undermines REDD's potential feasibility and sustainability.*[4]

One of the remote highlands peoples and places that we decided to test this hypothesis on was in Josephstaal in Madang Province. As the last stop on a missionary supply plane route, I was greeted by representatives from thirty-five or so clans in traditional attire at the remote airstrip. The subjective sense I had in arriving was very much that I somehow was part of a cargo cult delivery. The mix of beaming smiles coupled with curiosity and group huddling implied that light-skinned visitors were not all that commonplace in Josephstaal. Both that day, and at the next day's meeting, the predominance of grass skirted and scantily clad women was notable. They were not shy in making their presence felt and in speaking up. The men meanwhile appeared desultory and embarrassed to speak in comparison.

I tell this story because the worst of our small social science team's assumptions about the dynamic between private sector interests, government, and clans proved to be true. The implications of this experience are directly relevant to REDD. This is not only for PNG, but in the many countries in Africa, Latin America, and Asia where social capital at community levels is limited, and where the interface prevailing between communities living in dire poverty, logging companies, and corrupt government "facilitators" is subject to almost no regulatory oversight, and where legal recourse is limited. In fact, the joint work of Asociación Interétnica de Desarrollo de la Selva Peruana (AIDESEP) and Federación Nativa del río Madrea de Dios y Afluyentes (FENAMED) with support from Northern NGO partners in the Peruvian Amazon substantiates this concern through detailed analysis from another context (see Llanos and Feather 2011).

The thirty-five or so clan representatives I passed three hours in heated discussion with the following day had absolutely *no idea* what the Korean logging firm, Kosmo & Co. (Brown and Holzknecht 1993), who they had formally signed over their 100,000 hectares of forest to, was contractually obligated to do for them in return. They kept repeating that there would be "spinoffs", as if incanting this statement enough times would make it happen. But when queried as to the nature of the spinoffs, the delivery date of the spinoffs, the value of the spinoffs, the value of the 100,000 hectares of forest they had willfully signed over, they did not know. There were no details. There was only a visibly cocky twenty-something with a (true story) Michael Jordan hat and Michael Jackson T-shirt on, who had, apparently, been a lead facilitator for the company in the agreement with the local clans.

Yet, when it became clear in the course of our discussion that neither he nor the clans understood *what* they'd agreed to, and that this could pose a problem (and not a minor one at that), one excited woman in grass-skirt attire asked me straight out through my translator: So then what do *you* think we should do? I answered that I didn't know.

Should the clans renege on the company? Should they try and renegotiate terms with the company? Should they tell the government they were hoodwinked and swindled and hope that the government of PNG, renowned as it was for less than fully scrupulous behavior, would come to their rescue? Should the clans just grin and bear it, and hope for the best? Should they take the company to court and sue for false pretenses? And how exactly would a disorganized group of thirty-five clans with limited social capital and questionable leadership make an effective decision, one way or the other?

After thinking on it a bit, I suggested that the clans should (1) determine what was legally binding in the agreement, and (2) see if they could extricate themselves from their obligation to cede the 100,000 hectares of forest to the logging company,[5] as it appeared from reading the agreement that while they were relinquishing 100,000 hectares of primary forest to the company, there was no specificity as to what the company was providing in terms of spinoffs. Rather, the company would decide what it would deliver based on its assessment of the situation, considering the possibility for schools, clinics, roads, etc. The terms seemed so outrageous, that perhaps the judiciary system, corrupt as it was, could nonetheless consider their case favorably. But frankly, this was simply optimistic speculation.

I believe it is easy to demonstrate that local peoples today have as little *informed* sense as to the potential relevance to them of engaging in REDD as was the case in Josephstaal, Papua New Guinea twenty years ago. In both Josephstaal and REDD contexts today, contractual options are being presented to social-capital-challenged forest peoples to make sophisticated decisions about livelihood and land use options that would bedevil the brightest Harvard Business School MBA graduates. Too much of REDD is tainted by the same uncertainties and vagueness as prevailed in PNG in 1992. In fact, the situation is similar in respect to forest sector corruption (Greenpeace 2012), where the legacy of community rip-offs endures, creating wariness for REDD in the future (Dooley 2010).

Babon et al. (2012) provide data and analysis on how difficult it is for the public to be objective about REDD. They show how REDD issues have been "framed" in PNG since 2006 by the media. They conclude that "difficulties faced by journalists in accessing and verifying information has allowed REDD+ discourses to be politically driven", leading certain organizations to influence how REDD is perceived. In this regard, the mechanisms of carbon trading, talk of quick and easy money, and the potential for con men and scams to flourish has prevailed.

Substitute the Congo, Indonesia, and any number of central American countries, and the conclusion could be the same. Information and objective analysis have not proliferated about deforestation and its avoidance. Emphasis has been on establishing carbon trading as the means to mobilize much needed finance to *then* avoid deforestation and forest degradation while also conserving biodiversity and, hopefully, alleviating poverty.

However, for REDD to actually deliver on its potential for carbon dioxide (CO_2) emissions reductions, a *virtually unprecedented* international, cross-cultural, and cross-sectoral implementation regime has to emerge (Corbera and Schroeder 2011). It is one in which policy expertise, development actions, market forces, and a broad cross-section of actors in society align themselves towards a common long-term goal (Angelsen and Atmadja, 2008), and yet manage to remain flexible to the considerable ambiguities and experimental nature of the road ahead (Hajek et al. 2011). How likely is this "virtually unprecedented" implementation regime to be realized in practice given the pilot phase track record to date? And what will be the indicators that demonstrate that it has done so?

As of 2013, based on the past seven years of experience, it is unlikely this cooperation will materialize. Arguably, there has been less cooperation within the predominant international framework for helping REDD emerge, the UNFCCC which gave birth to the Kyoto Protocol and carbon markets as the means to combat climate change, than through programs outside the UNFCCC – e.g. UN-REDD, the World Bank's FCPF and Forest Investment Program (FIP), the Governors' Climate and Forests Task Force (GCF), and an evolving number of private partnership initiatives. This has led some to suggest that these may in the end prove more relevant than the stagnating efforts of the UNFCCC (Niles 2011).

On the other hand, the principal partners of UN-REDD continue to develop a framework that could turn into an eventual compliance system involving nation states, sub-national jurisdictions, and the private sector. Many of the highest profile international biodiversity conservation and development NGOs have been loudly backing efforts to link the fates of reduced emissions from deforestation with those of both enhanced biodiversity values and improved human welfare in remote forests of the developing world under REDD+. Private sector technical expertise, along with finance, has begun investing. And finally, in terms of social issues, the commitment to achieving free, prior, and informed consent (FPIC) has resounded loudly in international fora where policies and programs are being shaped. So too, the imperative to seriously address safeguards has also been indicative of parallel efforts to the UNFCCC process. Given this, why would anyone be so skeptical about the prospects of REDD from a feasibility standpoint?

Progress over the past twenty years on thorny issues such as participation, gender balance, and recognition of indigenous people and their rights through processes like the UN Declaration on the Rights of Indigenous Peoples (UNDRIP) and International Labour Organization (ILO) 169 has in fact been notable. When these elements are combined with the ever improving MRV tools being developed by partners outside the UNFCCC, the growing awareness of governments that policy reforms addressing tenure rights and property are needed to enable the inherent value bundled in carbon assets to be capitalized upon, the actual piloting of projects coupled to the rolling out of awareness raising and stakeholder convening at national and sub-national levels through UN-REDD sponsored processes, REDD prospects would appear promising.

Yet, it is arguable as to whether these elements are combining to create a whole that is greater than the sum of the parts, and that is in fact improving REDD's prospects. This is because the social dimensions of REDD really are not being squarely addressed, all public relations aside.

The basic controversy

In REDD, a polarized controversy has emerged pitting technocrats against skeptics and populists.[6] At issue is whether REDD is a viable program, or a sham.

On each side, a lot of fairly simple truths about REDD, development opportunities and costs for the poor, and the morality of action or inaction (and particularly what action at that), has been debated among different groups of people. A third thread that is relevant, but not central to the REDD debate and has been largely avoided, concerns whether the costs of mitigating the deforestation component of anthropogenic climate change (ACC) outweigh the projected benefits based on consensus based science, which most agree must, for coherency, be grounded in the Intergovernmental Panel on Climate Change (IPCC) process.[7]

Critiques of REDD itself are already numerous, with Lohmann's analyses (2006, 2008, 2011a, 2011b) most noteworthy. Of special distinction is the daily REDD-Monitor.com updating all things REDD, as well as mongabay.com. Both sites, along with the Global Justice Ecology Project, Madagascar Environmental Justice Network, and others like it, present a stream of critical thinking on REDD. Additionally, farmer organizations and indigenous peoples' (IPs) groups, social and university linked legal and ecological justice movements, other NGOs, and independent bloggers, radio, and filmmakers (see Miller 2012) are participants in the critique.

While justification for REDD comes from claims from proponents that REDD will reduce the 17–25 percent of GHG emissions in the global carbon account coming from deforestation, protect priority biodiversity, and even help secure livelihoods and poor forest peoples in the process, the proposed means for achieving this are not convincing, while the suggestion of a "triple win" is appealing, implementing REDD on its current pathway is highly dubious for the following reasons:

1. The prospects for sustainable carbon finance remain vague at best (Pereira 2012).
2. The accuracy of MRV tools is questioned.
3. "There is currently little clarity on what MRV systems covering social and developmental issues in REDD+ could look like" (REDD-Net 2010).

4. The lack of clarity about how the specific FPIC safeguard adopted under Articles 9, 19, and 28 of the non-binding UNDRIP (2007) is to be handled by the overarching UNFCCC program.
5. The current TMAs proposed to address social issues in REDD will in all probability fail given what is understood of their provenance and verified track record.

With the present and future livelihoods of 1.6 billion forest dependent peoples potentially at stake under REDD (IUCN 2012), if REDD proponents get it wrong, the implications for the poorest of the poor are significant. This is why, concurrently, donor funding for climate *adaptation* and *resiliency* is expanding – should mitigation not work, a back-up plan is needed.

While REDD continues to be projected, nonetheless, as offering beguiling prospects in international fora, particularly the parallel efforts of UN-REDD, FCPF, private investors, along with BINGOs[8] as Niles (2011) notes, on closer inspection considerable doubt as to REDD's feasibility lingers. How REDD will work in practice without impinging on the rights of local peoples, while facilitating development and not just stagnation and unduly high risk, is speculative.

Due to these risks, why people at local levels where deforestation is to be mitigated would participate in REDD is unclear. Why would they invest their time and opportunity costs for REDD incentives, when they still must meet subsistence needs as well as maintaining the prospects for future development (see Gregersen et al. 2010)? Only in heavily subsidized cases, or where private property rights are clear, and where they have confidence in receiving REDD benefits, would such investment be judged as rational.

For all those involved in establishing REDD MRV systems for voluntary market activities, facilitating transactions and payment for ecosystem services employing those systems to enable the creation of carbon asset value for its transacting, bringing buyers and sellers of carbon offset credits together, or advocating for or establishing legislation for the transacting of carbon offsets, the justification for continued enthusiasm in REDD is clear. If played correctly, there is either concessionary ODA funding to be had through REDD programming, or else perhaps money to be made for those who are early participants in voluntary systems, which stand the chance of evolving into compliance systems.

For those required to modify or give up wholesale current land use systems and subsistence practices, the lure of the REDD narrative, again, is less clear. While a basic premise underpinning REDD is that rational land users will opt to maximize revenue streams and will hence participate in REDD, there is no evidence that the probability of potential rewards outweighs those of risks of failure.

In REDD, revenues that in most instances have yet to be generated are to be shared through mechanisms that do not yet exist. This is to occur in countries where public sector corruption is very high, and where public confidence in the public sector is low to non-existent.

Accounting for revenues to be shared is also highly speculative. Revenue sharing depends on accounting calculations pertaining to baselines, proof of permanence of carbon assets over time, international market prices over which forest peoples have no leverage, along with little to no understanding about. This is not a scenario to inspire confidence.

For poor farmers living on the edge, speculating on livelihoods in the hopes of generating the latest "spinoffs", this time through an indecisive program known as REDD, is risky business for subsistence farmers. With prospective revenues to be determined, prospective enhancements to tenure security to be determined, prospective benefit-sharing mechanisms to be determined, and protective environmental governance at all levels for the time being simply imposed on poor rural farmers, why in the world, save in odd cases where these issues may be better clarified, would farmers invest in REDD?

Because of the extent of apparent uncertainty, it is fair to question how it is possible for project developers to successfully approach the non-binding principle of FPIC. In theory, all the main participants in REDD support FPIC principles.

For example, the UNFCCC is responsible for a number of components in REDD including the national strategy or action plan, establishing reference levels, adequate safeguards, and a national monitoring system (FCPF 2011). With FPIC considered as a safeguard, their position is clear.

Both the FCPF and UN-REDD review national Readiness Preparation Proposals (R-PPs). Safeguards and FPIC figure in both, though the FCPF understands the "C" in FPIC to refer to consultation and not consent. The many voluntary initiatives being launched through private sector finance, which often seek Verified Carbon Standard (VCS) and Climate, Community and Biodiversity Alliance (CCBA) certification and validation, also highlight obtaining FPIC as a central activity and outcome.

Yet, if local peoples often do not understand the basic nature of the REDD offering being presented to them, let alone the technical details pertaining to measurement and accounting specifics, and remain dubious about the mechanics of governance and benefit-sharing mechanisms, can FPIC actually be obtained? If so, can it possibly be credible?

There are many legitimate questions that can be asked of REDD proponents in the manner in which participation, and hence FPIC, are being approached. The questions are at the heart of the potential social feasibility of REDD and, thus, its overall feasibility and justification for international support. The sense from reviewing project identification notes (PINs) and project development documents (PDDs) is that coverage of issues pertaining to FPIC is, in general, highly superficial.

For example:

• What should the threshold criteria for FPIC actually be in situations where people have little understanding about what they may really be consenting to when it comes to revenue and other benefit streams where carbon values are difficult to predict accurately?
• Where people may have not undertaken a satisfactory (or any) cost–benefit analysis (CBA) of what their participation in a REDD scheme would mean in

terms of opportunity costs that are not realistically discounted over time, can credible FPIC realistically be obtained?

• How much capacity building will be needed to enable indigenous peoples, or functionally illiterate communities in the gamut of REDD-eligible countries, to fairly assess the content of agreements offered to them, including the investment implications of pursuing REDD in lieu of current management options, so that credible FPIC decisions are reached?

Without the tools to pursue credible due diligence, which is implicit in the FPIC compact, FPIC risks becoming little more than a ploy for REDD proponents to push programming through, particularly as criteria to define it remain under debate.

Premises

There are many premises held on the part of REDD planners that often go unstated. Ten of my own premises underpinning this book address the following:

1. *Local peoples living in forest communities are assets and not threats.* Local peoples not only deserve to be safeguarded from any negative impacts that could be engendered by REDD. They deserve to be central players in how any AD programs are designed and implemented if they are to work. This goes beyond expert analysis of social issues, to creating space at the decision-making table for forest peoples, and assuring they have the requisite capacities to participate. While this normative premise is consistent with the position of numerous rights-based NGOs such as the Forest Peoples Programme (Colchester 2007), The Rainforest Foundation (2012), and Rights and Resources Initiative (RRI 2012), it must go beyond rights imperatives to extend to feasibility if it is to positively influence practice.

2. *Current best practice is not good enough.* Whether best practice has become "tyrannical" as some have argued has become the case for participation (Cooke and Kothari 2001), or is nothing more than a panacea for unverified conventional wisdom for lack of better data, the mix of best practice and "lessons learned" have come to occupy a prominent perch in development and conservation literature. Where best practice is premised on perpetual learning, the opportunity to rethink the desirability for a new paradigm or strategic approach to complex problems such as REDD is virtually precluded through the type of group mindset produced by what Kahneman (2011) characterizes as "our ability to ignore our ignorance". "Best Practice" in REDD furthers our ability to ignore our own ignorance, as it is premised to begin with on the notion of success – in results even if not empirically verified, in methods even if not empirically verified, etc. This is not good enough in the REDD context, where the layers of social complexity outstrip any self satisfaction over current "Best Practice" methods, and "Lessons Learned", to deliver what is actually needed to achieve feasibility requirements.

3. *Carbon offset markets and poverty alleviation do not necessarily mix well.* The premise of REDD has been that millions of forest peoples stand to benefit from well designed AD programs as significant GHG mitigation is achieved. Yet as Michael Grubb, who came up with much of the original concept for favoring emissions trading so that "it would facilitate transfers from North to South to help enable clean development in the latter" (Newell and Paterson 2010), subsequently noted:

> Having created a market-based mechanism to cut carbon a lot of people seem to expect it to behave in a non-market way and deliver poverty alleviation, deliver sustainable development co-benefits, but fundamentally you create a market, it's behaving the way markets do, it chases the most cost effective things, where they can make the most profits, and I think that anyone who didn't expect a market instrument to behave in that way didn't understand what they were doing.
>
> (Grubb 2011)

As Munden put it (2011), this model has proven itself "hopeless", with little progress to show for REDD+ in terms of impacts over time and scale (Kanak and Henderson 2012).

4. *Using artificially low opportunity costs underestimates real costs and is inequitable.* Premising REDD project level activities on payments for ecosystem services (PES) that simply compensate the opportunity costs for very poor farmers raises ethical objections and is enough to justify seeking out another basis of payments (Karsenty and Ongolo 2012). That is because, at present, very poor farmers are unable to value their resources in broader international contexts, as the example of Josephstaal, PNG circa 1992 illustrated. In a similar vein, systematically underestimating full institutional and transaction costs to enable local stakeholders to negotiate equitable outcomes in REDD in an "informed" manner, to validate FPIC requirements, overstates potential profits to be made in REDD carbon transactions were FPIC seriously attended to. The McKinsey "cost-curves" have been roundly criticized in this regard (Dyer and Counsell 2010). It also underestimates the risks that disaffected partners to any "REDD carbon PES" may materialize at later dates.

5. *Northern countries are studiously avoiding addressing global climate change.* Science and recent GCC data suggest that GCC is in fact accelerating beyond prior worst case scenarios. Rather than being causal of extreme weather events, GCC is seen as exacerbating individual, seemingly anomalous events, from flooding to drought, to Arctic melt.

This inability to face up to the GCC challenge for what it is has led the North, one might hypothesize further, to perceive REDD as low-hanging climate change fruit. While less costly then attempting structural transformation of Northern economies along with those of China and India for example, the real costs of implementing REDD properly far exceed those first postulated by Stern (2006), Eliasch (2008), and others.

By implementing REDD, Northern seriousness and leadership (particularly European) has arguably been demonstrated. When poverty alleviation and biodiversity conservation incentives (the "Plus" in REDD) are added on, all at *far less cost* than approaching GCC comprehensively would require in terms of structural economic change, the allure of REDD becomes undeniable, particularly in the aftermath of the lingering global recession. Even if REDD is successful, which is doubtful on the current pathway, REDD remains a diversion from the principal challenge of climate change – inducing reductions in global carbon consumption, not simply offsets.

6. *The influence of the sunken cost fallacy leads to irrational behavior.* Because so much financial and personal investment has gone into REDD over the past seven years, abandoning it on its current trajectory would be difficult for proponents. When stakeholders stick to a sunken cost fallacy (Trivers 2011) – we have invested so much, we cannot afford to change, we will now "double down" – continuing to go straight when it is a left turn that is needed undermines all hope for the mission. In REDD's case, the goal is arguably worthy of holding on to, yet the means for getting there clearly needs major rethinking. Sinking further investment into the current approach is succumbing to the sunken cost fallacy.

7. *Participatory parity must become a requirement.* To achieve a new social contract in REDD, which is the only way in which widespread stakeholder engagement can likely be achieved internationally and in forested countries, far greater parity in involvement and decision making must be established. If it is not, it is hard to see how REDD can work.

    Ribot (2010) states that:

    > participation and alms do not constitute democracy or enfranchisement. For REDD+ to be an instrument of justice (as well as conservation) it must be affirmative—including material and representational "participatory parity" as its yardstick (Fraser 2008). If REDD only attempts neutrality it will deepen inequalities. To become a positive force for forest dependent populations, REDD must be guided by (and monitored through) clear standards for 1) democratic representation of local populations in all REDD decisions (meaning the discretionary power to make significant and meaningful choices), and 2) access to benefits (meaning local control over access to markets and forest resources).

8. *Trying to associate REDD with achievement of the Millennium Development Goals (MDGs) does not mean abandoning a feasibility lens.* The premise that the MDGs justify moving forward with REDD is appealing to planners. Much like biodiversity conservation or poverty alleviation, no one can contradict the allure of striving to achieve the MDGs. The problem, however, gets back to the means, which Easterly (2006), Moyo (2009), and Banerjee and Duflo (2011) have challenged on empirical grounds.

For REDD to succeed, far greater attention to the details of social feasibility will be needed. Simply stating that REDD will promote the MDGs is not enough.

9. *All risks are not equal.* At present, risk sharing in REDD is clearly out of balance. Customary forest right holders are being offered opportunities to speculate on their present "fragile livelihood security" for unclear prospects. Those offering incentives – intergovernmental organizations and multilateral banks and governments through the UNFCCC supported process; private sector groups and BINGOs through VCM initiatives – are all participating through soft investments, be they bilateral funding, shareholder investment, or proprietary investment bank supported agreements. The social contracts that need negotiating at international, national, and local levels *must* recognize and accommodate for these unique, differentiated risks. At present, this is not recognized beyond the fuzziest terms.

10. *"Blended" or "shared value" and "impact investing" are becoming all the rage.* Blended value is a facet of corporate social responsibility that in principle places social impact at the core of the value proposition. These private sector-driven investment mechanisms may be the next great elephant-in-the-room for REDD, as the financing for avoiding deforestation at global scale does need to come from *somewhere.*

While easy to reject on ideological grounds for many current REDD critics, if the terms of engagement for blended or shared value[9] and impact investing could be *negotiated* into new social contracts framing REDD governance arrangements and *incentive* structures, REDD may be redeemable. I know a good number of Pygmies, PNG villagers, and forest resource users across the global South who would be prepared to negotiate these terms in good faith. However, whether phrased "triple bottom line", "triple win", blended value or impact investing, to work in REDD, the terms will require negotiation. Top-down, virtuous policy making is not enough.

Climate change urgency

Part of the rationale for REDD involves urgency: the world pumped about 564 million more tons (512 million metric tons) of carbon into the air in 2010 than it did in 2009, an increase of 6 percent, leading IPCC scientists to debate whether they should be focusing on prior worst case scenarios from GCC, or develop new worst cases (Borenstein 2011).

REDD is one of a number of presumable responses to the urgency that GCC presents. Yet, it is unclear if avoiding deforestation through REDD represents a viable step for tackling this major subset of man-made contributions to GCC. While climate change indicators appear ever more urgent to address (Helm 2012), successfully mitigating GCC through REDD in impoverished developing countries is a contentious topic. Making out whether REDD is innovative and strategic, or simply a horrible idea, is hotly debated.

As climate change is both a scientific and a moral issue, it demands an approach that sees it more as a social phenomenon, as an "idea" that can be interpreted differently by different cultures and by our different sets of beliefs, values, and concerns (Hulme 2009). Since it can mean so many different things to different peoples, any solution to it will have to be extraordinarily clever. To date, the global approach to climate change first through the Kyoto Protocol and the Clean Development Mechanism (CDM), and now through REDD, has proven anything but up to the task.

As Karsenty (2012) and Verchot (2012) suggest, the likelihood of a regulatory REDD system now emerging in the embers of Durban and Rio+20 is ever more doubtful, and the ability for this occurring through a voluntary REDD+ mechanism is illusory at best. For some, Durban in fact appeared to be little more than an attempt to save the carbon market, versus the climate (see Dooley and Horner 2012; Parekh 2011). Without a credible REDD architecture, much current policy work and Readiness Phase programming efforts will largely be for naught.

As thousands of indigenous and local peoples are becoming aware that REDD is arriving, and stands to significantly impact their lives, perhaps for the better, perhaps for the worse, a twenty-first century battle is shaping up between those with local rights, and those distant from the front lines arguing for idealistic global concerns that often are beyond the immediate interest of local peoples. The battle involves customary rights holders versus nation states, big international conservation and development NGOs, intergovernmental organizations, international financial institutions, technical intermediaries and market makers, and select Northern countries supporting REDD. REDD offers an opportunity for students of stakeholder analysis to examine an extraordinarily rich stakeholder landscape.

Much of REDD is predicated on arcane technical issues pertaining to establishing credible MRV tools. These are used to measure and report on a spectrum of indicators from establishing historical baselines, to measuring carbon sequestered in real time, to assessing people's participation and FPIC, to mitigation activities. Much depends on how well nation states develop Readiness Preparation Proposals (R-PPs) that reflect stakeholder participation, so that any accusations of top-down planning and manipulation can be averted, and FPIC substantiated. This aspect of REDD remains thorny for planners.

Evaluating whether the value proposition proposed by REDD proponents is worthy of public and private sector support across the global community remains at the heart of the REDD debates. No single evaluation approach to consider REDD and its prospects is fully appropriate, as the program has never existed, is still being piloted, and nothing as ambitious at the scales envisioned, and the coordination required, has been attempted before.

Yet, the task is not impossible. There are in fact many lessons from the past, from economic development, biodiversity conservation, and broader social development programming that are pertinent to objectively appreciating REDD and helping it fulfill its prospects. That said, capitalization on experience and lessons learned often proves elusive in development and conservation programming, of which REDD is a subset of the two.

What can (or cannot) be learned from past experience?

Counterfactual scenarios of what would happen if REDD is not embarked on, as currently proposed, can be attempted. With the need to look at full costs and benefits to forests, people, and climate, this exercise is highly speculative. It has been used, however, to justify, moving forward, as analyses have suggested that many forest countries will be susceptible to deforestation due to expanding economic activity, expanding agricultural pressures from subsistence farmers as well as foreign demand for land. These differ from region to region. With subsistence agriculture seen by some to be a principal driver, while cattle ranching in the Amazon and large-scale agriculture and intensive logging in Southeast Asia are key.

Another more unorthodox counterfactual approach can also be employed in considering REDD – what would have occurred in the past to development and biodiversity conservation programming had things been approached differently from a methodological perspective? Would impacts and lessons learned have been more robust? Would the basis for embarking on REDD be far more secure than it is at present?

In the many cases where development and conservation results and impacts have fallen short, and where REDD appears to be perched for repeating many of these same mistakes again, assessment of *why* the earlier mistakes happened in the first place, whether they could have been avoided, what the implications for other techniques employed would have been, etc., could nonetheless help in more objectively assessing what is required in REDD. Understanding where REDD is likely to repeat the same mistakes as that of its principal programmatic forebears from which it draws directly for methodological guidance, what is usually termed "best practice" is key to assessing where pitfalls were faced, and how prospects can be improved. Assessing what *was* is therefore important to appreciate what *is*, and what *could be* in REDD.

On its current trajectory, there is not much evidence to suggest that success will be achieved in REDD. Clearly, to tackle climate change produced from tropical deforestation in the challenging political and remote geographic areas it occurs is no minor issue. Yet for the time being, REDD is a project of great vision, in which the whole shows no signs of being greater than the sum of the parts.

Part of the reason for why this is so has to do with the nature of inspiration behind the people driving REDD. A sense of enduring irrational exuberance about the possibility for forest carbon becoming a tradeable commodity that can provide financial incentives to mitigate deforestation, what has become the core premise in REDD, is a more accurate characterization of where things currently stand.

Arguing as an entrepreneur, Lou Munden (2011) suggests that irrational exuberance "is how things get done" and "is the most fundamentally important element of risk taking that we have", leading to inventions like the iPad and flights to the moon. To enable this irrational exuberance to reach fruition, for an analyst like Munden, the issue is not about challenging whether it should exist or not. Rather, the challenge is in figuring out how it can be productively channeled in the right direction. Here, by Munden's own account, the model for REDD proposed has not only fallen short, but is "hopeless", and "manifestly dysfunctional" as currently constructed (ibid.).

This manifest dysfunctionality is not only, however, because of the implausible model for carbon finance offered that continues to dog REDD in 2013. Additionally, it is because the strategy and approach for dealing with social issues, what one expert workshop entitled "the social dimensions of REDD" (Russell 2011), is equally dysfunctional as well. Lessons that should have been learned over the past thirty years from international development and biodiversity conservation about the role of people in planning, implementation, oversight and the like, resist *systematically* entering into formulation of strategy. Even best practice which standard setters and REDD project developers employ avoids considering what needs to be done for REDD *to work for people*, to achieve what I will refer to throughout this book as "social feasibility". Instead, the focus is on how social dimensions, particularly ex post social impact assessment, can best be addressed to work for planners and, presumably, donors and investors. In doing so, prospects for the program are further undermined as design issues are not prioritized.

Instead of social feasibility driving REDD such that planning and finance can be "productively channeled", as Munden suggests that the irrational exuberance driving REDD must be, a recycling of the same top-down planning mistakes is being repeated. This is leaving local communities and indigenous peoples marginalized on the planning periphery yet again, almost unbelievably, in lockstep with mistakes of the past. Were this simply an academic exercise, the implications would be less startling, but this is not the case with REDD.

Figuring out how to appreciate the experience from the past and present, so as to channel energies productively toward the future in avoiding deforestation, is thus the central focus of this book.

So while I agree with the many critics who already have noted that REDD will almost certainly fail to meet its deforestation mitigation objective, this failure need not be inevitable. Even under the market incentive scenario based on voluntary emission reductions (VERs) or "carbon offsets", which those critics of REDD reject outright as implausible and worse, if REDD were to approach the issue differently its prospects for success would be significantly enhanced from what is proposed at present. This would not guarantee success; it simply would make the probability of achieving it *higher* if a feasibility framework drove design and adaptive management.

On its current trajectory, REDD will neither enhance social welfare through poverty alleviation, nor conserve biodiversity in its quest for what euphemistically is known as "the triple win" for climate change, biodiversity, and peoples' social welfare. Here again, the means to enhance REDD's prospects rest on a clear methodological break from the present and past.

The rationale for this break, and what it may consist of, is explored in later chapters. Some may well label what is offered as little more than a panacea, while others might suggest it represents little more than the introduction of yet another utopian normative framework with scant empirical evidence to support it.

It is nothing of the kind. A new "social contract" in REDD defining a new social and political arrangement is needed. To work, this social contract must be the product of a multi-stakeholder negotiation process that is designed to meet one overarching standard – that it is technically and socially feasible for generating AD. That this be embedded under a governance framework that is adaptively managed is self-evident – it would not be feasible otherwise.

The UNFCCC actually provides conceptual scope for such guidance,[10] thereby opening, in theory, an administrative pathway at international levels. Whether the will can be deployed is another story. While academics can argue convincingly for why new social contracts are needed to address climate change, not simply revised or "tweaked" existing social contracts (O'Brien et al. 2009), getting decision makers *to act* in situations where clear programming opportunities exist in advance of ever expanding natural forest deforestation and its impact on GCC is clearly another issue.

The approach is consistent with recent analysis that suggests that adaptive governance of social–ecological systems is about successfully connecting actors and institutions at multiple organizational levels to enable ecosystem stewardship in the face of uncertainty (Boyd and Folke 2012). It is consistent with those arguing for emphasis of the social contract as reflecting the "consent of the people" (Weale 2004) to enhance resilience in adaptation to climate change (O'Brien et al. 2009),

---

Box 1.1 Calculating social welfare

But it is, of course, desirable that the choice between different social arrangements for the solution of economic problems should be carried out in broader terms than this and that the total effect of these arrangements in all spheres of life should be taken into account. As Frank H. Knight has so often emphasized, problems of welfare economics must ultimately dissolve into a study of aesthetics and morals.

(Coase 1960)

Box 1.2 Framework requirements for a social contract between stakeholders to REDD

- negotiated in good faith between parties
- best information is provided to parties to meet FPIC
- governance arrangements and criteria for decision making are at the top of the agenda
- social feasibility from the perspective of all parties guides the process: defining what each party can accept in an *informed manner* as "good enough" to move forward.

and resonates as well with Miliband's (2006) position for an "environmental contract" that clarifies the rights and responsibilities of government, businesses, and individuals towards the environment, whether or not "personal carbon trading" is considered part of the ultimate contract negotiated.[11] While Miliband argues (2006) that "[p]eople feel powerless in the face of threats such as climate change that require collaboration between individuals, businesses and governments", the same is true at local levels for REDD.

Moving towards a new social contract may begin with small steps, first by establishing more effective connections between people and institutions such that REDD initiatives are designed that can work. Premises and assumptions must be verified. Policies and actions must be adaptively managed as needed. At present, action on these levels is lacking.

To do so, a highly flexible resiliency framework, in which local institutions are as *important* as formal institutions that customarily hold sway over policy, is needed. This proposal is consistent with adaptive governance thinking, while possibly extending it a step beyond current focus (see Stockholm Resilience Center 2011).

My hope is that this approach will provide the basis for effectively engaging a broader community of stakeholders in REDD than is now the case. For the time being, those currently most stuck in the technocratic thickets of MRV and carbon finance issues are driving the process, to the detriment of climate change, biodiversity, and people.

Pushback from REDD proponents on feasibility?

The inattention to feasibility as a program design driver in REDD is linked to historically weak impact evaluation approaches in development and conservation over the past decades. That is, billions of dollars have been spent, and continue to be spent, on programming based more on conventional wisdoms than on any empirically verifiable data. Reasoned analysis that builds on lessons learned for what has determined *why* complex programs and individual projects were designed as they were, and often failed, is lacking. This partially has to do with very weak demands on the part of donors to demonstrate feasibility in the design phase, versus pinning hopes, irrational exuberance if you will, on highly improbable events and outcomes playing out in many of the world's most complicated places for programs and projects to succeed.

The costs of this failure in design method are best seen in a host of proxy indicators – biodiversity loss rates inside protected areas as well as adjacent areas to even those parks with highest level of protected status;[12] apparent redoubling of Amazonian deforestation rates,[13] with the world losing 6.4 million hectares between 2000 and 2005 alone;[14] rural poverty indicators exacerbating despite many irrefutable social welfare gains in urban centers, etc.

Again, by fundamentally involving people, all problems will not be solved if necessary technical components do not figure in planning and implementation. To be clear: it is not about "social dimensions" alone.

But, a broader strategy that credibly accounts for the social welfare of those who must be involved for projects to succeed will eliminate much of the repetition of oversights and miscalculations that have characterized ineffective development and conservation programming for decades now.[15] Participation and compliance oversight of promises by donor agencies, conservation NGOs or private sector implementers will not solve all of REDD's challenges. But done well, they will enhance the likelihood of success.

More specifically, I argue that defining the limits of social feasibility should be the product of a collective exercise. Consensus on the definition of what constitutes feasibility in the context of REDD, as well as the threshold for measuring its attainment, must be reached. Stakeholders should be able to agree on criteria for retrospectively evaluating whether results and impacts of REDD projects are in synch with the original design assumptions and methods employed to generate anticipated results. This should be a normal, business as usual procedure. Currently it is not.

Avoiding deforestation first at local levels, and then replicating up to broader scales so that REDD impacts in the aggregate can make a difference, will need to be based on the dynamism and capabilities of local peoples living in remote communities as equal partners at the negotiating table with governments and potential donors. NGOs and technical experts will have a key support role to play. If they continue to *dominate* the process, REDD will fail.

If a new social contract can be forged which enables stakeholder teams at national levels to actually work together with common purpose, AD could be achieved. The prospects for biodiversity conservation and poverty alleviation may also be enhanced.

There are grounds to believe there could be a convergence of factors that lead to more rational decision making on approaching REDD. While this has proven elusive, four factors could come into play:

- Unacceptable risk exposure for major players in REDD marketplaces leading to greater due diligence and up front demands for demonstration of feasibility, a factor playing in both funds or market scenarios.
- Rapidly worsening climate change indicators that coincide with high profile climate events in the developed Northern countries.
- Exacerbating droughts and flooding in developing countries creating hordes of new environmental refugees, and livelihood insecurity threats moving to the North (see Leighton Schwarz and Notini 1994).
- The recurrence of heretofore low probability climate events in both Northern and Southern countries with huge attendant social and economic costs.

On the latter, National Aeronautics and Space Administration (NASA) climate scientist James Hansen has suggested (2012) that

[t]hese weather events are not simply an example of what climate change could bring. They are caused by climate change. The odds that natural

variability created these extremes are minuscule, vanishingly small. To count on those odds would be like quitting your job and playing the lottery every morning to pay the bills.

At some point in the not so distant future, a tipping point will likely be reached. People in both rich and poor countries, mature and emerging economies, will become concerned enough with recurrent low probability, disastrous climate events which they may fear are becoming the norm. These will likely be causally attributed to ACC. If so, this will likely provide the political impetus for new approaches to addressing climate change, with REDD being but one programmatic beneficiary.

Housing bubbles and REDD

As the bursting of the US and European housing bubbles illustrated, both homeowners and lenders were complicit in poor analysis of risks associated with legally signing on to mortgage obligations. Neither side, along with government regulators, was responsible in pursuing proper due diligence. Everyone from United States Government (USG) federal regulators and their oversight committees, to investment banks, to banks offering home mortgages, to home buyers (qualified or not) was adhering to a bright-sided projection of future events that proved false. The problems that prevailed in the collateralized debt obligations (CDOs) markets that were responsible for US$542 billion in write-downs at US financial institutions from 2008–2009 alone were caused by a combination of poorly constructed CDOs, irresponsible underwriting practices, and flawed credit rating procedures (Barnett-Hart 2009; Schweigert 2012).

While not perfectly foreshadowing the evolution of REDD, lessons are to be learned from recent US experience particularly. Poignantly, some argue that carbon offsets "are inherently unregulatable, for unalterable scientific and logical reasons. Instead of reducing climate risk, they increase it and conceal it, along the way reinforcing environmental and social abuses of multiple kinds" (Lohmann 2009).

It is no stretch to suggest that at a *much* smaller scale, and involving different classes of assets, REDD is suffering from analogous design flaws that are predictable. The construction of feasible REDD mechanisms through offset markets is questionable, because the quality of the standards and regulatory best practice for technical and social MRV indicators is also highly questionable.

Despite the visible flaws, the standard setters for voluntary carbon projects, which represent the cutting edge of implementation practice on the ground, are in an equivalent position to the ratings agencies in the housing bubble. In that situation, agencies neglected examining the value of the securities serving as collateral underpinning the CDOs. In the REDD case, it is possible that the value of carbon assets is being oversold, as the value of social feasibility is essentially discounted to near zero by assuming it is being achieved indirectly through the deployment of other best practice tools.

Why planners may be optimistic about participation

If the Josephstaal example from 1992 represents a worst case in community engagement on the participation spectrum, is there any assurance that REDD will create a participatory process that will generate better results for communities? Is it because the notion "*pro-poor*" is being stapled to REDD from the get-go that things will work out better than they did in the "spinoffs" that PNG clans signed off on with Korean, Malaysian, Indonesian, and other logging companies over the past twenty-five years?

A recent Greenpeace review (Winn 2012) noted that millions of hectares of customary land had been stolen for logging through the same re-spun "spinoff" process in business since the early 1990s. While seemingly unthinkable, could the same type of "theft" be enabled under REDD as some onlookers focused on the "land grab" implications of REDD (Rayner et al. 2010; Sharma 2012; Lang 2012f) suggest? And from the other side of the world, clearly there are a number of Amazonian communities who also believe this to be so (see Llanos and Feather 2011).

If the feedback from community organizations, indigenous peoples, and NGOs in diverse countries such as Indonesia, El Salvador, Honduras, Panama, Mexico, and the DRC that is cited in chapters to follow is indicative, the rhetoric of FPIC is far exceeding substantive content for the time being. If REDD projects purportedly of benefit to the global community and national interests are not consented to by communities, what does this mean? And if the standard employed by the donors to REDD enables "consultation" to substitute for "consent", as has been the case for many World Bank projects (Goodland 2004), the door is opened to broad interpretation of what FPIC means.

The fact that the US (see United States Government 2010) supports an interpretation of FPIC where "consultation" suffices, complicates FPIC for all stakeholders concerned: "the US understands the concept of "free, prior and informed consent" or "FPIC" to call for a process of meaningful consultation with tribal leaders, but not necessarily the agreement of those leaders, before the actions addressed in those consultations are taken" (United States Government 2011). The USG bases its judgment on the view that "the US does not believe there is an international consensus in favor of a definition of FPIC that requires the agreement of indigenous peoples" (IFC 2011).

Concerning consultation, one obvious question that is worth exploring is whether REDD objectives will be feasible to achieve in a framework where consultation substitutes for consent. There is no empirical basis from lessons learned in development for believing that consultation will be sufficient in REDD. So can consultation create the conditions in which "win–win–win" outcomes for climate, biodiversity, and poverty alleviation can be achieved when we know from history that consultative processes are almost uniformly based on checklist procedures, and are weak? Or is REDD a grand pipe dream that at the end of the day will do little to mitigate deforestation and its contribution to global climate change, or make communities any better off through the consultative process?

For nearly eight years now the global community interested in climate change, deforestation, biodiversity conservation, indigenous peoples, innovative financing mechanisms, payment for ecosystem services, human rights, rural development, and other related matters of concern has been riding a roller coaster ride of tremendous optimism, apparent pessimism, rekindled cautious optimism, and more recent polarized debate, around REDD. This directly involves tens of millions of people from diverse levels and interests from around the world, and arguably billions more indirectly in terms of GCC impacts. This roller coaster ride alone makes REDD an intriguing issue.

As stakeholders vested from the most powerful to the most marginalized political levels of international society and the global economy are finding interests converging in space and time over REDD, REDD is a contemporary international issue whose success or failure carries great implications. If it fails, one imagines it will be difficult to generate enthusiasm to tackle big, complex problems that require international coordination in the future.

Box 1.3 Is REDD perched to succumb to the "thinking big" syndrome?

Cohen and Easterly (2009) frame the heated debate between those who "think big" to end poverty through economic growth, versus those who argue for a more modest, detail-oriented, bottom-up approach. The latter group suggests that a radical rethinking of the way to fight poverty is needed that focuses on identifying the right policy levers to push by listening to the poor, and building programs bottom-up, versus working deductively from the top-down (Banarjee and Duflo 2011). The thought has implications for REDD as well.

Banarjee and Duflo note that the lack of the right kind of information and a legacy of government failures make people mistrust government. Extreme poverty makes it necessary to give away services at well below true market prices. Since people often do not understand their exact rights, they cannot effectively demand or monitor performance. They suggest that this is a reason why government, NGO, and international organizational programs often do not work. Instead of looking to sabotage by a specific group as the cause, they see the problem residing in lack of attention to detail in design. They suggest that if the whole system was badly conceived to begin with, and no one has taken the trouble to fix it, then of course poverty will not be alleviated.

Not paying enough attention to the details in people's lives in the design process, lack of local awareness to rights, and design flaws are issues of great relevance in REDD too. Will REDD be another example of the "kind of magical recipe" (Forum Barcelona 2004) that discourages us through its failure?

If REDD is to become anything other than a repetition of grandiose program concepts gone amuck in five to ten years' time, it is best to quickly rethink the basic assumptions that underpin it. These assumptions are similar to those that have prevailed and driven development and biodiversity conservation for decades, all with questionable results and impacts.

If these assumptions are challenged with an open mind, REDD stands a fair chance of being redeemable. If on the other hand the global community continues down the present path, the insidious, superficial simplicity of the REDD proposition will lead us to more marginalized forest communities, accelerated deforestation, and unmitigated climate change.

The choice is ours as to whether we wish to tackle hard questions about the strategies, methods, and tools we employ in REDD, or if we wish to delude ourselves into believing that what is being proposed will be good enough. If it is the former, a healthy dose of optimism will still be warranted. If it is the latter, we shall continue to fail, and the failure will be collective.

Winners and losers

Conventional wisdom has it that there will be all winners and no losers in REDD. But is this true?[16] If REDD is part of a snowballing neoliberal agenda sweeping the conservation world, as many now argue is occurring, and of which REDD is clearly a subset, who stands to benefit and lose?

McShane et al. (2011) note that it seems clear that REDD will involve both gains and losses with respect to multiple values and from a variety of points of view: "Simply put, REDD involves trade-offs." Whether these trade-offs can be handled convincingly such that diverse stakeholders buy into the vision and program remains far from clear.

Peskett and Brodnig (2011) point out that as a new form of property which is associated with potentially large financial benefits, carbon stored or sequestered in tropical forests presents considerable opportunities and risks for poor people, demanding greater clarification in terms, policies, and law. Local people have never thought in terms of carbon, yet are being told that their forests will from hereon out principally be monetarily valued on the basis of units of carbon. How will this not upend holistic forest use that benefits all segments of forest communities? How will structures be put in place such that equivalencies in gains will offset any lost functions that community forest users may incur?

Another conventional wisdom in REDD that is derived from best practice in biodiversity conservation is that people are implicitly treated as *threats*, as opposed to assets. This implicit framing is indicated by the marginal role in decision making that communities play in REDD. This puts them in the position of *loser* simply by virtue of them being forced into situations where negotiation is not part of the social contract they are signing up for in REDD.

Vira (2012) challenges threat data to support the use of threat analysis in biodiversity conservation, as does Carter et al. (2012) in producing data that suggests that one of the orthodoxies underpinning biodiversity conservation for decades – that human–top predator "land sharing" is not possible – is apparently contradicted in the case of Chitwan National Park in Nepal. If people can in fact get along with top predators in congested landscapes, why cannot people be given a chance to become assets in REDD, instead of being set up to be losers?

To date, the global community has not succeeded in defining clear and acceptable conceptual and physical boundaries for REDD. This opens the door for ambiguity and loss of access to resources.

In the majority of landscapes where REDD will take place, property rights are contested between community institutions, the state, and private sector interests who may have been granted leasehold arrangements by the state in the absence of local consent. This constrains all aspects of REDD design and implementation. The effectiveness of monitoring, sanctions, rule setting for conflict management, and the adequacy of fit between resource management rules and prevailing conditions on the ground are susceptible to weak stakeholder buy-in resulting from ambiguous property rights. Where the state can override any customary claims, customary rights holders stand to lose.

Here consideration of the normative framework that Ostrom (1990) devised for management of "common pool" (or "common property") resources is clearly applicable in REDD. In the many situations where overlapping tenure claims exist, Ostrom's seven design principles for governing forest resources still remains to be clarified under REDD regimes.

Most troubling, the global community has not put forth a remotely credible strategy that will induce buy-in from the key stakeholder group with leverage over the end result of REDD programming – local peoples living in REDD-targeted forests, in forest communities dependent on those forests for their livelihoods. While deforestation is *not* simply an issue involving aberrant behavior of local peoples, but involves behaviors and incentive structures for economic actors and consumers far from the forest frontier, the success or failure of REDD or any other program addressing deforestation will begin and end with local forest peoples. They can help create winners in REDD, but they also can bear the principal weight of loss.

REDD's evolution amidst controversy

REDD is potentially important because it is one of the few attempts by the international community to actively rein in GHGs, in this case those attributable to deforestation that account for some 18 percent of overall GHG emissions and up to 25 percent of GHGs attributable to human agency leading to ACC. The question, however, is can REDD, as proposed, actually succeed? Moreover, is it the best approach to addressing deforestation and forest degradation contributions to climate change?

The basic idea underpinning REDD is deceptively simple: developing countries that are willing and able to reduce their deforestation rate set to a historical reference period will receive financial compensation. Transfers will be based and awarded either on foregone opportunity costs or on the value of carbon market prices.

Swingland's (2003) early compilation on the promise of market-based approaches linking carbon markets and biodiversity conservation provided logic and inspiration for breaking out of the cycle of ineffective and unsustainably financed biodiversity conservation projects reliant on public sector funding. This was before the notion of "double" or "triple win" scenarios was widely fashionable and broadly advocated for.

Yet despite the subsequent advance of UNFCCC and UN-REDD initiatives at international and national levels from 2007 on, lack of consensus continues to prevail as to whether and how it is *feasible* to create AD programs based on a payment for ecosystems logic. Policy constraints and practical realities relating to ownership of forests and the carbon within them, and obtaining the consent of parties needing to change resource management systems, remain problematic. Thus while ten years on the conceptual appeal of the Swingland (2003) edition retains theoretical allure for proponents, the failure to operationalize a mechanism at the UNFCCC level creates ambiguity and constraints. This leaves this subset of climate change activities in limbo at international levels. And while "cap and trade" was a principal issue in President Obama's run for the White House in 2008, in the US presidential and vice presidential debates in 2012, climate change itself merited no attention during the debates.

Part of the lack of consensus has to do with the premise that GHG emissions resulting from deforestation can be mitigated through a mechanism that places a value on standing forests, as opposed to forests that are cut down, and that employs a market mechanism to generate the finance needed to avoid future deforestation and forest degradation. This has been vehemently attacked over the past seven years by a range of academics, NGOs, and community activists.[17]

There are potentially plausible financing and programming approaches to avoiding deforestation other than REDD, though none in fact are risk free. Whether it is through markets, funds, hybrids of the two, or as recently proposed, "sustained investment" (Karsenty et al. 2012a), a whole different level of engagement when it comes to planning, decision making, oversight, and adaptive management will be needed. These too will need to abide by a higher standard than has historically prevailed in the development and conservation communities if they are to work.

This has rationalized the break off of VCM proponents, albeit that the US$147 million market in 2011 (Peters-Stanley et al. 2012) was still a far cry from what could be extrapolated for the potential US$5 trillion value for PES markets that Sukhdev suggested (2012) could result if full cost accounting for the value of ecosystem services was considered.

In 2012 meanwhile, the proportion of REDD projects in the overall forest carbon market fell 62 percent from 2010 to 2011, indicating that "projects came to terms with the unexpected complexities and costs of newly available methodologies; decreased demand from recession-constrained European buyers; and the intricacies of tenure, community building and evolving policy environments that characterized global challenges to REDD project implementation and finance in 2011" (Peters-Stanley et al. 2012).

At the same time, even *with* the existing demands placed on REDD project developers, compliance mechanisms remain distant yet still envisioned, while voluntary market activity is slowing. While proponents had believed that REDD represents a strategy that could be quickly implemented, and that could serve as a "bridge" for reducing near-term emissions while buying time to move into a fully fledged global low-carbon economy (Viana 2009; see Newell et al. 2012), the future is unclear at best. While REDD still remains theoretically appealing for many, many factors will need to go right from hereon out for this optimistic vision to be fulfilled.

In what remains of the Readiness Phase, a better understanding of how deforestation drivers (see mongabay.com 2012) can be addressed through different approaches to environmental governance, capacity building strategies, incentives, and benefit sharing arrangements would be useful. Presently, preponderant effort is placed on carbon finance and and the technical aspects of REDD – MRV, emissions reductions (ERs), and reference emissions levels (RELs). Social dimensions are categorically subsidiary, albeit important on a rhetorical level. This approach all but precludes attaining social feasibility.

Why REDD is seen as a solution to deforestation and forest degradation

Despite its many warts, REDD, and its recent expanded and far more complex version, REDD+, is seen as offering a large-scale GHG reductions program for viably addressing the 15–25 percent of GHG emissions (with 17 percent or 20 percent usually most appearing in the literature) emanating from the forest sector that are linked to deforestation and forest degradation. Yet, it is in fact far from clear that credible REDD policies and practices, with credibility characterized by Centre for International Forestry Research's (CIFOR's) 3Es – effective, efficient, and equitable – have been developed since CIFOR researchers began laying out the fundamental challenges (Kanninen et al. 2007; Angelsen 2008, 2009). The empirical data suggests we remain significantly distant as a global community from demonstration of CIFOR's 3Es in REDD.

REDD is premised on the assumption that the limiting factor in mitigating climate change and conserving biodiversity is *funding* on the finance side, in addition to the MRV tools needed to enable functional market transactions to flow. The idea is simple: secure the financing mechanisms, tighten the methodologies and ever-improve the measurement tools, and forests and biodiversity can be saved through AD, reduced forest degradation, enhanced proactive carbon sequestration, etc. Local peoples can receive "co-benefits" in the process – the "Plus" in REDD – if funding is in place, as the principal issues are technical and are being resolved. Meanwhile, biodiversity as another co-benefit will be conserved. Perhaps this is true. But, more likely, it is a gross oversimplification.

Taking into account deforestation drivers

It is difficult to design REDD projects to be feasible if there is fundamental disagreement over deforestation drivers, as appears to be the case when it comes to

REDD. As there are difficulties to establish clear links between underlying factors and deforestation or degradation patterns (Angelsen 2008), this complicates the efficacy of REDD planning.

While the proximate causes of deforestation and forest degradation are often easier to monitor and quantify because they relate more to specific deforestation and degradation events on the ground, underlying drivers of deforestation factor into local deforestation too. These, however, are more difficult to capture and plan for in specific projects. Thus, it is likely that projects may overemphasize the importance of proximate causes to deforestation in design.

While seductive, the *major* social challenges to REDD reside at institutional levels versus the technical MRV levels where much current emphasis is being placed. These involve the role of community based organizations (CBOs or CSOs) versus state agencies or BINGOs in conservation (see Hockley and Razafindralambo 2006; Hockley and Andriamarovololona 2007).

The challenges that face the developers of MRV tools are also not to be underestimated. The carbon flux in and out of ecosystems, forests, soils, etc. remains incompletely understood, highly variable, is dependent on a wide range of factors, thus difficult to assess accurately (EcoNexus 2012), and has complicated UNFCCC negotiations. It has also rationalized what anecdotal evidence suggests is a seeming funding bias for MRV in REDD.

In turn, many REDD critics have argued that this investment in MRV only makes sense in the context of carbon trading, and has less bearing on AD. Critics would prefer to see more emphasis placed on stakeholder engagement, governance and decision making, local capacity building, judicial reform, adequate oversight measures for REDD, adequate benefit-sharing mechanisms, and of course all the policy measures required to enact the above.

To move in this direction, multi-stakeholder negotiation processes are needed. The troubled UNFCCC process does not directly represent the interests of indigenous peoples and local communities, despite the stated UNFCCC principles that "policies and measures to protect the climate system against human-induced change should be appropriate for the specific conditions of each Party and should be integrated with national development programmes, taking into account that economic development is essential for adopting measures to address climate change" (UNFCCC 1992). Activities like REDD that fall under it should therefore not be disassociated from economic development at country levels. Yet, the types of negotiation processes that could lead to acceptable development outcomes for rural peoples have largely been absent.

Social feasibility: the key for moving forward

Failure in avoiding deforestation need not be inevitable. The probability for success in avoiding deforestation can be improved. This can be achieved through a revised REDD, or through other better adapted mechanisms.

The strategy that I propose to improve these prospects is based on systematic application of feasibility principles and frameworks. In particular, these must be directed to the social dimensions of the task at hand, as this is where the weak underbelly of REDD is most glaring.

There are four normative pillars of social feasibility that must be established for REDD to work sustainably. Each pillar is associated with a minimum threshold:

- *Tenure.* Working to establish a sufficient degree of tenure clarity through knowledge and agreements to rationalize multi-stakeholder investment.
- *FPIC.* Creating conditions for investing credible effort in satisfying genuine consent (and not simply superficial consultation).
- *Accountability.* Engaging civil society in oversight of REDD decision making, revenue flows, and national level programming to limit corruption.
- *Negotiated benefit sharing.* Negotiating clear, comprehensible benefit-sharing mechanisms for AD, and not simply carbon rights.

Absent these pillars, neither AD nor REDD will achieve sustainability. For these pillars to be prioritized, new social contracts between principal stakeholders to REDD at national levels must be negotiated that consider the multiplicity of land use needs, beyond simply the theoretical value of carbon. To be achieved, there must be international consensus on the need for making this a priority.

This type of strategy is consistent with Ostrom's findings: that the principle of subsidiarity be applied to successfully manage common pool resources (Ostrom 1990). As most AD activities will take place on lands in which overlapping property claims exist between customary rights holders, the state, and perhaps private sector interests such as logging and mining companies, these are common pool resources, regardless of any *de jure* title claims. Unless various rights holders agree on management principles, sustainability will prove impossible. Those living closest to the resources with the greatest leverage over sustainable use will more often than not have inordinate leverage over long-term prospects for the resources. This can become an asset in implementing REDD, as opposed to a threat as currently approached.

Social feasibility addresses those issues pertaining to resources and their sustainable use. These involve, among others, governance over resources (legally recognized and customary rights, decision-making over use, etc.), the capacity for analysis of situations such that rights can be invoked and optimal resource use decisions taken, adherence to human rights principles (e.g. UNDRIP), and *credible* application of FPIC, which for the time being remains more aspirational than palpable.

The call for enhancing the social feasibility of REDD echoes diverse concerns from a range of authors who have been working on social and methodological issues fundamental to success in REDD. The International Work Group for Indigenous Affairs and Asia Indigenous Peoples Pact (IWGIA and AIPP 2011), Forest Peoples Programme (2007), Rights and Resources Initiative (2008, 2011), Blom et al. (2010), The Munden Project (2011), and Richards' contribution to social impact assessment (SIA) in the CCBA (Richards and Panfil 2011), to name just a few, are all sources I have capitalized upon in broadening the framework for social feasibility in REDD.

Rights and Resources Initiative's (2012) point that despite the enactment of laws (of variable quality) to recognize tenure rights, their implementation has commonly been weak, and that rights must work in practice and not just on paper. This stimulated my thinking for outside-the-box solutions to approaching REDD.

In focusing on REDD in Latin America, Hall (2012) meanwhile raises doubts about whether sufficient account is being taken of the complex social, economic, cultural, and governance dimensions involved, advocating a comprehensive "social development" approach to REDD planning.

The consequences of indifference to social feasibility have been important. Deteriorating biodiversity and human welfare indicators in developing country contexts, along with development projects with oftentimes equivocal results have resulted. If the World Bank's suggestion that one out of two of development projects fail (see Kremen et al. 2000), clearly there is something wrong with the approaches employed to date, with infeasibility a component.

With weak or non-existent impact evaluation the cause of great recent concern in biodiversity conservation (O'Neill and Muir 2010; Brown 2010; Sunderland et al. 2012; USAID 2012), as well as weak economic development indicators more broadly (Cohen and Easterly 2009), weak M&E in biodiversity conservation even more so than in development has hampered social feasibility, as projects keep repeating the same design mistakes. While this realization has created great controversy since Chapin's (2004) watershed critique of conservation, it has done little to diminish funding flows, as big conservation NGOs have yet to be held accountable for performance by foundations, government donors or individual members (Brown 2010).

Meanwhile, reliance on expert opinion and anecdote for all things social has, despite the billions of dollars spent in conservation, been the norm. The paucity of public domain information on social welfare impact indicators is indicative of the disconnect between program funding, public information and knowledge, and data driven adaptive management. Reliance on reputation and positive public perception has prevailed.

In conservation for example, Duffy (2010) has convincingly written of how conservationists customarily argue that conservation practice is necessary and ultimately good for local people and biodiversity values given frequent linkage to ecotourism benefits. Yet Duffy (2010) disputes this conventional wisdom, suggesting that conservation has driven millions from their homes, altered otherwise sustainable resource management practices developed over hundreds of years if not millennia, and created animosity between the customary stewards of wildlife and the state, leading to avoidable externalities on the environment.

Paying attention to what the real subsistence, livelihood, and belief system costs and benefits of conservation are would enable better practices for people and wildlife to be developed. This echoes Neumann's (1992) earlier ethnography of conflict between people and wildlife in Tanzania, which is part of the broader literature on peasant resistance to state control of natural resources (Scott 1977; Peluso 1992). This lesson has yet to be assimilated into mainstream conservation, and the potential implications for REDD are equally serious.

As further billions of dollars of investment are in the pipeline to be spent to meet conservation and development goals, and now climate change mitigation through REDD, legitimate questions can be raised about whether the assumptions as to how this will be achieved are justified. And as millions of people will be impacted, and multiple environmental values including biodiversity, ecosystem services, and now GHG emissions and climate change are in the balance, the question is not academic.

Facile acceptance of the assumption that best practice will be good enough as long as financing is in place, will prove fatal to people's production and resource management systems, climate, and the environment.

Framing REDD

Di Gregorio and Davidson (2008) note that "a frame acts as a conceptual lens that brings certain aspects of reality into focus, emphasising a certain way to understand an issue, while shifting others into the background". Framing is important to understand in terms of how issues get presented to the public for a variety of purposes. These purposes may be supported by scientific data and analysis, which themselves may have been framed by scientific findings, scientific controversy, change in science, science reports; or they may exhibit political objectives that highlight unsubstantiated, theoretical, optimistic views that may or may not prove able to promote social welfare at the end of the day.

The feasibility of whether AD and REDD are successful or not begins with the manner in which the challenges are framed. This framing impacts all aspects of design and implementation of programs, including assumptions about what is adequate in terms of policies, tools, methods, and approaches (PTMAs).

To date, the framing of the carbon market dimension of REDD appears to have taken precedence over what it will take to actually avoid deforestation and mitigate forest degradation effectively and sustainably. The issues are not necessarily the same. The impacts from the manner in which AD and its presumed response, REDD, have been framed are still being played out as policy debates, REDD pilot activities, and the first round of any positive or negative impacts from what has transpired over the past six or seven years begin to trickle in.

The lax framing of issues has enabled tremendous leeway in interpretation about what represents "good enough" in terms of PTMAs. This has led to developing standards and best practice that is anything but good enough. These standards and best practices will end up imperiling REDD, as well as the AD mission unless rethought. So too, biodiversity conservation and poverty alleviation as co-benefits also are threatened by overly weak standards and best practice that relies on providing broad choice to users to interpret what is "good enough".

The debate over what REDD is and should be includes not only academics, NGOs, and peasant and farmer activist organizations who have led protests against REDD and the vision of proponents since its inception in 2005. Now, even the Governors' Climate and Forests Task Force (GCF) which links the US State of California, once the world's fifth largest economy and now ranked ninth,[18] and a leading proponent of carbon offsetting with the States of Chiapas in Mexico and Acre in Brazil, is feeling an imperative to "reframe ongoing efforts in a manner that will build on previous work" (GCF Task Force 2012). If the previous work was bearing fruit, there would be no need for a reframing. It, however, as the GCF notes below, is not.

That the GCF wishes to reframe how REDD is approached in programming it supports is potentially laudable, as stated in its September 25–27, 2012, Chiapas meeting's objectives.[19] Yet, in fact, efforts are needed to reframe REDD at international and national levels *across the REDD* landscape. It is not just about building "on previous work". What is needed urgently is a rethinking of basic premises and assumptions by critically reviewing past failures and successes where and if merited.

Box 1.4 Governor's Climate and Forests Task Force, Chiapas, Mexico, September 25–27, 2012, Session 2(b) Reframing REDD+

REDD+ is at an important crossroads. The early vision of an international REDD+ mechanism that would deliver large-scale compensation for successful efforts by tropical forest countries to reduce greenhouse gas emissions from deforestation and forest degradation has not materialized. Many political leaders who made courageous and politically risky decisions to put REDD+ into practice are frustrated by the lack of deeper financial commitments to REDD+. Some of the early supporters of REDD+ have left office in the wake of new elections, and their successors are wondering what to do with nascent REDD+ programs. Indigenous peoples and traditional forest communities have participated in numerous dialogues on REDD+ and have been approached by project developers proposing obscure carbon deals, but tangible benefits for their communities are virtually non-existent. Farmers and livestock producers have seen few benefits for the steps they have taken to forgo deforestation and reduce their emissions. Given the many challenges facing REDD+, the continued lack of progress in developing a global framework, and the general fragmentation of climate policy, it is important to step back and reassess the prospects for REDD+ and, more importantly, to reframe ongoing efforts in a manner that will build on previous work, promote a broader policy agenda and corresponding institutional frameworks for low emissions development, and provide more effective ways to deliver funding to activities on the ground, where it is most needed.

How best practice language enables feasibility to be bypassed

Participatory rural appraisal (PRA) did more thirty years ago to transform how planners interface with rural people in project planning processes than anything before or since. Yet, much more arguably has come to be expected of what PRA can in reality deliver. REDD is just the latest area where use of PRA diverts attention about what really is needed to properly address social dimensions. But it is not the only factor in this diversion. PRA is not built to solve all problems.

The comprehensive, apparently benign technocrat language of development and conservation has created an intuitively appealing, politically correct smokescreen. By referring to "best practice", "theory of change", "PRA", and "FPIC" that are presented as part of "guidance" by the lead VCM certification bodies for REDD (see Richards and Panfil 2011; Richards 2011), the CCBA and the VCS, a solid intellectual artifice that comprehensively covers all necessary bases is established. With an alliance comprising some of the biggest names in biodiversity conservation backing the standard setter – including Conservation International (CI), CARE, Rainforest Alliance, The Nature Conservancy (TNC), and the Wildlife Conservation Society (WCS) – on the surface it would seem that the social dimensions of REDD are in good hands.

While facilitating certification by industry-recognized standard setters and their chosen audit, validation, and verification agencies clearly raises confidence levels for funding agencies and investors that proposed projects are viable based on the core guidance they provide, the guidance ironically does not provide explicit information on feasibility. Rather, this is inferential, and assumed to be a product of a long-term "learning by doing" effort framed by theory of change and the specific TMAs of best practice. The "learning by doing" strategy is premised on the assumption that, eventually, regulatory problems in carbon markets will become manageable.

There is absolutely no empirical basis for believing that this guidance provided by the CCBA to address the social dimensions of REDD will actually work any more so than it does in biodiversity conservation. The guidance is based largely on experience, rather than data and analysis, that has been presented as verified by the biodiversity conservation and development industry; i.e. it is primarily anecdotal, and appears to suffer from the kind of "weak or vague causal maps and articulation of aims" that United States Agency for International Development (USAID) (2011) for one is attempting to rectify in its programming.

Despite many billions of dollars spent in biodiversity conservation and development projects over the years, verifiable information is sorely lacking. Instead of reducing the perception or reality of biased measurement or reporting due to conflict of interest or other factors, it appears that the guidance shaping CCBA auditor processes is perpetuating vague causal mapping which may be linked to biased measurement and reporting due to potential conflict of interest or other factors.

Box 1.5 Appreciating REDD as a problem of
environmental governance and framing

Thompson et al. (2011) argue convincingly that

> REDD+ is more than an impartial container for the various tools
> and actors concerned with addressing anthropogenic climate change.
> Instead, even as it takes shape, REDD+ is already functioning as a form
> of governance, a particular framing of the problem of climate change
> and its solutions that validates and legitimizes specific tools, actors and
> solutions while marginalizing others.

They state that the manner in which REDD is framed raises important
questions about how to critically evaluate REDD+ programs and their
associated tools and stakeholders in a way that encourages the most effective
and equitable pursuit of its goals. They question a variety of premises: What
is to be governed? What tools are used for this governance? What forms
of objects are to be governed? And what are the forms of environmental,
economic, and social knowledge that are considered legitimate under this
framework. They note that indigenous peoples' participation in REDD+
is embedded in a framework that attempts to bring about environmental
governance by aligning the interests of a wide range of stakeholders to
bring about desired environmental outcomes. Understanding this alignment
is key to the authors, as it illustrates that in fact the alignment "has been
incomplete, suggesting an emerging crisis of governance within REDD+
that will compromise future project and policy goals, and thus the well-being
of many stakeholders".

  This emerging crisis can and should be avoided.

The challenge to framing REDD thus involves identifying *who* is participating
in framing, as well as *what* is being framed.

In Box 1.5, Thompson et al. (2011) note the process must begin with an
assessment of premises of *which stakeholders* are involved in framing the problem
to be explored in REDD. They ask what information and knowledge is to be
admitted in analysis, what tools are to be employed, and thus how "alignment"
between stakeholder participation and sought after outcomes is framed. They
suggest that human welfare of the poor is dependent on this.

This can be a perfectly valid springboard for considering how REDD has been
framed to date, as well as what will be needed from hereon out to coherently frame
avoided deforestation and forest degradation, not simply REDD or REDD+.

A new social contract is needed

For the dream of REDD to become more plausible, let alone come true, a lot more effort needs to be put into taking the concrete steps needed to address a range of feasibility issues, with social feasibility the most pressing. A first step in this process is admission that *a new REDD social contract* is needed. The framework for this social contract is presented below.

The social contract is premised equally on principles of morality, good science (natural and social), and common sense. At present, too much emphasis is inappropriately placed on apparent, unrealistic assumptions about how local people will participate in REDD because of the supposed economic utility that REDD will generate. This assumption is highly questionable.

Based on current information we have as a global community about a range of factors key to the success of REDD, the slope to success is indeed slippery. For example, carbon markets, price signals, the ability to devise credible revenue or benefit-sharing mechanisms, the ability to resolve outstanding conflicts due to resource tenure ambiguity or even compliance with prior social responsibility contracts (SRCs) in the forest sector between logging companies and local peoples, are all issues that can undermine REDD from succeeding if poorly approached.

The only way for REDD to successfully move forward and enhance the probability for success will be to create a credible framework for making AD and carbon sequestration attractive for local people taking on added risks and responsibilities. For this, a new social contract that brings together citizens, nation states, and a range of international stakeholders is needed. For the time being, the dial is still stuck on top-down planning processes and presumed capacities that have already begun to show their inability to create a satisfactory enabling environment for REDD.

Intellectual inspiration

The book has been influenced by numerous thinkers on both the policy and practice side of development and conservation, and REDD. Theoreticians and practitioners from participatory development and its critique, common property research, critique of neoliberal approaches to biodiversity conservation, economic anthropology, tropical forestry and forestry economics, rights and resource based advocacy, indigenous peoples' movements, along with philosophy and psychology have all influenced the analysis and theory of change proposed here.

The overriding inspiration, however, is less from a particular scholar or school of thought than from a manner of thinking about situations and problems. Put simply, *Redeeming REDD* has been most inspired by thinkers who have been willing to challenge conventional and received wisdoms by examining underpinning premises and assumptions. I believe that REDD, like biodiversity conservation and, to an extent, "development" more broadly, have suffered and continue to suffer from a lack of critical reflection. Orthodoxy, however well intentioned, has stifled innovation.

In this manner, incisive challenges to conventional wisdom that have changed subsequent thinking about complex topics ranging from the evolution of ethnic groups, to social science research methods to the nature of the problem in REDD include: Southall's (1976) analysis of the evolution of distinct ethnic groups in the south Sudan; R. Schutt's (2009) influential social science methodological research text for students which introduces key notions such as illogical reasoning and selective observation where people prematurely jump to conclusions or argue on the basis of invalid assumptions, an apparent prevailing phenomenon in REDD; and finally Thomson et al.'s (2011) clear analysis of why environmental governance, and not the technical gadgetry of REDD, will have fundamental leverage over the success and failure of current and future REDD initiatives.

On a different plane, but fundamental to my main arguments, is the current introspective trend in biodiversity conservation practice to consider the relationship between empirical evidence and reporting of conservation impacts. As mentioned, it is increasingly recognized that most planning decisions are largely not based upon evidence, but overly so upon anecdotal sources (O'Neill and Muir 2010). Furthermore, very little evidence has actually been collected on the *consequences* of past practice, so that current and future decisions cannot be based upon the empirical experience of what does or does not work. Rather, it is more about conventional wisdom about the efficacy of what is referred to as "best practice".

I argue that best practice guidance has unwittingly become a factor in negatively impacting millions of people in conservation (Dowie 2009; Duffy 2010), and will impact millions more in REDD. As this best practice has never been *comparatively assessed* to determine which methods are most successful or not, this leaves industry thought-leaders free to opine on experience and conventional wisdom (Brown 2010). This strategy minimizes objective analysis from data generated within the industry on social impact in particular, enabling donors and members of conservation organizations freedom to continue to provide funding on the basis of reputation, I argue, more than empirical track record.

Because biodiversity conservationists are among the principle proponents of REDD+, given the prominent position of biodiversity conservation as a co-benefit of the REDD+ agenda, appreciating conservation's enormous intellectual and methodological impact on REDD is the first step in strategically approaching any reframing of REDD. It is also key in assessing why, from a stakeholder analytical perspective, it will be difficult to introduce the logic for any changes to the prevailing paradigm driving REDD, even if it is arguably based on a degree of illogical reasoning, as the latter is defined by R. Schutt (2009).

In terms of biodiversity conservation practice, Ferraro and Pattanayak (2006) ask some basic questions. These also have implications for achieving the core REDD mission, along with the co-benefit side of the "Plus". That these questions still beg for answers in 2013 is indicative of basic methodological issues that remain to be resolved in conservation and, hence, in REDD as well.

Ferraro and Pattanayak (2006) ask:

- What is the most effective way to slow deforestation?
- How can we reduce poaching of protected species in low income nations?
- Does conservation education lead to changes in behaviors that affect biodiversity?

While REDD planners may suggest that the ongoing pilot programs are in fact answering these questions, I will argue in this book that evidence is unconvincing to support this.

Two more strategic questions are worth thinking about from the outset: *If the issue is reducing deforestation, is REDD the best mechanism? Or will it be more effective to enable communities to take the lead in field situations, where currently they remain in a traditional, subservient role to other planners and decision makers in REDD?*

Other questions pertain to enabling conditions and feasibility of avoiding deforestation and sequestering carbon:

- Has the UNFCCC process become too stultified to move forward with *any* feasible approach to avoiding deforestation?
- Is the opposition to REDD too widespread and representative of diverse stakeholders that it has achieved "critical mass"?
- Has the legacy of the credit default swap debacle in the US and other countries poisoned the waters for REDD carbon markets?
- Will the combination of corruption and disregard for human and indigenous peoples' rights in a number of the potential flagship REDD countries preclude any reasoned hope for success?
- Is California in its financial state really going to become the leading beacon in REDD in places like Chiapas, Mexico, when broad *campesino* opposition to the process and objectives is voiced, and consent appears impossible to obtain?

The two preconditions to success

Many factors will determine success or failure in REDD, or through some other mechanism that promotes AD in countries where forests are either highly degraded or subject to near-term risk, or where forests have already been degraded and carbon sequestration through tree planting is an option. Yet, there are really two preconditions that *must be met* for AD through REDD or any other mechanism to be achieved. These currently do not figure in UNFCCC negotiations, or high level policy dialogues, save in the most oblique manner.

The two preconditions are:

1. A credible social contract between stakeholders to avoid deforestation and sustain forest management must be established through negotiation.
2. REDD projects must reflect social feasibility.

These preconditions do not include other technical feasibility issues that will also impact on success (such as approaches and tools employed in MRV) because these are already prioritized by thought-leaders and funders of REDD. As noted, social feasibility and a credible social contract have never been a priority. Well implemented, they both hold leverage over the destiny of REDD.

By underscoring social feasibility, instead of the current preoccupation of policy makers with social safeguards, these preconditions are consistent with the views of one International Union of Forest Research Organization (IUFRO) Global Forest Expert Panel member (Vira 2012) who stated that "the language of safeguards is perhaps a little defensive. It kind of accepts that there's going to be a negative social consequence and we've got to protect against it. You might actually want a more positive embracing of the social objectives."

I agree.

This book is about what "a more positive embracing of social objectives" would consist of in REDD. The argument is that this embrace would need to be both fundamental and transformational in its scope, as opposed to relying on the "learning by doing", technocratic "tweaking approach" that currently prevails. If implemented, the chances for avoiding deforestation would be significantly enhanced. If maintained along its current trajectory, REDD will resoundingly fail. In this respect, "redeeming REDD" is highly conditional.

2   Theses and theory of change

This chapter lays out the major theses I present in the book, along with what I feel is needed as a theory of change approach to enable REDD overall to succeed. It examines how REDD has become what it is, and explores if change for the better is possible, and what it would comprise if so.

Current gaps in REDD

The most glaring gaps, or questions of strategy, pertain to issues that involve the *social dimensions of REDD*:

- The absence of feasibility, and most particularly social feasibility, as a decision-making criterion both for project design in REDD and in validation of the project development document (PDD) process by the principal standard setter for social issues in REDD, the CCBA.
- Reliance upon information, education, and communication (IEC) strategies to predominate in stakeholder convening meetings in lieu of requisite capacity building to enable analysis and credible FPIC.
- In the absence of convincing evidence, reliance nonetheless on "best practice" from development and, particularly, on BINGOs as principal service providers for addressing the social dimensions in REDD.

Thesis #1: establishing a new social contract is imperative

Avoiding deforestation has merit on multiple grounds. As of 2013, the proposed means for achieving this remain unconvincing through REDD. This is because involvement and support from key stakeholder groups on the ground in forested countries – communities, activists, NGOs – is either lacking or confrontational with REDD objectives. This leads to confusion, indecision, and risk that undermines REDD. These risks apply to donors, governments, investors, and of course, local peoples.

To right this course, a policy and plan for establishing new "social contracts" between stakeholders at international, national, and local levels will be needed.

Up to now, no such social contract pertaining to REDD exists. Rather, REDD is a program designed and directed by agents external to forest communities who have legitimate stakes in the status and future of forests, but not necessarily more so than the peoples living in them who have rights as well.

While sounding implausible on the surface, the institutional framework for establishing such social contracts in REDD exists at each level. What is lacking is the recognition of the necessity, the will to invest in what is required for these contracts to be established, and a plan for doing so.

Once this framework for establishing social contracts is established at different levels, specific PTMAs that address social feasibility issues can be developed by capitalizing on the weaknesses and strengths of antecedents and current practice. Here, thought-leaders will prove instrumental in helping objectively assessing what is known, what requires further proof of concept, and how best to move forward.

If an integrated package of TMAs is employed that focuses on promoting technical and social feasibility, the probability of avoiding deforestation will be much improved. These TMAs require policies that support the logic for development of TMAs that are more appropriate for the circumstances facing those most directly involved in REDD on the ground.

Thesis #2: empowering local people is key to REDD working

People living locally with natural resources of global import that are being either deforested or degraded, forests and biodiversity in particular, must become central planners, decision makers, and managers in the destiny of these natural resources. If they do, the prospects for conservation will be significantly enhanced. If they do not, prospects for failure are likely guaranteed. To do so, the real costs for capacity building to achieve minimal threshold levels must be objectively factored in. This is regardless of the financing mechanisms for combating deforestation that the global community promotes.

Data suggests that forests managed by local or indigenous communities for the production of goods and services can be equally (if not more) effective in maintaining forest cover than those managed under solely protection objectives through national forest sector support agencies (Nepstad et al. 2006; Bray et al. 2008; Ellis and Porter-Bolland 2008; Porter-Bolland et al. 2011). This brings into question the necessity, and logic, for carbon markets where principal governance and accounting authority over forests is placed in the hands of national governments and their departments and directorates, or, in the case of proposals for "nested" governance responsibilities, with sub-national entities.

In fact, whether it is REDD, REDD+, biodiversity conservation, or bilateral development assistance programs, local people need to become far more central actors in practice than they ever have been to date if success is to be achieved. The two principal reasons for why there may be reason to believe that a new day is dawning for placing people central in REDD, development and biodiversity conservation are the following:

1. Civil society in Northern countries is increasingly demanding greater accountability of what taxpayer investments are achieving, with the *possibility* that this could impact on donors.
2. Remote peoples with access to cell phones and internet technologies are generating greater de facto transparency, are asserting rights, and demanding accountability of elected officials in ways that even twenty years ago would have been unimaginable.

Both elements, if leveraged, could play into a favorable evolution for REDD. Using climate change as a rationale for urgency, REDD and REDD+ risk heightening the rhetoric that mitigation objectives can be achieved in the absence of feasible design principles being employed. This will simply perpetuate mistakes in policy and practice already made, further undermining what otherwise may represent an opportunity to put people first in one component of mitigating ACC where local groups *can* make a difference if empowered – combating deforestation and forest degradation at local levels.

The framework for a theory of change

Theories of change are highly detailed and are, by definition, the product of a participatory process. The outline or framework for developing a theory of change is presented in Figure 2.1.

For a theory of change framework to be useful in 2013 for REDD, there must be a willingness among those driving REDD programs to rethink the current theory in which policies supporting market incentives, technical best practice, and safeguards are packaged to enable AD outcomes. In the reformulated theory of

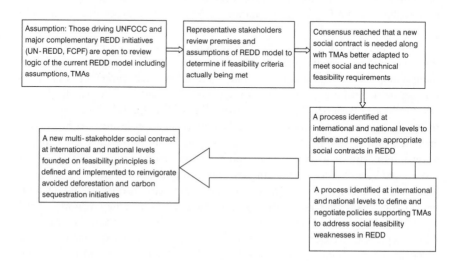

Figure 2.1 Framework for a revitalized theory of change for REDD

change framework proposed here, focus is on review of premises and assumptions, negotiating new social contracts at international, national, and local levels, and development of better adapted tools to enable socially feasible outcomes to be achieved.

Rights and REDD

While REDD is currently focused on technical approaches to planning and implementation, the success of AD and carbon sequestration is directly tied to the quality of rights to resources stakeholders enjoy, governance arrangements, capacities facilitating stakeholder participation, and the ability of right holders to withhold consent from REDD+ programming moving forward if situations are inappropriate.

The most important right that needs to be enforced is for the millions of local peoples directly impacted by REDD to have access to the information, and to have the time and a minimum level of skills, to think through what is being offered in REDD. In order to have the right to consent, they must be able to receive and *process* relevant information optimally. Their lives depend on it.

For REDD to work under any of the financing mechanism scenarios, one precondition must first be met: policies and mechanisms for sharing benefits and risks underpinning REDD must be credible in the eyes of (a) principal stewards, and (b) financial participants of AD. If the principal stewards are not convinced, then the "permanence" of forests will become insecure. If financial backers are not convinced, be it through funds or markets, requisite funding flows will not be made available. Figure 2.2 depicts this.

The precondition for success in REDD is at present not close to being satisfied. Intriguingly, while development theorists can ask "how much would need to 'work' to justify continuing to provide official development assistance (ODA) – should it be 5 percent, 10 percent, 30 percent or less? Should aid be provided if less than half of it can only be shown to have worked?" (Riddell 2009), these questions are inappropriate for REDD. It is less about quantity than quality, which is to be determined by the quality of the social contract binding stakeholder common interests, and the approaches used to capitalize upon the potential inherent to the social contract.

At present, the potential inherent to the social contract is minimal, because key stakeholders that need to be part of the contract are not convinced. That is, both people responsible for the destiny of forest resources that release or otherwise sequester carbon, who have the most direct leverage over REDD outcomes on the ground, together with those responsible for funding the actions to encourage appropriate behaviors, require reassurance from each other that the conditions are mutually recognized for what is required for the investments to *verifiably work*. And here, governments too are key for establishing enabling policies.

For this reason, all stakeholders to REDD *must* be concerned with the architecture and mechanisms put in place that enable, demonstrate, and validate that REDD is actually, as opposed to rhetorically, working, and is sustainable over

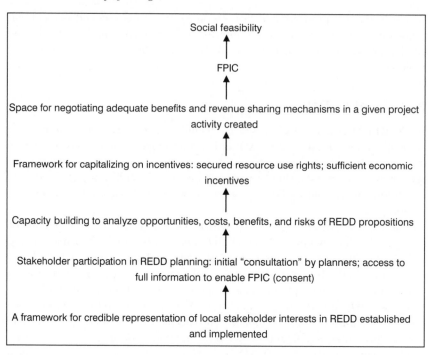

Social feasibility

↑

FPIC

↑

Space for negotiating adequate benefits and revenue sharing mechanisms in a given project activity created

↑

Framework for capitalizing on incentives: secured resource use rights; sufficient economic incentives

↑

Capacity building to analyze opportunities, costs, benefits, and risks of REDD propositions

↑

Stakeholder participation in REDD planning: initial "consultation" by planners; access to full information to enable FPIC (consent)

↑

A framework for credible representation of local stakeholder interests in REDD established and implemented

Figure 2.2 Components and steps to achieving social feasibility in REDD at local levels

the time frames necessary to justify continued investment. This is at the heart of what needs agreeing to in the social contract. Absent this, which is currently the case, investments in REDD for mitigating climate change make no sense from the perspective of either private investors or local peoples asked to modify resource use patterns and livelihood security.

The hype over sophisticated tools like geographic information systems (GISs) and the latest in MRV methodologies that have taken the fore in conservation and now REDD planning can assume a more reasonable support role. In putting the brouhaha over these tools in proper perspective, we must, as a development and conservation community, figure out how *once and for all* to place people first in planning, decision making, and administration of said efforts. Whether it is conservation or REDD, both are at heart *social* activities.

While the same development and conservation community has been circling its horses around empowerment for at least three decades now since Robert Chambers introduced the notion of people-first driven development, not much beyond the rare success story is there to show for it. Public participation is not a panacea for either development or REDD; far from it. It is simply a *first precondition* that must be met, one that has become routine in development speak for at least twenty years now. That said, the sense and limits of what public (or "popular", or "community") participation means is more often than not left up to the discretion of the individual speaker.

A key first step in empowering people will be to eliminate the knee jerk relativity that has come in simplistic application of terminology such as "participation", "community", "empowerment", etc., and demand some consensus so that both the public and practitioners are clear on what specifically is being referred to when we employ these terms.

The basic REDD appeal

In the beginning there was RED. Then REDD. Now REDD+.[1] A REDD++ that values soil carbon cannot be ruled out either.

RED(D) was conceptually appealing at the outset: improve the prospects for impacting GCC by focusing on tropical forested country contributions to GHG emissions by avoiding deforestation. The basic idea was simple: developing countries with high deforestation rates would be willing and able to reduce their deforestation rate if they received financial compensation, making standing forest more valuable than forests that were cut. Transfers would be based on foregone opportunity costs and on the value of carbon market prices. Technical measures for establishing historical deforestation reference levels would couple with measurement techniques to facilitate valuation and market transactions.

REDD was a reaction to the inability of the 1997 global climate agreement, the Kyoto Protocol, to develop policies related to deforestation and forest degradation under the mechanism. This left developing countries out of climate change mitigation activities under the CDM which was established to enable "investments in Southern countries to be credited to Northern ones", in which offsetting Northern emissions along with paying for energy and emissions saving projects in the global South ("Clean Development") would be the policy focus (Newell and Paterson 2010). While the Kyoto Protocol had been the only global pact legally requiring the developed world to reduce emissions as of 2012, the prospects for its being provided a new legal mandate preventing it from dying were unclear, constraining global emissions reductions further.

Deforestation and forest degradation were excluded from Kyoto because of the apparent complexity of measurements and monitoring for the diverse ecosystems and land use changes. In reaction, a bloc of forest countries created what became known as the "Coalition for Rainforest Nations", which included Papua New Guinea, Costa Rica, the DRC and other forest nations. At the eleventh UNFCCC Conference of the Parties (COP) in 2005, the Coalition for Rainforest Nations submitted a request to the COP to consider "reducing emissions from deforestation in developing countries". The matter was referred to the designated technical expertise under the COP – the Subsidiary Body for Scientific and Technical Advice (SBSTA), thus setting in motion the REDD process.

Subsequently in 2007 at the UNFCCC COP 13 in Bali, an agreement was reached on "the urgent need to take further meaningful action to reduce emissions from deforestation and forest degradation". The framework for REDD-Plus (REDD+) was clarified under the UNFCCC "Bali Action Plan" which called for: "Policy approaches and positive incentives on issues relating

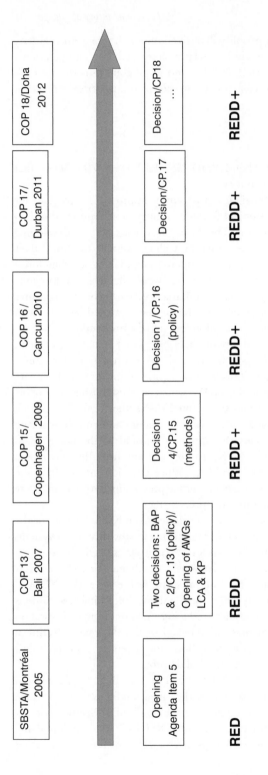

Figure 2.3 Major milestones in REDD evolution under the UNFCCC. Adapted from Paz-Rivera (2011); UNFCCC (2011, 2012b)

Key: AWGs – Ad Hoc Working Groups; BAP – Bali Action Plan; COP – Conference of the Parties; KP – Kyoto Protocol; LCA – Long-term Cooperative Action; SBSTA – Subsidiary Body for Scientific and Technological Advice.

to reducing emissions from deforestation and forest degradation in developing countries; and the role of conservation, sustainable management of forests and enhancement of forest carbon stocks in developing countries" (UNFCCC 2008). The Coalition of Rainforest Nations has driven the urgency of the REDD process at the nation state level ever since, finding opportunistic synergies with Northern CO_2 emitting countries, including both non-profit and private sector stakeholders as well.

With the addition of the "Plus", which was affixed onto REDD to create income generating opportunities for poor, rural, forest peoples who have few viable development options – i.e. to create a "Pro-Poor REDD" (see IUCN 2012), conserve biodiversity, and sequester additional carbon by enhancing forest stocks – the conceptual appeal of REDD became truly compelling. Understandably, who could be opposed to such a proposition?

Complications in implementing the initial REDD vision

While a December 2009 deadline was set for reaching an agreement on the specifics of implementation of an international REDD mechanism at COP 15 in Copenhagen, anticipating that implementation arrangements for the short and medium term would be specified, this proved overly optimistic. As of 2013 these details are still being worked out, with no definitive architecture for REDD in place. The devil has proven very much tied to the missing details.

When it comes to defining governance arrangements, much still remains to be clarified. Full stakeholder participation in REDD planning and decision-making processes, transparency in reporting, benefit-sharing mechanisms, and how ambiguous resource tenure and rights to carbon require clarification.

Contextually, the market price of carbon continues to tumble, with no clear basement to the price of carbon in sight. This undermines element one of the REDD model, as neither compliance requirements nor strong demand are on the horizon for REDD in 2013. Overall in the compliance markets, projections of a supply glut from EU ETS carbon run through 2020,[2] while in regard to voluntary markets specifically, the Caisse des Dépôts, Climate Division, recently provided the following projections: 30 million Mt CO2e absorptive capacity in the voluntary market from all sectors combined; 100 Mt CO2e/year of potential supply from REDD+ projects alone (Karsenty 2013), suggesting oversupply from REDD forest carbon as well.

So long as carbon-friendly government policies for setting a cap to rationalize demand for carbon as a tradeable commodity remain off the table in the key emitting countries – the US, China, India, etc. – the broader context shaping awareness and demand for any REDD carbon will be compromised, leaving initiatives to philanthropists, some corporate social responsibility investors, and corporations wishing to associate themselves with the brands of conservation organizations and the enunciated REDD+ cause. With the politics of debt and deficits in the US coupled to teetering economies across much of Europe, leadership from those countries traditionally turned to is, for the time being, questionable.

The ideal for what REDD could be at first glance is seemingly virtuous, elegant, and plausible: package global climate and environmental concerns in a way that national and local entities can work out solutions in partnership with international partners. Enable developing countries to take credit for this, while "nesting" accounting in such a way that local resource stewards in principle receive a share of net benefits (Marshall 2007). The solutions will benefit global stakeholders in terms of emissions reductions, while contributing to forest and biodiversity conservation, *and* sustainable development[3] locally. Clearly, if successful, REDD will be worthy of "WOW" acclamations.

On a more basic nuts and bolts level, there is no clarity of how outstanding technical requirements for establishing baselines and objectively measuring ERs in countries where corruption is rife will be accomplished. Governance structures through which payments to resource stewards managing forests and carbon can be believed in are inexistent, or else not widely believed in. Meanwhile, few have faith in the ability of international and national leaders to deliver through the UNFCCC framework that has been created. The obvious conclusion: a major reboot in REDD strategy is warranted. A simple tweaking of the current plan is not enough.

While much has been made of the technical advantages that nested MRV approaches meant to facilitate the implementation of projects within national accounting frameworks potentially offer (Chagas et al. 2011), nesting does little to answer feasibility pertaining to governance arrangements. Even with nesting, details as to how informed local analysis and decision-making that are key to the FPIC safeguard standard can be enabled are lacking.

In REDD, certain types of risk continue being underplayed. Inadvertently or not, underestimation of risk supports a major experiment that inevitably is subsidized on the backs of poor forest peoples. The need to believe in REDD on the part of its major proponents – Northern governments: big, mainstream conservation NGOs;[4] an array of private sector technocrats mediating early stage REDD programming; and Southern governments who legally claim the right to transact carbon credits[5] – continues to overwhelm objectivity as regards the nuts and bolts of REDD implementation.

Systemic challenges

The WWF's Forest & Climate Initiative argues that *systemic* challenges to REDD at the level of scale, integration, governance, and sustainable finance (Chatterton 2012) need to be resolved. The suggestion is that pilots of REDD+ projects at scale in three major rainforest blocks around the planet may help to provide lessons on how to address the challenges identified.

Specifically WWF notes that REDD+ projects are too small to influence climate change, operating merely at hundreds of thousands of hectares. They are not methodologically integrated. They usually focus on one methodology that addresses a particular type of threat, whereas threats are multiple in forest landscapes, requiring that low carbon development strategies simultaneously address multiple deforestation drivers.

REDD+ projects tend to establish their own governance institutions and arrangements that risk being in addition to existing forest management structures. This may undermine existing institutions. Government and official development assistance cannot address the problem, as priorities are geared to identifying mechanisms to take investment to scale that meets UNFCCC and forests' governance system safeguards to, presumably, unleash funding flows.

Other sources of implausibility in REDD planning exist too. These include:

- top-down planning and decision making remains cloaked in bottom-up rhetoric;
- the assumption that REDD, like other development challenges, is primarily technical in nature and is amenable to technically engineered MRV solutions;
- the leap of faith required to accept the counterfactual premise underpinning REDD MRV systems – that it is possible to accurately predict what will happen in the future with and without a REDD intervention;
- the premise that carbon trading can work based on MRV systems, underpinned by credible RELs that can be developed in oftentimes corrupt, data-poor countries (and that are rated low on the TI Corruption Perceptions Index) where data for historical deforestation rates and deforestation drivers is questionable;
- the apparent delusion that social issues are essentially subsidiary to technical issues in REDD, and that these primarily concern the need to safeguard rights and minimize damage, versus maximizing the legitimate development benefits for citizens in developing countries (as *both* are the issue).

Psychological explanations for REDD

There are a number of psychological approaches to considering REDD that further buttress the belief that implausibility currently exceeds credibility. These involve the effects of expert bias on REDD, overly optimistic outlooks, and based on criteria established by one Princeton philosopher, "bullshit" (Frankfurt 2005).

Unrealistic "bright-sidedness"[6] has prevailed in REDD from the outset. This bright-sidedness has been founded on a technocratic conceptual framework that weaves principles, standards, and best practices. These have been developed by technical agencies specializing in standard setting and the range of MRV components.

The proposition appears straightforward: baselines under REDD are established, measurements of ERs and sequestration is accounted for at local, sub-national, and national levels. The range of activities to either avoid deforestation or degradation, or sequester carbon and eventually transact it in units through voluntary or compliance markets, is adaptively managed such that the whole system becomes efficient and sustainable. Safeguards are instituted to assure no harm to IPs and others. Market incentives elicit AD and other carbon and biodiversity-friendly behaviors.

NGO and scientific thought-leaders, Northern governments at national and state levels, ODA donors, financial backers (investment bankers, corporations, and individuals), carbon traders, foundations, and private sector firms developing methodologies and projects are the principal stakeholder bloc currently advancing this bright-sided vision. As a variegated bloc, the incentive to embellish or overstate the viability of REDD is clear when one considers that the potential stakes of the value of global environmental services could reach the multiple trillions of dollars (Sukhdev 2012). While REDD is targeting a small but important subset of these environmental services, clearly the imagination of vested stakeholders may understandably focus on the positive side of outcomes while minimizing downside potential. For their part, Harvey and Dickson (2010) anticipate that REDD markets could be valued at US$15–30 billion annually.

For some thinkers, this situation could create the conditions for exaggeration, self-delusion, or simply bullshit to overwhelm objectivity. To others, insinuation of door openings either through the Mafia's proven, insidious role in the forest sector, or more generalized and widespread corruption in most REDD participating countries, has been key (see REDD-Monitor 2008; Reardon and Hooper 2012; Standing 2012) and demands focused attention.

Bullshit, Frankfurt (2005) notes, is an inevitable byproduct of public life "where people are frequently impelled—whether by their own propensities or by the demands of others—to speak extensively about matters of which they are to some degree ignorant". Since, according to Frankfurt, bullshitters ignore truth instead of acknowledging and subverting it, bullshit is a greater enemy of truth than lies.

In REDD, because of the pilot nature of programming in the Readiness Phase, the door is open to Frankfurtian bullshit. Claims or assumptions surrounding finance, stakeholder participation, consent, methods to design projects and safeguard peoples' interests are subject to Frankfurt's criteria for what constitutes bullshit.

For example, to obtain funding and validation that standards have been adhered to, REDD participants appear to be having to "speak about matters of which they are to some degree ignorant". Were this a theoretical matter that did not directly implicate people and their livelihoods, the basis of concern over those who may arguably be ignoring the truth while employing linguistic semantics about what this is, would not be a major issue.

If and when REDD is agreed to have fully faltered, which appears increasingly likely barring a paradigm shift, millions of impoverished forest peoples' lives will be impacted. Not only is this pernicious given the stakes for forest peoples, it is, to build respectively on the work of Kahneman (2011), Ehrenreich (2009), and Frankfurt (2005), "exaggeratedly optimistic", overly "bright-sided", and at the end of the day as reference to Frankfurt's analysis suggests, "bullshit".[7]

Kahnemann (2011) might ask if REDD has not fallen victim to uncritical acceptance and the exaggerated suggestion that the multitude of improbable events required for REDD to work will come true. He likely would suggest that a "halo effect" for REDD had been created by both early influential reports such as that of Stern (2006), Eliasch (2008), McKinsey & Company (2011), and the consistent bright-sidedness of the highly respected Union of Concerned Scientists in its advocacy: "Reducing deforestation is a practical, cost-effective way to help avert global warming—and it also makes important contributions to saving biodiversity and supporting sustainable development."[8] Conversely, he could also ask if critics of REDD have not fallen victim on the opposite side to group think bias, where those who spoke early in criticism of REDD created a weight of opinion that unleashed a lemming effect among followers repeating essentially the same set of criticisms, albeit with mixed impact to date.

REDD would probably present a classic example of self-deception to Trivers (2011). The arguments if presented to an objective, disinterested party would seem to indicate that the chances of success are low, despite the ostensible laudable objective. From a social evolutionary perspective, the advantage of self-deceit is clear – so much the better to delude others and secure sustainable financing for one's own programs.

Furthermore, with decreasing public finance, any comparative advantage to forward one's personal needs will lead to survival success. Hence, REDD offers a wonderful opportunity in a survival of the fittest world. Conditionality demands from donors on big conservation NGOs to demonstrate proof of principle prior to releasing funds is limited. Conventional wisdom and best practice as to methods leverages continual funding streams under the pretext of urgency.

Ehrenreich (2009) in turn could argue that REDD has fallen victim to the cultural foible of positive thinking that is predominant in America, inarguably a major thought-leadership locus for the promotion of REDD.[9]

Bright-sided thinking in REDD is characterized by neglect of all the "negative people" and their negative commentary in favor of more positive people and positive strategies for REDD to work. Positive thinking will enable the complex, for some implausible, requirements for successful REDD to align. Thought-leaders say it will, so it will. Critics who suggest things will not work in REDD, as historically has been the case in conservation, are simply discounted given the urgency of the mission. Discounting may take the form of total ignorance of opposition, to repudiation of a critic's credentials, to relativization of claims by arguing that it is still too early, that the data is not yet in to declare failure.

In a more academic line, REDD appears to fit the terms of development critiques that consider "increasingly far-fetched, abstract and artificial images of a world order" (van den Berg and Quarles van Ufford 2005). REDD, and the expanded REDD+ variant as formulated at sub-national levels, assumes and depends upon an order and integration at different scales that most sober eyes would well consider far-fetched.

From a forest sector, field practitioner's standpoint, Poynton (2012) is more directly pessimistic:

> At all the COPs and everything, it's ridiculous that all these meetings since Bali, there are huge long lists of the mitigating things we've got to do to make REDD work. It's chicken's entrails towards Venus. Now there's no chance. You're talking about human, natural systems, and you've got to do this, that, hold your mouth to the left, dance on one leg, point your finger up in the air while pointing your bum to the ground. These things aren't going to work.

The argument can be made that the narrative for how proponents see REDD working is based on a selective story for REDD that suits economic, environmental, philosophical, and political purposes. This line of thought is aligned with Lang's suggestion (2011a) that "bullshit is a better word than myth for 'describing the arguments put forward in favor of carbon trading'. The argument is that a myth is a 'fiction or half-truth, especially one that forms part of an ideology'", whereas since the truth is irrelevant for the promulgator of bullshit, it can either be true or false, as its purpose, per Frankfurt, is purely to promote the bullshitter's agenda.

Best practice and REDD

The need to understand the issues driving behaviors of people and groups leads to the mapping of stakeholders and their interests, also known as "stakeholder analysis". Best practice around stakeholder analysis and engagement is fundamental to REDD's prospects. At present, it is deficient on numerous levels.

Absent sophisticated understanding of how stakeholders actually use resources key to REDD, maintain or claim rights to those resources, interact with one another, etc., it is difficult to imagine that the foremost objective of REDD – carbon permanence – will be achieved. All MRV will be beside the point, as the probability for achieving carbon sequestration through conservation, reforestation or afforestation risks being undermined by disaffected stakeholders.

The manner in which best practice in REDD as currently promoted *does* attend to social and institutional issues is familiar to professionals in the field. This creates a sense of credibility through "consultation" and "participatory development" approaches relying on PRA and rapid rural appraisal techniques that have been the principal tools to interface with community level groups among development and biodiversity conservation professionals since the 1980s.

While useful for basic needs assessment, and collecting perspectives in large group settings, consultation and participatory development as has been practiced has not led to transformational breakthroughs with measurable results and impacts to support effectiveness or sustainability in either development or conservation. Nor do best practices tackle complex governance problems, or address situations where informed analysis over land use options and trade-offs is incorporated.

While REDD proponents are acting *as if* lessons are in hand and REDD methodologies are proven, successfully piloted, and adequate to guide REDD, the empirical grounds for accepting this are at best tenuous. At worst, they are disingenuous. As a highly placed senior manager at a major conservation organization put it off-the-record by email about being hopeful about REDD+:

> We are all hopeful about REDD+. And of course we all know it will be a disaster. It would be a miracle if some of the $4.5 billion were to drop from the tables of the consultants and managers like you and me to the tables of "the poor". History shows it won't.[10]

Box 2.1 Grounds for hope

As Rights and Resources Initiative puts it (Pearce 2012):

> [T]here is hope – derived largely from local communities and progressive actors. The local custodians of the world's remaining natural resources are becoming difficult to ignore. The recognition in 2011 of the importance of forest communities in maintaining vital carbon sinks is only one example. A rise of popular politics asserting more control over local resources is challenging business-as-usual and leading to political changes at the national level, which, in turn, is exerting an influence internationally.

Outline for a solution

The justification for enhancing the prospects for avoiding deforestation builds on the thoughts of many people. If and where there is any originality, it is in focusing on several basic principles to frame a new approach to design and implementation in REDD that has a significantly higher probability for succeeding than the current course.

The solution I suggest is based on a combination of factors. These include empirical experience in development and biodiversity conservation, and lessons learned through results and impact reporting in those two fields. It also includes common sense and a dose of morality.

The suggestion is unavoidably burdened by its being normative; empirical data to show that it will work in practice does not exist since it has never been tested. In this respect, it is no different than REDD itself.

As the current approach to REDD relies heavily on highly flexible, highly subjective theory of change logic at project levels, that provide five to ten years for working out methodological issues, and up to twenty-five to thirty years to assess success overall (see Richards and Panfil 2011), any theory of change framework that can tighten up performance reporting requirements over the short term merits consideration.

Furthermore, the best conservation practice upon which current strategies employed are based has not empirically proven itself workable. As Kareiva et al. (2012) admit:

> By its own measures, conservation is failing. Biodiversity on Earth continues its rapid decline. We continue to lose forests in Africa, Asia, and Latin America... This move requires conservation to embrace marginalized and demonized groups and to embrace a priority that has been anathema to us for more than 100 years: economic development for all... But we need to acknowledge that a conservation that is only about fences, limits, and

far away places only a few can actually experience is a losing proposition. Protecting biodiversity for its own sake has not worked.

If true, those looking for guidance from conservation experience and best practice must take the approaches and tools offered by the biodiversity conservation community with a large pinch of salt. There is no data that demonstrates any causal relationship between conservation practice and successful results and impacts for social indicators that will be pivotal in REDD success.

While a handful of biodiversity conservation thought-leaders are admitting this, the ripple effects across the conservation and REDD communities is just beginning to be felt.[11] While the solution proposed involving better economic valuation of resources (Swingland 2003; Kareiva et al. 2012) is useful, it does not get at all the social welfare issues required for either conservation or REDD to work.

REDD has yet to figure out how to embrace social welfare *operationally*. The preoccupation with the technical elements of carbon valuation, coupled to apparent simplistic analysis of property rights and the requirements for negotiating workable tenure agreements to enable AD or more proactive afforestation or reforestation activities, persists. Until *full cost accounting for social welfare for impacts* is introduced, REDD prospects will be seriously hampered. At the same time, the emphasis on community safeguards, while laudable, is morally problematic if this leads to rationalizing that development is not a priority, and justifies maintaining people in status quo conditions should market-based approaches to poverty alleviation fail.

Communities can demand-drive REDD

Research and scholarship in recent years from community forest management suggest strongly that grassroots, community level organizations have a significantly important role to play in how REDD evolves (Agrawal and Angelsen 2009).The data on the gamut of community forest management (CFM) projects is clear – communities that are given enough support can contribute to reduced deforestation and increased forest carbon stocks.

The framework offered here builds upon this scholarship, and is based on the following principles:

1. Feasibility must be the driver of all program and project design. Currently it is not. Social feasibility in particular must be stressed.
2. To be effective and sustainable, those most directly responsible *on the ground* for stewardship of forest resources in AD, activities to mitigate forest degradation, and any other proactive activities pertaining to the "Plus" in REDD+ must *negotiate* the conditions for their effective participation. These conditions must be favorable to elicit sustained buy-in. This cannot be assumed. It will also involve capacity building and adaptive management.
3. Market incentives, funds, and hybrids of the two can be *considered so long as the offerings are feasible* for local stakeholders to buy into, especially when wholesale behavior change impacting livelihood security is expected.

4. Avoiding deforestation will also require policies that impact more distant deforestation drivers beyond those verifiably attributable to unsustainable resource management practices of the poor. These will involve demand-side drivers influencing consumption patterns, governance over resource use and flows, and factors influencing corruption at different levels.

If focus can be maintained on these four principles leading to the design of program-level activities, the contribution of deforestation to climate change can be reduced. This proposition is very different, however, from one where a market incentive-based framework that few understand prevails in the face of enduring resistance or outright rejection.

Transacting of carbon is not the pivotal issue for reducing deforestation, as performance measurements over the past ten years from the principal markets where carbon has been traded already suggest. If a compliance market for REDD offset credits is to be pursued, its only chance of success is if it is wedded to a rigorous social feasibility agenda. But this will dramatically change current REDD-enabling environment programming.

Rationale for empowering local managers

Gorenflo et al. (2012) note that analysis of satellite data shows that indigenous lands occupying one-fifth of the Brazilian Amazon (which is five times the area under protection in Brazilian parks) currently represents the most important barrier to Amazon deforestation (see also Nepstad et al. 2006). Moreover, biodiversity is equal to, if not higher, in areas with more indigenous presence than areas with less, recognizing that these are all areas where population densities are low to begin with.

Since "the inhibitory effect of indigenous lands on deforestation was strong after centuries of contact with the national society and was not correlated with indigenous population density" (Nepstad et al. 2006), this indicates an inherent conservationist ethic among Amazonian IPs. This suggests that IPs' practices have significant sustainability implications in AD and, therefore, REDD, certainly in Amazonia, but potentially elsewhere as well.

This recognition among biodiversity conservationists of the role IPs and other local peoples may play in conservation is welcome. It complements the position long posited by numerous social science researchers (see Anderson and Grove 1987; Homewood and Rodgers 1991; Pretty et al. 1995; Fairhead and Leach 1996) that local people can play a *lead role* in biodiversity conservation. I suggest this should now be extended to include REDD.

Social scientists have recognized that people are embedded in ecosystems, have co-evolved with ecosystems, and have helped shape much of the renowned biodiverse, forested landscapes around the world. People are in large part *responsible as agents for biodiversity richness and ecosystem function* in some of the world's greatest landscapes, and are not simply coincidental bystanders.

Far from being wilderness areas or so-called wildlands, the great biodiversity values found in these landscapes have been shaped by the hands of peoples – hunter-gatherers, pastoralists, and agriculturalists in particular.

This perspective is in contrast to still persistent, conventional wisdom that people are the root of the problem, often need to be removed from these landscapes, or else need to have their resource management systems modified if forests and biodiversity are to be saved. Suggesting that it is because of low population densities that people and wildlife co-evolved underestimates active decision-making processes these peoples used to shape their environments sustainably. As conservationist biologists have themselves noted, traditional practices are often sustainable and can serve as the basis for conservation today (Gurung et al. 2006). The same is of course potentially true for REDD too, though the implication that economic benefits could be significant could arguably have a perverse impact if not well synchronized with customary values in places such as the Sacred Himalayan Landscape where strong customary practice still prevails.

Based on early indicators, REDD nonetheless appears to be falling into a similar knee-jerk trap that on an operational level treats local peoples as threats, while rhetorically strongly espousing respect for IPs and traditional and indigenous knowledge (TIK). The inconsistency between attitudes and actions noted decades ago by LaPiere (1934) appears worsening with susceptibility for media manipulation on all sides not to be underestimated (see Lopez de Victoria 2012).

By creating methodologies that enable the MRV of carbon based mainly on assumptions about proximate deforestation drivers, this does little to assure that the fundamental social issues that historically have plagued conservation will be met. This is particularly true for attribution of deforestation to proximate causes – usually unsustainable farming practices, charcoal making and timber extraction, etc. – that are automatically deemed the problem, and are easier to address than other potentially more important drivers that are indirect.

Yet here, of course, the argument could easily be made that underlying drivers such as lack of viable agricultural technologies, lack of alternative energy sources, and generalized poverty are causal of deforestation as well as any other indirect causes such as demand for illicit wood products from an emerging economy like India or China.

MacKinnon (2011) suggests that it is not just poor people who drive habitat and biodiversity loss: "[t]he global trade in wildlife, for instance, is big business and so profitable that it is often run by organized crime syndicates that also specialize in drug running and human trafficking". When coupled to the acknowledged role that the Mafia play in "conflict resources" such as trade in timber or minerals for arms (Global Witness 2006), it is clear that REDD strategies and policy must clearly transcend the fixation on local resource users as primary deforestation drivers if it is to succeed.

Yet, are REDD projects being designed to address these important drivers?

Capitalizing on communities: subsidiarity, democratic representation, tenure, and empowerment through capacity building

Subsidiarity (Schumacher 1973; Ostrom 1990), democratic representation of local peoples linked to adequate revenue sharing, market access (Ribot 2010), secured resource use rights if not legal tenure (Bruce 2012), and empowerment of local people through enhanced analytical, decision making, and negotiation skills (Brown et al. 2008) are all *social dimensions* fundamental to the success of REDD.

The assumptions underpinning this focus are threefold:

1. The legacy of positive experience in community driven resource management is broad enough to encourage belief that local peoples are *as competent* as any other stakeholder to drive combating deforestation activities, via REDD or any other mechanism, provided that they receive requisite support.
2. If communities do not drive the process, REDD will inevitably *not* be socially feasible, and transparency and accountability will also be weak.
3. To work, local stakeholders will need capacity building and working relationships through multi-stakeholder partnerships, as simple fund transfers to communities will likely not lead to satisfying results in most cases.

The rationale for this focus is straightforward. Communities own and administer over 377 million hectares of forest in developing countries, representing 22 percent of total developing country forests, three times more than that owned by industry or individuals (White et al. 2004). This creates an excellent, largely untapped structural opportunity that planners can capitalize upon.

Subsidiarity principles can serve as the basis for scaling up REDD programming to enhance social feasibility and, with it, the chances for success. The key now is to broadly pilot how to go about this, as this has not yet been done during the REDD's Readiness Phase.

As a first step, incentives must be provided to sovereign states to abandon the convservation model wherein inefficient state agencies, often technically advised by BINGOs, receive the lion's share of REDD resources in the Readiness Phase, assume authority, delegate responsibilities, and oversee projects that become unaccountable for actual performance justifying continued funding. This is a flawed model for environmental governance in REDD that needs changing.

The only way this will occur is if a critical mass builds at international levels for rethinking the current REDD paradigm, and moving towards a new social contract that focuses on local stakeholders, governments, and responsible impact investors in REDD (see Figure 2.1). These three must form the core for developing social contracts that will enable financial flows to be built up on feasibility principles, as opposed to theory. Facilitation from the UNFCCC, UN-REDD, and FCPF, along with technical assistance from international research centers and BINGOs would be *supportive* as opposed to determinative as it is today.

Strategy for approaching social feasibility in REDD

For social feasibility to be met, consensus must first be reached among policy makers, donors, and project developers, about (a) lessons that have been learned about participatory development broadly that REDD is drawing on,

(b) empirical evidence of potential community capacities in forest management that REDD could draw on, (c) empirical evidence of civil society capacities for oversight and advocacy in the forest sector critical for REDD oversight, and (d) analytical, decision making and negotiation needs among local stakeholder groups for FPIC and effective revenue-sharing mechanisms to be reached.

While issues pertaining to rights and capacities of IPs and communities are far from the only issues of concern in REDD, the current approaches to addressing what some social scientists have reasonably called "the social dimensions of REDD" (Russell 2011) are ineffective.

In many respects, while the international community vested in REDD has deemed that the pilot phase of REDD activities is nearing a close with implementation in hand by 2014, the fundamental understandings about deforestation drivers, policies to induce appropriate behavior changes, and appropriate approaches (strategies, tools, and methods) to do so remain questionable despite pressures for scaling up of activities in the implementation phase.

Better TMAs

A first step is to develop TMAs that are more effective at eliciting social feasibility than those now employed. Communities must themselves objectively analyze ecological, economic, and social factors impinging on FPIC in REDD. Under current TMAs this is not achieved. Without this, the REDD community will continue to rely on participatory approaches that are useful for ranking perceived community needs for delivering compensatory project benefits, but are not well adapted for more complex analytical tasks where extensive technical information is provided and internal institutional capacity to make informed and representative decisions is vital. Few TMAs have been developed explicitly for this purpose.

One possible toolkit that can serve as a reference for what may be applicable at scale in REDD is the Community Options Analysis and Investment Toolkit (COAIT), developed during seven years of funding from USAID and WWF under the Central Africa Regional Program for the Environment (CARPE) by Innovative Resources Management, a non-profit organization the author founded and managed (Brown et al. 2008). An adapted version of it would likely be of great value in REDD. This is because the manner in which community perspectives on land-use planning in biodiversity conservation zoning exercises were handled by COAIT appears to fill many of the requirements for community participation and articulation of vision under REDD, all the more so if FPIC is a serious concern.

Lessons from integrated conservation and development projects

The lack of evidence that the respective goals of conservation and development can be jointly reached (Agrawal and Redford 2006) perpetuates strategic and methodological ambiguity in conservation. And while recognition of the thorn that poverty and social issues pose is unequivocal among conservationists, there is still *no consensus* among practitioners in 2013 about how squarely poverty and other social issues should be addressed for conservation to be successful.

This lack of consensus will plague REDD too. How can conservationists credibly now mount poverty alleviation components to REDD, when they have failed miserably over the past thirty years at livelihood related co-benefits through integrated conservation and development projects (ICDPs) and their predecessors?

REDD must deal with the same sets of governance and landscape level management challenges that the past decades of biodiversity conservation projects have faced. The added dimension is that incentives must now be concretized in a far more convincing manner than proved to be the case in the era where ICDPs were first designed and implemented (Brown and Wyckoff-Baird 1992).

In practice, feasibility was ignored in the design of ICDPs, with design based more on external judgments for how incentives should be structured. This is in contrast to what was needed to create viable linkages between conservation and development with communities, which is consistent with Hughes and Flinton's (2001) point a decade ago that "the use of development tools to achieve conservation objectives – was neither understood by implementing counterparts in national and provincial government, nor sufficiently integral to ICDP design and practice".

It is presumed by REDD project developers that, by using PRA methods in REDD, projects will escape the problems faced in ICDPs. But PRA *was* usually employed in ICDPs as well, with the results being what they are. The implementation of sub-national REDD projects thus pits REDD against some of the same obstacles that have dogged ICDPs.

Firstly, forest dependent communities will likely demand short-term livelihood improvements that are not always compatible with the long-term objectives of forest conservation (Brown 2004). With agricultural practice often expected to be curtailed in conservation concessions where REDD is put into practice, it is unconvincing that offering employment opportunities will necessarily suffice as a trade-off for altered production systems.

Secondly, the benefits of conservation are largely global while the cost of conservation is mainly incurred by forest-dwelling communities that must forgo exploitation for the sake of conservation (Kremen et al. 2000). These issues are particularly important given the current global trend towards the decentralization and/or deconcentration of natural resource management in the tropics (Larson and Ribot 2008; Batterbury and Fernando 2006).

While ICDPs were criticized as ineffective as originally designed in linking incentives to improved conservation outcomes (Wells et al. 2004; Garnett et al. 2007; Weber et al. 2011), it is unclear whether either ICDP design or project implementation was ever adequate to begin with. In REDD, as in ICDPs, the basic challenge of balancing development with sustainable resource management remains the same (Brown 2010b). Much therefore is still to be learned from past failure.

Blom et al. (2010) also points to the necessity of learning from the difficult lessons presented in ICDPs, as "issues of equity will likely pit these sub-national projects against some of the same challenges that have dogged ICDPs". This leads to the conclusion that REDD project developers stand to learn a great deal from the lessons generated by this experience. Contextualizing the challenge further, Blom et al. (2010) remind practitioners of the various cycles of big new ideas that eventually proved disappointing, the legacy of which REDD cannot escape.

Market access for non-timber forest products (NTFPs), ICDPs, forest certification, and community-based natural resource management (CBNRM) were all once believed to be the new way forward for tropical forest conservation. Each of these approaches has turned out to be based on impracticable assumptions when applied in the field and have not met the high expectations set for them. Will REDD be the next on this list?

Yet, if ICDPs were so unsuccessful as has been argued, what exactly will REDD project developers be doing this time around such that outcomes will prove different? Learning from poorly designed ICDPs will not bear much fruit unless methods are reconsdered and radically revised.

Until there is consensus on whether the problem with ICDPs was with the theory, project design, implementation and adaptive management, or some combination of these, learning from ICDPs in REDD will be handicapped. Experts will continue to talk around the issue of *integration* of conservation and development, as if it cannot happen or should not be tried. Clearly if REDD projects, similarly to ICDPs, are not approached through a feasibility framework, there is no way they can work. The problem though resides less with the concept, than with the manner in which people fudge around with what needs to be done.

Without understanding how to fundamentally break from the failure of ICDPs, it is almost impossible to envision REDD succeeding, as the problems of integrating development to avoid deforestation are equally complex and unavoidable.

And while sub-national REDD projects may be required to undergo third party methodology and project design validation, and ERs must be verified to ensure unbiased reporting (Wertz-Kanounnikoff and Angelson 2009) (thereby offering a potential method to improve oversight into REDD that ICDPs did not benefit from), this does not address the fundamental flaw in either ICDPs or REDD projects.

While this third-party verification safeguard represents a step up from ICDPs which did not demand any such outside scrutiny, it does not address any inherent problems with the standards used in assessment of project development documents, the quality of the information and analysis provided, and whether validation and verification processes by auditors are lax (see Swedish Society for Nature Conservation 2013).

Lessons from CBNRM

CBNRM is known primarily as being a project-based approach to conservation where communities are provided management responsibilities and, at times, authority over key decision-making processes. The manner in which CBNRM is distinguished from ICDPs, and the learning it too represents for REDD, is summarized as follows:

[CBNRM] activities took many of the best elements of the ICDPs for their design, and emphasized the importance of land and resource tenure, ownership of natural resources, decision-making authority, and the governance structures that made those elements possible. This approach has been quite successful in areas where the value of the resources is high enough to compensate the communities for immediate foregone benefits. This was particularly true in Southern Africa where CBNRM focused on sport hunting which provided the greatest source of revenue for the communities and community members. In others areas, such as the Sahel, programs focused on trees, forest products, and forest tenure, in the belief that if communities benefit, they will manage sustainably.

(Hecht et al. 2008)

The experience of over twenty years of piloting CBNRM approaches is pertinent to appreciating the challenges facing REDD. While CBNRM is widespread, it has yet to become the cornerstone for conservation programming as a growing number of practitioners feel is justified in a range of contexts when the empirical record and potential logistical advantages are considered.

Establishing the policy and practical framework for scaling up the potential of CBNRM continues, however, to be elusive. Similar questions pertaining to the relation between tenure rights and resource use remain unresolved, and the inability to articulate tenure and management authority has undervalued CBNRM's potential.

For example, Hockley and Andriamarovololona (2007) underline that current community-based natural resource arrangements in one of the globe's premier countries practicing CBNRM, Madagascar, through the Transfer of Management (*Transfert de Gestion*) program that nominally provides communities with incentives to conserve, has in practice not accomplished what was aspired to.

There is a large gap between the community's perceptions of *Transfert de Gestion* and those of external stakeholders, with the community's vision more closely matching that inevitably required to produce a sustainable solution. Community perceptions were grounded in the "oral contract", established by mediators during the process of public education (*sensibilisation*) and negotiation. This oral contract had stressed development assistance in return for abandoning forest clearance. There remains a large gap between this oral contract, and the official contract signed by community-based associations (COBAs). The problems with contract stability therefore resulted in part from asynchronies of power and information during contract negotiation, which led external agencies to drive too hard a bargain, while promising assistance that either did not arrive, or fell far short of expectations.

REDD may well amplify similar "asynchronies of power" that Hockley and Andriamarovololona (2007) refer to. Those implementing *Transfert de Gestion* in Madagascar have thought of CBNRM as providing a win–win scenario, assuming that COBAs would be self-sustaining after community institutions have been established. Yet communities have received very little support from external stakeholders prior to contract signings and thereafter (Josserand 2001).

By expecting COBAs to be self-sufficient, while also making no provision for the external evaluation of their management, external stakeholders to Malagasy CBNRM sought to gain the positive externalities associated with successful COBA management, but "at zero cost to external agencies" (Hockley and Andriamarovololona 2007). The same appears to be the case in REDD, where benefit-sharing mechanisms remain unclear, and up-front investment in establishing the policies together with TMA conditions for achieving feasibility are unconvincing; project developers' preoccupation to keep institutional and transaction costs to a minimum so as to induce investors of the potential margins to be made from REDD investments appear to undermine the appropriateness and feasibility of their own offering. As these cost savings can only come from subsidies provided by local stakeholders from receiving less capacity building and thus being *less capable* to analyze REDD and provide informed consent, this in turn impacts on the long-term feasibility of the enterprise.

Similarly to Malagasy CBNRM where the system "has adopted a naive view of Community-Based Natural Resource Management, assuming a pure win–win scenario where the interests of communities and external agencies are perfectly congruent" (Hockley and Andriamarovololona 2007), this same win–win simplicity has been exhibited in REDD from the outset. Unless it is rethought through, this will undermine REDD outcomes and sustainability as it has hampered CBNRM.

Ultimately it will be about bargaining zones and not carbon rights

Merlet and Bastiaensen (2012) argue that situations where land is at stake, such as that posed by REDD, are characterized by the superimposition of multiple normative orders. Each order claims rights to the same land. Even though there is little explicit legislation or policy when it comes to carbon rights, which is understandable given ambiguity over forest tenure rights to begin with, this evolving asset class creates a further degree of complexity due to the abstract nature of what REDD appears to be prioritizing the most.

The commodification and capitalization of carbon rights is complex in practice due foremost to the potential multiplicity of claimants to carbon rights. As there is more than one "legitimate" or legal owner of land, in practice it is often the state or private sector title holders that claims prerogative to land, based on formal title; customary rights are put into a subordinate position. Thus, there will be multiple legitimate or legal claimants to carbon credit rights as well.

Karsenty et al. (2012a) debunk the "rhetoric" of the "carbon rights" aspect of REDD, arguing that the concept is "useless and misleading", as it promotes rent seeking which will disappoint those seeking it in any case. As remunerations in REDD can become "disconnected from the active contribution to the production of emission reductions, which is a public good by nature", private investors will be profiting from public goods emanating from common pool resources, creating contradictions from the outset.

Because checks and balances on land claims either do not exist or are rarely enforced due to the power of the state and the weak capacities of local rights holders, norms supporting state prerogatives over land predominate in developing countries. In REDD, these claims and prerogatives underpin the potential feasibility of what is envisioned in the "green economy", as both private investors or NGOs working in partnership with investment bank financial backing both work through governments to gain access to what frequently are common pool forest resources. These resources often are the subject of lingering dispute over attribution and use, with the state traditionally exerting its powers of eminent domain to attribute land use as it chooses. In countries where public administrative corruption is rife, this poses immediate problems of legitimacy now for REDD and those establishing carbon concessions, just as it may have for those previously obtaining rights for logging concessions in the past.

The normative orders that must be considered in REDD thus include the state and local jurisdictions as the "management right holders", IPs enjoying customary rights that are respected to varying degrees constitutionally or through forest codes in different countries, NGOs or corporate entities seeking to create REDD conservation concessions, international corporate investors in land and other resources, and financial investors who may be working with NGOs or other private sector groups. For Merlet and Bastiaensen (2012), it is the power relations and social struggles going on inside the various political arenas related to land, and now carbon, that ultimately will count. The key to moving forward, they believe, is to strengthen the bargaining capacity of weaker actors within these political arenas when it comes to land. This will also hold true for carbon.

Bargaining zones

In 1990 an eight-country workshop in Queen Elizabeth Park in Uganda was organized by the author on Buffer Zone Management in Africa.[12] The justification for the workshop at the time was that there appeared to be inconsistency as to what buffer zones were, and what the rights were of peoples residing in newly gazetted buffer zones attached to national parks. Conflicts were brewing in these contested spaces. With their apparent proliferation on the horizon, a need for clarification appeared imperative. If "REDD project areas" are substituted for protected areas as the geographical unit of analysis, the issues covered in 1990 on buffer zones are highly germane to REDD today.

The key to come out of the 1990 workshop was an emphasis on stakeholder analysis, particularly the inclusion of the perspective of marginalized peoples living inside, and on the peripheries of, protected areas (PAs). The most salient point made was a comment from a public administrator from northeast Uganda on buffer zones.

For the administrator from remote Karimoja, buffer zones were "bargaining zones", places where stakeholder interests collided, requiring negotiation to reach resolution. For him, biodiversity stood no chance in buffer zones or adjacent

PAs in the absence of forthright stakeholder bargaining over access, use, and associated rights. His premise is little different from US President Obama's 2013 suggestion in Jerusalem that for peace between Israelis and Palestinians to be achieved, "part of what we're going to have to do is get out of some of the formulas and habits that have blocked progress for so long" and get back to the bargaining table.

The logic of this insight resonated first viscerally then intellectually among the diverse stakeholders present at that workshop. Workshop participants – NGOs, community representatives, and government representatives – all recognized that the administrator from Karimoja was right; it is inevitable that resource use in buffer zones must involve negotiation if sustainability is to be sought and obtained.

Yet somehow, serious negotiation has managed to elude biodiversity conservation in the twenty-plus years since that workshop. And now with REDD, this same mistake is being repeated, as planners driving the REDD process are not even setting up the possibility for bargaining to occur. Rather, the dictate of the state, the views of big conservation NGOs, along with a handful of private sector industry leaders has been predominant as to how resource management and land use should be approached, and how carbon trading can facilitate this, for all intents and purposes, top-down process.

Yet, while resource use can be mandated by state fiat, this may bear little relation to sustainable use if stakeholders either do not recognize the legitimacy of the state's authority or feel they cannot afford to. Across the developing world in landscapes where REDD is being tested, conflicts are brewing. These may have as much to do with the lack of opportunity to negotiate outcomes as they may have to do with the substance of REDD itself.

The real problem is, therefore, that local peoples are left off the invitation list when it comes to any bargaining in resource management decisions, be it in buffer zones or now in REDD. For buffer zones to be sustainably managed, it was acknowledged by nearly all participants from eight African countries, bargaining must occur. Everyone also understood that this vision was unrealistic in Africa circa 1990. Tellingly, it apparently remains as unrealistic in 2013, and undermines prospects for coherent environmental governance in African countries' approach to REDD. While the situation is somewhat different in countries like Brazil which has more progressive legislated rights for empowering IPs, many Latin American countries and Asian countries face similar issues and conflicting stakeholder views that have no apparent forum for structured resolution.

We are still waiting for bargaining to occur in buffer zones when it comes to biodiversity conservation. Local stakeholders also remain waiting for bargaining to begin in REDD. This bargaining can begin with identification of the landscapes where REDD will occur, and the criteria for deciding strategies, priorities, and analysis of methodological options. If in even five years local peoples are still waiting, the prospects for REDD will surely be no better than they are for the long-term sustainability of buffer zones outside national parks, whose "fortress conservation" model is of questionable sustainability in a world of global recession.

What can economic anthropology contribute to understanding REDD?

Economic anthropology has potential significant contributions to make to the analysis of REDD. Oddly, economic anthropology has yet to squeeze its way into the conversation about the significance of REDD for remote peoples with technologically simple production systems and relatively limited access and participation in market economies. Nor has it been turned to for operational guidance, to help in the design of methods that may be appropriate to rural peoples facing difficult to comprehend, yet important trade-offs presented in REDD.

As REDD is premised on the creation of a new asset class – carbon offset credits – that can be transacted on behalf of remote rural, subsistence based peoples, with allocation of benefits determined by government agents and intermediaries transacting the carbon – the question is whether this can lead to positive net benefits. Economic anthropology can assist in assessing trade-offs and values, as well as interpreting stakeholder behaviors, as the methods of formally assessing benefits from the perspective of a business may not fully capture values that a culturalist perspective may (see Sahlins 1974).

The past century of debate over the nature of human economic behavior has raised intriguing questions about human nature and social process that are relevant to REDD.

For instance, is economic behavior rational so that maximizing profit or other utility is always predictable? Is economic behavior, rather, embedded in social relations, and is otherwise unintelligible absent this broader context? Or is it some hybrid of the two, and is pivotal to advancing abstract agendas like REDD?

Arguing that human economic behavior is anything *other* than a hybrid is, I believe, reductionist, polemical, and simply ridiculous. In effect, this is what REDD planners ultimately do, basing the theory of change they propose in REDD on the foundation of economic maximization principles, with a twist of the *possibility* for added benefits like tenure reform and some capacity building perhaps thrown in. But as in Josephstaal, PNG circa 1992, or the Peruvian Amazon circa 2011 (see Llanos and Feather), details of the agreements for tenure reform, capacity building or *the REDD contracts themselves* remain vague. People are expected to buy into the economic incentive model, regardless of the details.

Several examples from Madagascar and the Sudan from the anthropological literature suggest why rural peoples in REDD countries may hesitate over providing FPIC in REDD. For Bloch (1975) for example, individuals could theoretically undertake acts for their own economic ends, but it was best to understand action in the context of the social systems in which people were embedded. Action that is independent from a broader social group's concerns and values was anathema to most people. Maximization of utility was not seen to be a driving force for either the Merina or Zafimamaniry, despite the glaring contrast in all aspects of their material and symbolic cultures that offer different incentive structures and constraints even forty years later.

Berger (2010) notes that Bloch ultimately compared Merina irrigated rice cultivators with Zafimaniry swidden farmers from the perspective of their property rights, kinship systems, and technologies of production and power, as well as by analyzing the respective symbolism of their tombs and houses. The fetishization of lands and tombs (that fascinated Bloch in the Merina kingdom) was a consequence of slavery, irrigation systems, and state-building through long-distance commerce (Berger 2010). It had nothing to do with individual decision making, beyond perhaps that of the king.

The point here is that to understand natural resource management practices of swidden agriculturalists like the Zafimanry, who are among those perceived as threats to remaining primary forests landscapes in Madagascar, a lot more than just assumptions about maximization of individual utility need to go into the planning mix. While Allnut et al.'s (2013) work in Masoala National Park shows that "[d]espite nominal protection, support and financing from national and international organizations for the last 15 years, the current deforestation rate in [the Masoala] national park study area is higher than in many forests that lack protection altogether," emphasizing the need for remote sensing tools to monitor anthropogenic forest changes, the breakdown of forest governance after the 2009 Presidential coup is even more notable. For Masoala and other parks like it to have a chance, through REDD+ or conservation projects, local people must perceive principal stake in conservation outcomes, as state agencies can all too frequently not be relied upon. Yet, just as biodiversity conservationists simplified IEC messages to meet with conservationists' vision for how Malagasy forest landscapes should be shaped, to frequent dismal conservation consequences (see Freudenberger 2010), REDD planners seem to be dumbing down the complexity of rural decision making even further.

In parts of Africa where sharecropping is fundamental to livelihood security for multiple parties as Robertson (1987) has described, *understanding how local people calculate win–win situations* on their terms is obviously of relevance to REDD. The PDD process, even if and when CCBA validated, does not demand that this type of analysis of local decision-making calculus be part of project justification.

It should. While providing useful flexibility and time, relying on abstract theories of change to provide insight onto the calculus of individual community members may inadvertently encourage project implementers to lose focus on what is important to local peoples in specific contexts. These play directly into feasibility over the short and medium term.

One final example: in looking at the evolution of the two major ethnic groups in what is now the South Sudan, the Nuer and Dinka, Southall (1976) needed to consider environment, linguistics, ethnicity, and religion, along with the cultural penchant for maximization of cattle herd size through both raiding and internal reproductive dynamics. To attempt to understand the evolution of Nuer and Dinka as peoples over the centuries, *all* these, and others, required understanding.

Now imagine hypothetically: to attempt to convince Nuer and Dinka to place greater emphasis on say tree planting or conserving carbon rich agricultural soils (were this eventually creditable under REDD) by offering material and cash incentives, and were REDD to attempt to do so, would likely miss the mark.

Basing policy on market incentives to reduce deforestation reduces contextual variables into overly simplistic categories. While anyone who has professionally experienced for extended periods the desperation of rural poverty in remote forests of the world will attest, farmers, pastoralists, and resource users generally will opt for pathways out of poverty if possible. Most poor people do not like being poor, as there is nothing romantic about it. But, escaping poverty will *not* necessarily be opted for, either, at the expense of social relationships that provide the principal safety net that rural people depend upon. The social safety net insurance that communities provide to individuals is important too. Economic anthropology could help planners to avoid simple mistakes by keeping these safety nets well in mind.

Premising programs like REDD primarily on the projected responsiveness of remote rural peoples to economic incentives risks diminishing the rich contextual fabric that rural peoples' decision making is based upon. West et al. (2006) speak to a similar simplification practice in protected area management, where "rich and nuanced social interactions connected to what natural scientists see as the environment are condensed to a few easily conveyable and representable issues or topics". The superficial consideration that REDD thought-leaders give to highly complex social and cultural institutional contexts in which REDD *must* become embedded if they are to be feasible, is reflected in the checklist approach to complex issues shaping resource use patterns that are treated in PDDs and national level R-PPs (see Chapter 5). The assumption that trickle-down economics will be enough to induce wholesale behavior changes, in a free, informed, consenting manner, even if local stakeholders participate in planning meetings and shake their heads collectively and affirmatively that they understand and consent, is simplistic and potentially counterproductive.

Other generic factors that individuals in any community setting would need to consider prior to being in the position to provide FPIC for a REDD endeavor, if provided the chance, would include fleshing out community level institutional factors:

- Will competition between community members lead to conflict?
- Are there adequate customary conflict management mechanisms in REDD landscapes to "de-risk" potential for conflict arising?
- Is maintaining community solidarity as important to me as any increased revenues generated from REDD that I may gain?

These issues, and the framework for understanding them, have been at the heart of economic anthropology for over sixty years now since the great LeClair and Schneider (1968)–Dalton (1971) debates which centered on the adequacy of conventional economic categories explaining how price influences social behaviors in peasant or modernizing economies.

These debates were all the rage in economic anthropology in the 1960s and 1970s between so-called formalists, substantivists, and Marxists. They have recently been rekindled in the aftermath of the 2008 financial crisis. Where "markets and money were left to find their way around the world without much political interference" (Hart et al. 2010) ,the questions that have driven economic anthropologists for decades have resurfaced as to the nature and purpose of economics in society.

Recent research from within economics itself (Galbraith 2012) has highlighted how "inequality was the heart of the financial crisis"[13] in 2008. This involved credit extension, derivatives markets, the near total collapse of the financial system and the global economy.

Taleb (2007, 2009) both before and after the crash explained how black swan events were a firestorm in the making, simply awaiting a spark. His dictum *"giving someone no map is better than giving them a wrong map"* (2009) is more than relevant for forest dwelling central or West Africans who may simply not be as hell-bent on maximizing profit motive as upper East Side Manhattanites, or their capital city elite counterparts in developing countries. This issue of *not* giving community level people *the wrong map*, which to the contrary, REDD appears well on the road to doing, is imperative to get a handle on sooner rather than later.

The influence of society and culture on individual behavior is thus not ephemeral. This is why the type of analysis conducted by Ostrom (1990) or Costanza (2012) on prospects for common pool resource management regimes *may* offer guidance for structuring operational platforms for avoiding deforestation and sequestering carbon in practice. So too, this can help in anchoring stakeholder negotiations towards new social contracts to more logically structure REDD, to provide the kind of roadmap that is needed when compared with current consultation processes.

To shape REDD programs on anything other than a hybrid balance between people wishing to better their individual lot, yet working under the umbrella of community social norms and influences, is counterproductive. Understanding how economic anthropological thinking has addressed the relationship between economic behavior and social and cultural institutions can help REDD thought-leaders understand how best to frame policies and incentives, along with the very TMAs employed, to make AD and mitigated forest degradation work. It also can help REDD planners and project developers avoid giving would-be beneficiaries wrong maps, or perhaps no map at all if more appropriate.

Social science expertise and process issues: engaging people in planning and decision making cannot be substituted for

Social scientists can help in framing the challenges that programs face in reducing deforestation and forest degradation. They can help in elucidating local understandings and models that are relevant to identifying how incentive packages can best be shaped to induce behavior changes in avoiding deforestation. But they cannot take the place of local participation in analysis and decision making. Sustainability depends upon it. Yet, too often social science expertise is assumed substitutable for local participation and voice. While complementary for advisory purposes, it is not equivalent to local participation.

If given the chance, local peoples can take the lead in defining how avoiding deforestation can work. In Cameroon and the DRC, under the US NGO Innovative Resources Management's (IRM's) work through WWF/US on the USAID-funded CARPE project, local communities were able to develop land use plans that represented what *they felt*, after protracted and in-depth analysis on their part, represented a basis for sustainable management as part of landscape level land use planning. IRM piloted and then replicated the COAIT,[14] an approach that was developed specifically because on-the-shelf participatory approaches did not address social feasibility issues at landscape levels for biodiversity conservation.

COAIT was successfully tested between 1998 and 2006 in large forested landscapes in Cameroon and the DRC. These were areas that communities were willing to invest years of step-by-step efforts in bottom-up planning on the assumption that *their collective vision* for how zoning should be approached could serve as a basis for project level land use planning (Bonis Charancle et al. 2009). While it is unclear if and how this information, based on two years of capacity building leading to land use planning in the case of the DRC, was capitalized upon in CARPE by WWF, the lesson learned was that *even the most remote multi-ethnic communities incorporating Pygmies and diverse ethnic groups* were prepared to work together in visioning, analyzing, and defining their view of a feasible land use plan for the future. There is no reason why REDD could not build on lessons from this experience.

With their current resource-use patterns and livelihoods often at stake in REDD, it is implausible to expect that abstract propositions for downstream benefit sharing that are conditional on changed local behaviors will necessarily lead to sustainability. The internal divisiveness within communities that creates rifts between average people and elites is a phenomenon that external experts can, at best, facilitate solutions for, but that only local peoples can resolve. The types of guarantees that people will require to elicit buy-in, and the role that local peoples feel they must play to enhance compliance, transparency, and accountability in REDD commitments cannot be mandated by external REDD planners. They must be negotiated to be viable.

Most important of all, communities will need to go through credible collective analytical exercises that are most appropriate from their viewpoint. This means that in addition to participation, local peoples will need to have the ability to assess whether they even *wish* to participate, under what conditions, and at what scale. They will need to objectively assess potential benefits, costs, trade-offs, and come to conclusions as to whether REDD propositions are feasible for them or not. For REDD to work, contrary to current best practice through UNFCCC and voluntary processes, they will need to identify what their negotiating positions are in terms of commitments, agreements, capacity building, percentage shares of future revenue streams, etc. These skills transcend prevailing participatory methodological best practice.

These too are *not* issues that social scientists, however well intentioned, can represent in place of local peoples. These are issues that must emerge from collective analysis and consensus seeking. In shortcutting this process, REDD planners, thought-leaders, and standard setters undermine REDD feasibility.

Under present REDD guidance, this type of participatory process is *not* part of the requirements that are placed on project developers. The demands on project developers are, as we will illustrate in later chapters, more basic. While simplicity is oftentimes good, in this case simplicity does not appear to add value.

It is relatively easy to mobilize communities for large stakeholder meetings, particularly if daily allowance incentives are involved. Identifying basic needs and general issues can be accomplished in this type of meeting. REDD programs have proven adept at this to date.

Yet, the complexity within and between communities is such that attaining rapid consensus and agreement about complex management issues that involve analysis of trade-offs between economic, ecological, and social issues is unlikely to be achieved via rapid appraisal processes. For REDD to be sustainable, community engagements must reflect cost–benefit and risk analysis of economic, ecological, and social issues. This has eluded REDD planning processes to date.

How past development failure has implications for REDD

Escobar (2001) provocatively asserted that "development" is nothing more than "discourse", without real positive impact on societies beyond enabling the rich to get richer and the poor to stay poor. Polemical or not, the legacy of disappointment to failure in development programming is *impossible* to avoid in any serious consideration of the future prospects for REDD.

Empty idealism is characterized by "disjunctures", or disconnections between "abstracted models of replicable management techniques" and the specific contexts in all their disordered, at times chaotic, webs of human relationships and messy political realities (see Quarles van Ufford and Giri 2003; van den Berg and Quarles van Ufford 2004). REDD is an excellent example of this abstraction and reification process, with the problem being that the likely costs for failed idealism will be severe, and will be principally borne by many of the world's poorest peoples with least political and economic clout.

The contexts in which REDD will be implemented will, more often than not, be influenced by historical precedents of development failure or else unkept promises. REDD and REDD+ will invariably be competing against these failures and disappointments, along with wariness of all types of intervention from external actors which will include state agencies, international NGOs, transnational coalitions backed by intergovernmental and multilateral donors in partnership with business interests – whomever (see Rew and Khan 2006).

While not insurmountable, this context is part and parcel of what any feasibility analysis for REDD in a particular national and sub-national context will need to tackle, starting with analysis of the feasibility of REDD in specific landscapes in all their ethnographic richness. The inherent challenges far surpass what two- or three-day well-intentioned PRA sessions are capable of delivering in terms of community analysis, meaningful consensus and hence, FPIC.

3 REDD's path to date

This chapter is an overview of REDD's path to date in terms of (a) UN (UNFCCC and UN-REDD) level programming, (b) major partner initiatives including the World Bank's FCPF and other donor technical support, (c) big conservation and development NGO activities, (d) the private sector, and (e) opponents of REDD. Major issues as they have evolved are touched upon to provide contextual information about how REDD has gotten to where it has, and where it is going from here (see Figure 2.3). These will be further developed in later chapters.

Climate change debates and REDD as one proposed solution

Deforestation and forest degradation account for about 17 percent of current global emissions. REDD is one response from the global community to ACC. As it can address 25 percent of the anthropogenic portion of the global GHG account, I believe that REDD has emerged as one of the key areas for action on mitigating climate change by virtue of its conceptual approbation by influential thought-leaders like Stern (2006) and Eliasch (2008). According to expert proponents, without REDD+ the goal of limiting the rise in global temperatures to 2°C above pre-industrial levels will be much harder, and substantially more expensive, to achieve (Gledhill et al. 2011).

Many scientists now believe that keeping the lid on a 2°C temperature rise may no longer be feasible. The Massachusetts Institute of Technology (MIT) Integrated Global System Model that has been used to make probabilistic projections of climate change from 1861 to 2100 projects a 90 percent probability that global temperatures will rise by 3.5–7.4°C (6.3–13.3 degrees Fahrenheit) in less than one hundred years, with even greater increases over land and the poles (Sokolov et al. 2009). Yet much debate continues among scientists over expected and actual observed warming, along with allowed emissions based on improved comprehension of the carbon cycle at different time scales and assumptions used in modeling (see Knutti and Plattner 2012). Alarm grew further in 2012 when one of the world's leading climate scientists, Fatih Birol of the International Energy Agency (IEA), projected that by 2100 temperatures will rise by 6 degrees centigrade (Eilpirin 2011).

To mitigate climate change, focus has been on the Kyoto Protocol, the main policy instrument of the UNFCCC that emerged from the Rio Earth summit in 1992. While the majority of nations have signed on to the Protocol to address climate change, it has proven ineffective as the principal global policy instrument to address the challenges.

In 2013 there is little doubt that the GCC regime has been inadequate. The current centerpieces for multilateral action against climate change are the UNFCCC, its associated Kyoto Protocol, the Copenhagen Accord, and the COP 17 Durban Platform for Enhanced Action ("Durban Platform"). The Kyoto Protocol includes firm commitments to curb emissions only from developed countries, but does not include the US and has no meaningful consequences for noncompliance; it has also come under unprecedented strain as Canada officially withdrew from the accord in December 2011 (Council on Foreign Relations 2012). Progress has proven glacial.

Perhaps in response, the decision to move forward with REDD has been influenced by the mix of stalled progress and perceived urgency posed by ACC, and the consensus among 97–98 percent of the world's scientists that ACC[1] is real, significant, and really must be dealt with (Anderegg et al. 2010).

Programmatically, the watershed moment for REDD first occurred in 2005, when a group of forest nations made a proposal within the UNFCCC that negotiations for a post-2012 agreement should include incentives to mitigate emissions from deforestation. The Bali Action Plan which was agreed on at the UNFCCC conference in Bali in 2007 began to formalize the process. It included provisions on *R*educed *E*missions from *D*eforestation and forest *D*egradation (REDD). These sought to provide an economic incentive structure that would reward activities for reducing emissions from sources of deforestation and forest degradation. In the course of subsequent UNFCCC negotiations in Cancun in December 2010, REDD proposals were broadened, culminating in an agreement on supporting expanded objectives under REDD+. As well as including activities which reduce emissions from deforestation and forest degradation, REDD+ now explicitly includes the conservation of forest carbon stocks, sustainable management of forests and the enhancement of forest carbon stocks (Gledhill et al. 2011).

Through the UNFCCC COP in Durban in 2011, and Rio+20 in 2012, international negotiators have failed to reach consensus.[2] While REDD and REDD+ are, in 2013, targeted for broad roll-out over the next few years, the final shape of REDD programming remains to be adequately defined.

In this regard, REDD "is", as it has been evolving conceptually and programmatically for a decade. Yet REDD "is not", in that lack of consensus over fundamental issues constrains scaling up from the pilot, Readiness Phase where these issues were to have been *convincingly* resolved.

Despite the fact that international negotiations leading to REDD compliance mechanisms have yet to be finalized, the progressive uptake in volumes and total values transacted in voluntary forest carbon markets from 2008–2010 has progressed (Diaz et al. 2011; Peters-Stanley et al. 2011), creating grounds for certain optimism, and explains why many voluntary market proponents feel that voluntary initiatives represent the vanguard for REDD.

With the absence of international consensus over the scope of UNFCCC-driven programming in REDD, an increasingly piqued critique of civil society groups from around the globe about REDD's implications has emerged.[3] So too, developing synergies between big conservation NGOs, financial institutions, philanthropists, and carbon offset design and trading intermediaries are leading to unexpected alliances between heretofore disparate stakeholders and interest groups.

Slowly, but perhaps surely, momentum is being generated for the emergence of a functioning, voluntary REDD offset marketplace that may well prove sustainable, as stakeholders await decisions on a compliance based marketplace. This voluntary market will certainly involve a corporate social responsibility (CSR) and philanthropic core, though Robinson's (2012) remarks for the implications for BINGOs of CSR programming to conserve biodiversity should be taken seriously for REDD as well.

REDD as a leading mitigation approach

In 2013, REDD represents one of the most prominent global efforts to combat ACC. While REDD alone cannot come close to fully mitigating ACC, *if* successful, it could serve as a key placeholder for what may turn out to be a palette of policy reforms requiring global coordination. For this reason alone, REDD is important.

Despite the overwhelming scientific consensus on ACC, a resistant, vocal, and politically strong group of citizens based primarily in the US, and backed by a small cohort of vocal scientist-deniers,[4] remains influential over *the absence* of serious discussion in the US over climate change policy, let alone enactment of policy itself.

Whether blind to the indicators of ACC, or denying human agency in climate change, by suggesting that climate change either reflects natural cycles, or even political hoax, can account for seemingly extreme climate events. That is: it is not as bad out there as Climategate conspirators would have the public believe.

This denial enables "policy makers, climate contrarians and skeptics, and those simply not paying attention to either actively deny it or to just look the other way, committing the planet to more and more change" (Gleick 2012). This denial has had, undeniably, a strong impact on the emergence and evolution of REDD, as a seemingly "low" cost program (politically and financially) which ostensibly generates win–win outcomes in both developed and developing countries.

As context, the US released 5,424,530 tonnes of CO_2 equivalent in 2009 (EIA 2009), which in 2012 was 26 percent above 1990 levels. Global emissions of carbon dioxide (CO_2) – the main cause of global warming – increased by 3 percent in 2011, reaching an all-time high of 34 billion tonnes in 2011, with China's share at 29 percent, followed by the United States (16 percent), the European Union (EU27) (11 percent), India (6 percent) the Russian Federation (5 percent), and Japan (4 percent) (Olivier et al. 2012). Despite its economic slowdown, the data indicates that US climate policy will be pivotal for GHG mitigation for years to come. The White House announced on November 25, 2009, that President Obama was offering a US target for reducing GHG emissions in the range of 17 percent below 2005 levels by 2020. How this target will play out in the post-2012 election aftermath is anyone's guess.

By 2010, optimism for addressing climate change was fully eviscerated by the prevailing political sentiment in the US. It was taken off the table for the foreseeable future by US President Obama who stated that "cap and trade was just one way of skinning the cat", that it is "a means not an end", and that he was "going to be looking for other means to address this problem" (Obama 2010).

In the aftermath of devastating superstorm Sandy on the northeastern portion of the US in late October 2012, it was clear that President Obama had yet to figure out how to skin the climate change cat. In a cnn.com lead story on October 31, 2012, professor of geosciences Michael Oppenheimer concluded that "climate change will probably increase storm intensity and size simultaneously, resulting in an intensification of storm surges". This article stood in contrast to the silence of climate change over three presidential debates in the month of September prior to Sandy. Despite the fact that 67 percent of Americans believe the earth is warming, political movement on the issue is frozen, with no thaw in sight despite the increasing frequency of so-called once in a lifetime climate events.

For climate change mitigation policies to work, political, economic, and technical strategies need to be well tailored to the fickle views of respective citizenries that remain steadfast in their need for either maintaining standards of material and social welfare in the case of the US and much of the EU, or achieving "acceptable" standards of living levels in the case of Brazil, Russia, India, and China (the BRIC countries). For the lattter, this is an entitlement, just as it was for Western (now "Northern") countries. With the aspirations of billions of people in play, any capping of development in heavily coal dependent countries with a growing fleet of vehicles is challenging. This applies not only to India and China, but to the US as well, insofar as coal production remains important in twenty-five states, with two, West Virginia and Pennsylvania, key "swing" states in national elections.[5]

As backdrop to the sense of urgency among thought-leaders in the global community, United Nations Environment Programme (UNEP) Executive Director Achim Steiner notes that the risks of unsustainable resource use are formidable: "[i]f current trends continue, if current patterns of production and consumption of natural resources prevail and cannot be reversed and 'decoupled,' then governments will preside over unprecedented levels of damage and degradation" (UNEP 2012).

This makes the logic for initiatives like REDD undeniable.[6] Yet there remains a troubling anomaly in REDD. As REDD is shooting for "offsets" – e.g. net zero reductions, net zero increases – it merely enables us to continue treading water, without cutting into the disturbing trends Steiner notes. For many, and considering its likely social costs, this renders REDD unviable.

What is REDD+?

In their compendium on REDD+, Angelsen et al. (2009) suggest that REDD+ is an umbrella term for local, national, and global actions that reduce emissions from deforestation (i.e. reducing forest area) and forest degradation (reducing carbon density), and enhance forest carbon stocks in developing countries (REDD+). It is perceived differently by different stakeholders which, arguably, represents part of the challenges it continues to face.

The "plus" (+) sign indicates *enhancement of forest carbon stock*, also referred to as *forest regeneration and rehabilitation, negative degradation, negative emissions, carbon uptake, carbon removal* or just *removals*, where "removals" refer to sequestration of carbon from the atmosphere and storage in forest carbon pools so as to increase the amount of carbon per hectare.

Yet as Angelsen et al. (2009) note, REDD+ is not only about climate change. Other goals, known as "co-benefits" (i.e. benefits in addition to reduced climate change) are also vitally important.

There are at least four types of co-benefits to consider. First, forest conservation, in addition to storing carbon, provides other environmental services, such as preserving biodiversity. Second, REDD+ actions (e.g. financial flows) and forest conservation might have socioeconomic benefits, such as reducing poverty, supporting livelihoods and stimulating economic development. Third, REDD+ actions may spark political change toward better governance, less corruption, and more respect for the rights of vulnerable groups. Fourth, REDD+ actions and forest conservation could boost the capacity of both forests and humans to adapt to climate change.

With so many simultaneous objectives on a biophysical level alone, REDD+ epitomizes the quest for "triple win" outcomes. Here, the role of markets, and the creation of carbon assets that can become securities for international trading, have become a lightning rod around which most attention in REDD debates has been focused. For many, this fourth element represents a "win" for private capital that some NGOs, community activists, academics, and practitioners find difficult to accept given the nature of the public goods in question (Karsenty et al. 2012a).

If we substitute the pursuit of avoided deforestation (AD) for conservation of biodiversity, a reasonable substitution given that the biggest conservation organizations in the world are committing significant resources in the pursuit of both objectives simultaneously, then REDD+ represents the closest thing that *attempts* to address Peter Kareiva's hypothesis "that conservation will achieve greater conservation gains if it supports plans and actions that optimize meeting multiple objectives (food security, energy, jobs, etc.) as opposed to single mindedly pursuing only conservation goals" (Kareiva 2012; quoted in Revkin 2012).

Sunderlin and Atmadja (2009) noted several years ago that REDD+ has generated interest as "a ground-breaking concept for saving tropical forests", with proponents believing that REDD+ funds will be an incentive to keep forests standing and to restore and perhaps even establish new forests. While for those less in favor of the idea, REDD+ repeats the same old story about throwing lots of cash at forests as a be-all and end-all solution to deforestation and degradation.

The controversy over REDD+ heated up further in late 2012. When prices in the principal market for carbon in Europe, the EU ETS, continued sinking due to a combination of oversupply of credits, unanticipated reductions in GHG emissions, unexpected increase in available green technologies for power generation, continued weak global economics and recession, unanticipated winter warming leading to reduced energy demands (Clark and Blas 2012; Business Green 2012). With successive years of once in a generation hurricanes battering Northeastern America (Irene and Sandy), REDD+ was one of the few policy tools which an important set of stakeholders remained moderately bullish about.

While the REDD agenda of AD is certainly ambitious, the addition of sustainable management of forests, conservation of forest carbon stocks, and enhancement of carbon stocks not only increases the scope of REDD tremendously, but also increases the complexity of the challenge. As Newell and Patterson (2010) put it about actions to reduce climate change, "we have the first instance of societies collectively seeking a dramatic transformation of the entire global economy". And given the lack of momentum generated for the green economy in the aftermath of Rio+20 in 2012, this transformation will likely occur in fits and starts, as opposed to one fell swoop.

Procedurally, nations wishing to take part in REDD+ must start by preparing to enter the mechanism (Phase 1, the so-called "preparatory" phase), notably by identifying causes of deforestation, establishing a carbon inventory, and explaining how they intend to model their future deforestation. Phase 1 with "fast-track" funding, comes from the FCPF managed by the World Bank and from the UN-REDD program, co-managed by the Food and Agriculture Organization of the United Nations (FAO), UNDP, and UNEP.

After this preparatory phase, nations can enter phase 2, which involves implementation of REDD+ policies and measures, again with dedicated REDD+ funding such as the FIP or through bilateral agreements. Lastly, phase 3 is the full-scale implementation phase, which no nation had reached by late in 2012, where efforts could be supported by dedicated forest carbon funds or possibly through access to the UN backed carbon market. As of 2013, however, negotiations on MRV for verifying emission reductions were still ongoing.[7]

REDD+ and the green economy

For UN-REDD, one of the pre-eminent drivers of REDD+ programming globally during the preparatory "Readiness Phase" of activities, the conceptual framework for where REDD+ fits in to environment and development concerns, is part and parcel of the green economy.[8]

REDD+ is portrayed as appearing to represent a wing of a transnational green economy centered on natural capital. Meanwhile, a significant portion of natural capital assets are found on lands of peoples with customary tenure claims. While these peoples have ironically been marginalized by major trends of the global economy up until now, the assumption is that these assets over which local peoples have de facto use rights and customary claims will become part of the core of the new green economy.

While one of the few points of conceptual clarity to come out of Rio+20 was the global commitment to the green economy, of which the commodification and trade in natural capital assets is pivotal, it is likely that this will come in tandem with mounting resistance from farmers movements, many smaller and national NGOs, academicians, and others who fundamentally oppose this.[9] While not opposed in principle to REDD+, COONAPIP in Honduras withdrew from the UN-REDD driven process in March 2013 due to perceived weak consultation and respect of FPIC principles (see Lang 2013).

Aguilar et al. (2012) in a harsh critique of the REDD+ program in El Salvador note that the REDD+ mechanism has stagnated and it is weakening under the UNFCCC multilateral process, both as a result of methodological shortcomings and the lack of funding for implementation. The latter has been exacerbated by the collapse of carbon prices (Clark and Blas 2012) and the reduction of trading volume in carbon markets, due to the absence of ambitious ER targets.

Why REDD is so politically expedient

It is reasonable to hypothesize that for the US, as well as the UK, Norway, and other Northern European countries along with Australia and Japan, REDD has become a politically expedient option. At present neither the US, India, or China can politically tolerate compliance cap-and-trade programs to reduce global emissions, let alone "nothing less than a reorganisation of society and technology that will leave most remaining fossil fuels safely underground" (Lohmann 2006), as the most aggressive activists of climate change propose.

While the EU Emissions Trading System (EU ETS) is committed to mitigating climate change through the auctioning of permits combined with free allocation of credits for CO_2, nitrous oxide, and petrofluorocarbons,[10] the reorganization that Lohmann (2006) suggests is totally off limits for the foreseeable future.

In comparison to ambitious societal reorganization, or even cap and trade, where the perceived externalities on business and jobs in a time of economic recession are seen as overly high by American voters, REDD is a program that average US citizens remain either unaware of or indifferent to, as any externalities from REDD will principally be felt in distant developing countries.[11]

Financial contributions from the US government to REDD, meanwhile, should they reach the hundreds of millions of dollars, would represent a fraction of what the type of "reorganisation of society and technology" needed as Lohmann (2006) suggests to keep fossil fuel in the ground and cut into, not just offset, emissions.

Hyperbole around payments under REDD

Some refer to PES, of which REDD is a distinct subset, as an "appropriation of nature" (Fairhead et al. 2012), that poses potential risks to local land rights and livelihoods (Larson et al. 2011). Others suggest an emerging crisis of governance within REDD+ as representing a fundamental weakness that will compromise future project and policy goals, along with the well-being of many REDD stakeholders (Thompson et al. 2011).

Concern is also voiced that should significant payments for provision of ecosystem services be made for forested land through REDD, this is likely to raise the value of forested land as an asset class, leading to speculation and implications for reduced food security among the poor (HLPE 2011).

Fueling the controversy beyond climate change issues, *incredible* figures have been alluded to for the potential economic value of ecosystem goods and services. Hyperbole aside, stories about the monies that REDD will in theory bring to rural peoples have created a tizzy in remote forests of the Congo, PNG, and elsewhere. When extrapolated across REDD-eligible countries, a lot of international buzz about REDD as a vehicle not just for poverty alleviation, but for establishing prosperity, have been in the works over the past seven years.

As noted, respected sources have publicly projected that ecosystem values and services could be pegged annually as high as US$4.5–5 trillion (see Sukhdev 2012). This has translated into information in orders of magnitude in the billions of dollars, reaching into the remote hinterlands of forested areas of developing countries, creating great expectations, confusion, and controversy in the process.

As the earlier anecdote from Josephstaal, PNG circa 1992, suggests poor rural peoples are often prepared to sell the clothes off their backs for little to nothing. If they can receive billions, or millions. or thousands. *or even hundreds* of US dollars for simply not cutting trees, of course on the surface they will in most cases accept on the spot. Whether this constitutes informed consent is another issue.

In 2013, despite the equivocation of carbon markets, REDD+ continues to be the principal mechanism now being promoted to capitalize upon this potential value of ecosystem services to be used to mitigate negative effects of ACC. Meanwhile, the valuation of ecosystem goods and services remains inexact.[12] If just a fraction of Sukhdev's (2012) claims for the value of environmental services were to be promoted through policies and programs being shaped through REDD, the expectations, hackles, and concerns among remote forest peoples, anti-REDD activist circles, as well as middle of the road development practitioners that have been raised to a pitch, may begin to be assuaged. This has yet to materialize.

Arguments pro and con for investing in climate change mitigation

While scientific consensus that human activity is the principal cause of climate change is broadly accepted, it is unclear whether REDD is an appropriate or feasible policy mechanism for addressing ACC. In recent years, a vibrant debate has been engaged among economists and policy wonks over the evidence and logic for climate change and its mitigation. Many of the elements of the debate are useful in considering whether a rational basis has been established for REDD as a mitigation strategy.

One group that firmly believes so is the Union of Concerned Scientists. In 2008 when great optimism prevailed, the Union argued that REDD offers a clear solution, stating that "REDD is clearly an inexpensive approach compared with emissions reductions in the energy sectors of industrialized countries" (Boucher 2008). It projected that costs per ton of reducing current CO_2 emissions from deforestation by half, under pessimistic scenarios with only opportunity costs, implementation, transaction, administration, and stabilization costs considered, would represent less than a third of the then current (mid-2008) capped carbon market prices. It argued that conservative estimates showed that US$5 billion in funding annually could reduce deforestation emissions in the year 2020 by over 20 percent; that US$20 billion could reduce them by 50 percent; and US$50 billion could result in a drop of 66 percent. As of late 2012 this analysis from 2008 was still on the Union of Concerned Scientists' website, leading to the conclusion that this remains its position.

It is amazing to consider this optimism over REDD prospects and its potential impact prevailed five years ago in 2008. Times indeed have changed, with many country commitments for funding REDD still pending follow-through.

In the popular press and blogosphere, and on the critical side of mitigation strategies for ACC, Manzi (2008, 2010a, 2010b) has argued that uncertainty must be central to the decision logic in addressing what to do, if anything, in mitigating climate change. Manzi contends that cost–benefit analysis (CBA) must, in *some* manner, be used to inform decision making. His viewpoint is counter to unconditional proponents of the "precautionary principle".[13] To Manzi, the potential for unlimited mitigation investment given less than 1 percent risk as implied by the "fat-tailed uncertainty" posed by worst-case ACC, or "Catastrophic Climate Change"(CCC) (Weizman 2007, 2011; Krugman 2010) is unjustifiable, as the potential mitigation costs may well exceed actual economic losses incurred from climate change at a 2°C rise in temperature.[14]

On the other side, Weizman and Krugman argue that given "how rapidly probabilities are declining and how rapidly damages are increasing that far exceed the worst case probability distributions of the principal scientific reference on climate change science (Weizman 2011), action must be taken." In this context, the argument is that adherence to standard cost–benefit analysis principles are thoroughly misguided, as they take little account of the magnitude of the uncertainties involved in extrapolating future climate change so far beyond past experience (Weizman 2011).[15] Thus, the implications of the worst case rationalize urgent action.

Manzi, on the other hand, argued in 2008 that the foremost authority in the world climate modeling community, the IPCC,[16] projected that even in the most extreme ACC scenarios, we should expect warming of about 4°C over roughly the next century. Notwithstanding Birol's recent upgrade of worst case estimates to a 6°C rise in temperature by 2100, this earlier estimate "should cause the world to have about 1 to 5 percent lower economic output than it would otherwise have", leading to a twenty-second century world that is "about 3 percent poorer than it otherwise would be (though still much richer per capita than today)" (Manzi 2008). These are costs that, to Manzi, do not necessarily justify mitigation investments that could dwarf them by several orders of magnitude.

Given the IPCC projections, Manzi concludes that there really is

> massive uncertainty (rather than mere risk) in our ability to predict the impacts of anthropogenic global warming. Recognizing this should lead to two actions: (1) improve the science to better-specify these extreme risks, and (2) hedge this uncertainty by making "insurance-type" investments today that would provide protection if an extreme AGW scenario ends up happening.
>
> (Manzi 2008)

This argument between Manzi, and Weizman and Krugman loops back to Al Gore's influence on climate debates and the push for cap and trade through the best seller and film *An Inconvenient Truth*. It is relevant to REDD not because of who is right or wrong; in fact a "reasonable person" could accept buying into either side of the argument, as judgment relies on values, principles, and how one weighs data and unknowns.

The issues frame the relevance of REDD as a policy response to climate change. First, they provide the overarching context for appreciating REDD as a mitigation response and second, because REDD represents, in microcosm, a *far* more moderate, empirical case for approaching ACC than the broader issues that Weizman, Krugman, and Gore on one side, and Manzi on the other, have been parrying.

For example, would Weizman, Krugman, and Gore support REDD as a mitigation response for the 17 percent of carbon emissions that deforestation is responsible for?[17] Would it be relevant enough? Would Manzi buy into REDD as one, albeit relatively minor "insurance investment", in terms of scale of impact that a CBA could support, given what we currently know and conjecture about costs? Oddly, this line of questioning is not one that has been at all pursued in the REDD literature.

While ACC/AGW is clearly occurring, what is less clear is whether we have evidence that is *remotely* sufficient to justify our belief that REDD is a credible response in a full cost accounting scenario. Therefore, two questions appear reasonable to pose:

- Should we not seek to understand the probability of success, failure, or risk of any single REDD project before embarking on it?
- Should we not seek to understand the same for REDD programming overall?

To work, REDD requires an aggregate of successful actions at multiple levels – local, sub-national, national, international. Succeeding at each of these levels will be difficult.

At local levels, the ability to aggregate actions on a sustainable basis requires a stretch of great faith that is not supported by empirical evidence from development and conservation experiences in remote regions. As Rayner et al. (2010) has made the point for forest sector issues broadly, whether the focus is climate change mitigation, human development, biodiversity conservation or trade, a more effective approach to coordination is needed for improvements to forest conditions and livelihoods to be achieved. REDD has not demonstrated the capacity for such coordination.

Acceptance of the precautionary principle to move forward with REDD on the basis of unacceptable climate change impacts is also problematic. If there are no grounds for confidence that REDD can be made to work to begin with, it is not good enough to justify moving forward because of the climate change imperative.

Furthermore, even if measurable outputs could be secured in the short term, the question of long-term sustainably is problematic. The solutions that are proposed may all be worthy of *piloting*, but can only be so if a rigorous monitoring and evaluation program exists for assessing their value prior to rolling out on a broad scale. At present this type of evaluation and learning appears more rhetorical than real. Meanwhile, many institutional interfaces and mechanisms that could be piloted in the Readiness Phase that a new social contract would necessitate are not being piloted. These are activities that require piloting and subsequent learning of lessons. Absent this, broad buy-in will be hampered.

Finally, there has not been any serious reflection about the high probability for risk of failure in REDD overall. Whether this equates to a 50 percent, 75 percent, or 99 percent likelihood is impossible for anyone to assign.

Analysis of whether the component pieces to success in REDD are plausible and worthy of support could be formally analyzed. Under the current theory of change framework that is applied, these experiences are loosely tracked and adaptively managed. There is no indication from the reading of early PDDs that have been CCBA approved, for example, that contingencies for *abandoning* projects because of thresholds not being met are part of a broader decision-making framework. At present, the theory of change approach presupposes a twenty-five-year time frame, with no indication that projects can be shut down because of invalid assumptions or analysis that certain social welfare costs for example are proving too high to bear, or that unforeseen risks pose unacceptably high threats to justify further implementation.

The REDD community is presently going down a pathway where, at best, it appears like the visually impaired are leading the blind. REDD programs and specific projects are not looking at what needs to be observed, and properly seen, to avoid critical mistakes. Oddly, a critical mass of informed people who understand the high probability for failure at this level are yet to stand up and demand a more reasoned approach.[18] This positions forest peoples to bear undue externalities.

Deforestation drivers

Deforestation is a problem with many causes. How well REDD is addressing this is highly debatable.

Tropical deforestation claimed roughly 13 million hectares of forest per year from 2000–2005, with a shift from poverty-driven to industry-driven deforestation, and geographic consolidation of where deforestation occurs (Butler 2009a). Recent scientific studies show that "large, commercial agriculture and timber enterprises are the principal agents of tropical deforestation, which is responsible for about 15 percent of global warming pollution worldwide "(Union of Concerned Scientists 2012).

Some analysts suggest that this means that poor people are less responsible than industry or corruption for deforestation (Koyunen and Yilmaz 2009). This would seem to be consistent with Dove's earlier point (1993) that forest peoples are poor because of the degradation caused by others. By this logic, this should mean that most policy levers should be directed towards industry, or corrupt public administrative practice, as a solution in mitigation strategies. But this does not appear to necessarily be the way that it is playing out in REDD at the project level.

Drivers refer to both root and proximate causes of deforestation and forest degradation. Considerable attention has been paid to drivers of deforestation and forest degradation, along with potentially applicable policy measures (see Kanninen et al. 2007; Chomitz et al. 2007). Identifying deforestation drivers are key to designing appropriate mitigation measures.

Some analysts argue (Rudel et al. 2005; Butler and Laurance 2008; mongabay. com 2012) that the fundamental drivers of tropical forest destruction have changed in recent years. Prior to the late 1980s, deforestation was generally caused by rapid human population growth in developing nations, in concert with government policies for rural development. These included agricultural loans, tax incentives, and road construction. Such initiatives, especially evident in countries such as Brazil and Indonesia, promoted large influxes of colonists into frontier areas and often caused dramatic forest loss.

More recently, however, some suggest that the impacts of rural peoples on tropical forests seem to be stabilizing (Wright and Muller-Landau 2006). Although many tropical nations still have considerable population growth, strong urbanization trends (except in Sub-Saharan Africa) mean that rural populations are growing more slowly, and are even declining in some areas, so that the impact of population pressures locally is highly variable.

Yet, the popularity of large-scale frontier colonization programs has apparently begun to wane in countries like Brazil, though recent policy changes legislated there may lead to setbacks in deforestation indicators in years to come (*The Economist* 2012). Some recent research (Kissinger et al. 2012) shows that while 80 percent of deforestation is now estimated to be directly driven by agriculture, there is tremendous regional and local variability as to causes of deforestation. Koyunen and Yilmaz (2009) suggest that corruption is more strongly correlated to deforestation than rural population growth.

In Latin America, commercial agriculture is the main direct driver, responsible for two-thirds of all cut forests, while in Africa and tropical Asia commercial agriculture and subsistence agriculture both account for one-third of deforestation, a significant regional difference. Mining, infrastructure, and urban expansion are important, but less prominent drivers worldwide. Clearly, it would be useful to determine to what extent corruption would factor into commercial extractive activities' impact on deforestation.

Kissinger et al. (2012) conclude that economic growth based on the export of primary commodities and an increasing demand for timber and agricultural products in a globalizing economy are critical indirect drivers. They also find that the decrease in forest quality through degradation is prevalent in over 70 percent of cases in Latin America and tropical and sub-tropical Asia. This is due to commercial timber extraction and logging activities, while in Africa, fuel wood collection, charcoal production, and, to a lesser extent, livestock grazing in forests are the main degradation drivers.

Despite the increased role that commercial activities and corruption play in deforestation and degradation, it appears from REDD policy and public relations materials that most attention in ODA programming and actual projects is directed to the role that local resource use plays in deforestation and forest degradation. This framing begs an obvious question: Is program and project design most efficient based on actual deforestation and degradation drivers?

Fairhead and Leach (2000) wrote on earlier biases against local peoples that have driven policies to mitigate deforestation. While long before the arrival of REDD, these earlier assumptions about unsustainable farmer practices appear to also have energized how REDD has been approached as a problem, as the conventional wisdom perceiving local peoples as threats has not been disabused.

In their analysis of deforestation in West Africa, Fairhead and Leach noted (2000) that claims of one-way deforestation had completely obscured what seemed to have in fact been a large increase in the area of the forest zone in recent centuries, leading to exaggerated estimates of deforestation at scale that, in turn, misled regional and global climatic modeling. And by exaggerating estimates of deforestation, in effect seeing farmers as the proximate cause of deforestation, it is "impossible to appreciate how farmers may have actually been enriching and managing their landscapes in sustainable ways". In turn, this obscures the historical experience of inhabitants and the origins of their claims to land, while justifying a cascade of corrective actions that may be inappropriately targeted.

For example, large swaths of common pool resources that are customarily owned are also claimed as state lands. Under the politically correct banner of GCC mitigation through REDD, these lands may be grabbed by the state or sub-national entities in regions or other jurisdictions.

Some communities argue (Comunidades de la Región Amador Hernández, Chiapas, México et al. 2012) that this is what has happened in Chiapas, Mexico, where the state government has negotiated a carbon offset agreement with the state of California under the latter's AB 32 legislation. This has received support from the Governors' Forest Conference (GFC), which in turn is supported by the Gordon and Betty Moore Foundation and other foundation partners, along with CI and Starbucks. While California's concerns for ACC mitigation are legitimate, these Mexican communities reject both their role and the overall approach.

This situation is consistent with Fairhead et al.'s point (2012) that "[i]t is the prerogative of First World conservationists (backed up by the power of the state), to determine whether land uses are compatible with their interests or suitable for the purposes of the buffer zones". That is, "First World conservationists" (BINGOs) together with the "power of the state" decides appropriateness and feasibility of land uses and management.

International NGOs *do* consult with communities as part of best practice. Yet, there are no standards whatsoever for consultation. The locus of decision-making power in the vast majority of cases resides with the state and international advisors, and partners from development cooperation agencies and BINGOs. Community perspectives are factored in as NGOs; the state, and foreign technical assistance agents determine what is most appropriate in local settings. As IPs and communities have begun to claim that displacements have occurred to accommodate REDD projects (Beymer-Farris and Bassett 2012), it is clearly imperative that macro level planning that justifies community displacement be transparent and credible.

Yet, if much of the understandings of local peoples' roles in deforestation in REDD-eligible contexts like West Africa are based on fundamental misunderstandings about land use, as Fairhead and Leach's (2000) analysis suggests, then what of the proposed solutions proffered through REDD policies that are based on similar assumptions? On another level, while it is clear that the interactions between changes in agricultural technologies and deforestation are fundamental to making REDD+ effective in the long term, these have not yet been given due consideration in national strategies (Pirard and Belna 2012).

As drivers continue to be incompletely understood and dubiously prioritized, appropriate solutions will remain elusive. In REDD, policy makers are preparing to catapult from the Readiness pilot phase into full implementation, *as if* solid enough understandings on deforestation drivers and adequate solutions are in hand, when they are not. Nor will they be much better developed even in two years' time.

International demand and deforestation drivers

From an investment standpoint, would attacking the 15 percent contribution from large commercial agriculture be policy dollars better spent than in focusing on small farmer contributions to deforestation? If these may either be exaggerated or else be infeasible to transform under the framework of options that REDD is currently promoting, a rethinking of strategic emphasis may well be in order.

One example of where this tack may be evolving is in Indonesia. The government has been put into the limelight over how agribusiness and forestry are principal drivers of deforestation, with the setback to agribusiness peatland conversion representing a victory for community and NGO advocates (REDD Monitor 2012). While the situation remains fluid in 2013, this may signal the onset of an important shift in policy and its implementation in Indonesia, and could serve as a model for how agribusiness-deforestation drivers could be approached elsewhere.

At another level, the absolute rise in population globally, together with the demand for resources – forest products, water, etc. – places serious pressures on tropical ecosystems, and is rising. This has tremendous impact on forests that absorb direct pressures from local resource use. The dynamic as to how this plays out in any particular forest is complex, and not at all well understood. It has also contributed to the urgency that REDD has exhibited on a policy level – e.g. the assumption that if policies can be put in place, then deforestation drivers can be better contained. The assumption is that best practice combined with new and improved incentive structures that better value forest resources, will induce AD and reduce forest degradation. This theory of change has yet to be tested or proved during the Readiness Phase.

On the international level, globalized financial markets and a worldwide commodity boom have in recent years created a highly attractive environment for the private sector. Under these conditions, large-scale agriculture – crops, livestock, and tree plantations – by corporations and wealthy landowners is increasingly emerging as the biggest direct cause of tropical deforestation (Butler and Laurance 2008). A surging demand for grains and edible oils on the consumption side has run up against an increasing demand for biofuels.

Some suggest that biofuel use could increase carbon emissions by increasing destruction of forests when displaced local farmers clear land. This carbon released from deforestation is linked to biofuels, and was projected to exceed carbon savings by 35 percent in 2011, rising to 60 percent in 2018 (Bowyer 2010).

Biofuel production is also linked to rising food prices and pressure on land. In effect, the high demand for biofuels is giving rise to harmful land investments (Bailey 2008; Zagema 2011), raising questions about the strategy proposed by Killeen et al. (2011) in their report for CI, for example. According to the Oxfam report on land grabbing (Zagema 2011), governments seem to have aligned themselves with investors in biofuels, welcoming them with low land prices and other incentives, and even helping to clear people from the land. From a biodiversity standpoint as well, concerns have been raised (Danielsen et al. 2008). Will the same be replicated under REDD for agricultural as well as forest carbon?

Triple win scenarios to reduce carbon emissions, conserve biodiversity, and promote economic development in developing countries through "sustainable biofuel" plantations of oil palm, jatropha, and eucalyptus are now under discussion. This involves explicitly linking cultivation of biofuel feedstocks with forest conservation and reforestation, and thus be creditworthy under REDD (Killeen et al. 2011), and is at the center of controversy over the definition of forest in REDD.

At present, monoculture plantations are REDD+ creditworthy under the UN definition of forests. Many activist NGOs are vehemently opposed to this, as this policy it is argued will lead to the cutting of primary forest through perverse AD "incentives". If agribusiness can sell primary forest, cut timber, and get REDD credits for sequestering carbon under biofuel or palm oil plantations, and at the same time disrupt local livelihoods regardless of any jobs created on plantations, this may represent more of a triple loss than gain.

Species extinction, protected areas, and REDD

The Millennium Ecosystem Assessment (MEA) (2005) documented how humans are continuing to fundamentally change Earth's diversity of life to meet growing demands for food, fresh water, timber, fiber, and fuel. More than 1.1 billion people live in the world's twenty-five biodiversity hotspots (Myers et al. 2002), and more than 45 percent of the 100,000 protected areas developed for biodiversity conservation have in excess of 30 percent of their actual land area under crop cultivation (Scherr and McNeely 2008) to meet food security needs.

The current species extinction rate is 1,000 times greater than the fossil record, and is projected as ten times less than the rate that may exist in fifty years (Millennium Ecosystem Assessment 2005). This is at a time when conservation dollars are increasingly limited (Groves et al. 2002). Meanwhile, human population will grow from six to eight billion by 2030, to potentially twelve billion by 2080 (Scherr and McNeely 2008).

At a global level, sixteen of the twenty-five biodiversity hotspots identified as conservation priorities by CI are found in areas where a fifth of the population is poor and malnourished (Barry and Taylor 2008). Currently, the average protected area coverage in those hotspots is 10 percent of the total 2.25 million km^2, with the figure dropping to 5 percent of 2.25 million km^2 if only IUCN management categories I–IV are considered (CI 2010). While good demographic data within that range of 112,478–225,000 km^2 are not available, coarse extrapolation from Myer et al.'s (2002) estimates of 1.1 billion people living in hotspots, or Niezen's (2003) suggestion of 300 million people occupying so-called "wild places", conservatively suggests that tens of millions of people are being impacted through PA programming globally.

The overlap between conservation and REDD programming is therefore of serious financial and practical interest to conservationists. It is also of life and death relevance to local communities and IPs.

While protecting and managing biodiversity well by 2020 "in the world's most outstanding places" is WWF's first proposed outcome of its conservation strategy (WWF 2013), this implies that well managed PAs will figure squarely in achieving their conservation management objective in the world's outstanding places. The stakes for the tens of millions of people impacted by PAs will thus only heighten over time, as both PA programming and REDD+ will be framing the basis for livelihood sustainability. Yet, as Kareiva et al. (2012) argue, the arsenal of PA strategies has failed to mitigate biodiversity loss, suggesting that this relates to conservationists' failure to adequately integrate economic development into its TMAs. While earlier conservationists' analysis argued that the overemphasis on monitoring and eco-regional analysis at small scales has come at the expense of strategic allocation of scarce conservation resources (Cleary 2005), this quest for more scientific rigor has not occurred in reference to social dimensions in conservation.

Regarding governance, about half of the 20,000 state PAs created in the last forty years overlap in complex arrangements with "indigenous" customary territories (Molnar et al. 2004), and do not account for local or "tribal" peoples. Accounting for them only increases this ratio. Anywhere between one and sixteen million people in Africa alone could become environmental refugees from protected areas *if* existing legislation were strictly applied (Geisler and de Sousa 2001), while in India up to four million people face potential eviction following amendments to PAs (Kothari 2004).

Given this trend, it is a legitimate question to ask conservation organizations how they intend to support *implementation* of the human rights conventions and policies promoting mitigation from displacement as pertains firstly to the Convention on Biological Diversity (CBD), and now to REDD. To date, REDD has not addressed the issue of the balance between human rights and global imperative to mitigate climate change, systematically or convincingly.

Poverty and REDD

When the poverty line is set at US$1.25/day, 25 percent of the developing world's population in 2005 is poor, with an extra 400 million people living in poverty compared to 1993 (Chen and Ravallion 2008). The number of poor has almost doubled in Sub-Saharan Africa over the past decade, from 214 million to over 390 million, with the share of the world's poor by this measure living in Africa rising from 11 percent in 1981 to 28 percent in 2005 (ibid.). Given the recent expansion of PAs in Africa to over 10 percent of the terrestrial surface of many countries (e.g. Rwanda, Madagascar, and others where the figure is closer to 20 percent), it is also legitimate to wonder whether this trend will alleviate or exacerbate poverty. This is because difficult socioeconomic conditions still prevail in remote areas where PAs predominate. Overall, the relationship between biodiversity as a mechanism for poverty alleviation remains ill understood (see Leisher et al. 2010).

Here Hockley and Andriamarovololona (2007) in their analysis of biodiversity conservation policy, ambiguous tenure rights, and Malagasy poverty in rural areas is instructive. As projected carbon pricing remains modest in REDD (from under US$2 to US$4/ton) it is unlikely that, after all institutional and transaction costs are accounted for, poverty alleviation will be *at all* dependable and transformative through REDD. Rather, if well implemented, REDD could be a modest supplement for rural livelihoods, though the full opportunity and transaction costs to local peoples merit far more serious scrutiny than has been the case to date. And where breakdowns in governance occur as has been the case in Masoala National Park in Madagascar since 2009 (see Allnutt et al. 2013), it is unclear if REDD can possibly compete with less savory incentives.

On another level, while the World Bank Institute for example has devoted a 250-page manual to REDD opportunity costs (2011), none of it considers the opportunity costs to society and culture from communities engaging, or not, in REDD programs. Clearly, local decision making does not occur in a vacuum.

Legal questions pertaining to ownership of carbon or other PES credits and entitlements very much remain to be resolved. Vhugen et al. (2012) state:

> This new and poorly-defined commodity [carbon] raises questions about who holds the legal rights to the benefits associated with REDD+ activities, or in other words, as defined in this paper, who holds the "carbon rights"... If those who predict that REDD+ will result in the transfer of substantial funds from developed to developing countries for forest management are

correct, then there is an urgent need to revise or find ways to adequately interpret national laws and policies in time to take advantage of the REDD+ mechanism that is ultimately adopted and to ensure that populations living in and near forests have the opportunity to fully and effectively participate in, and benefit from, REDD+.

The jury is thus out on whether the marketplace for REDD offset credits will contribute to poverty alleviation or not (Angelsen 2008). Very much will depend on how projects are designed, how safeguards are factored in, as well as what level of accountability the purchasers of carbon assets will demand in the marketplace for how far poverty alleviation needs to be demonstrated (see Angelsen et al. 2012).

What learning is being generated in the Readiness Phase?

During the Readiness Phase of REDD, between 40 and 80 percent of resources have been devoted to designing and setting up national monitoring systems (Simula 2010). This clearly limits allocations for the social dimensions of REDD. With the negotiating mandate of the UNFCCC structured to agree on the most effective and efficient way to prevent anthropogenic interference with the climate system (Bucki et al. 2012), there is much latitude for interpreting what will produce effectiveness and efficiency. Not surprisingly, the "social dimensions of REDD" (Russell 2011) appear as laggards in all aspects of REDD Readiness programming, necessitating dedicated workshops to advocate for why social soundness in REDD and social dimensions require more attention and better integration into programming.[19]

No matter what the rhetorical discourse concerning the primacy of safeguards and participation in REDD, for the time being, social dimensions in REDD are clearly not at the top of the priority list for agencies driving REDD programming. MRV and carbon finance are clearer priorities than are social dimensions. As it took countries such as Brazil and Europe several decades to develop operational forest monitoring systems, and other countries will need time to catch up (Bucki et al. 2012), there is very fair reason to assume that social dimensions in REDD will continue to be relegated to lowest tier priority. In this scenario, overall REDD feasibility will suffer.

Where conventional wisdom falls apart in REDD: Readiness Plan Idea Notes, Readiness Plans, voluntary standards, consultation, political capital

Readiness Plan Idea Notes

Readiness Plan Idea Notes (R-PINS) were prepared at national levels and were step one in a two step process where national "Readiness Plans" would be produced through support of the FCPF. The FCPF was launched by the World Bank in late 2007,[20] and as of October 2012 had mobilized US$429 million in funding through thirteen donors and "Carbon Fund Participants" including the EU, Agence Française de Développement (AFD), the US, and one notable NGO – TNC.

After completion in twenty-five countries, this opened the way for eligibility for World Bank funding of Readiness Plans (R-Ps) that would lead to detailed REDD strategies at national levels. The plans cover a number of fundamental conditions: national reference scenarios for deforestation, MRV systems, tenure issues, and other governance aspects, the manner in which forests laws are enacted, and factors pertaining to forest communities' role.

In an early analysis of first generation R-PINS in Panama, Guyana, Paraguay, the DRC, Liberia, Ghana, Laos, Nepal, and Vietnam (Dooley et al. 2008), it was found that human rights, land tenure, FPIC, and governance were dealt with superficially:[21] "The process has been rushed, implicitly directed towards a market based REDD and dominated by centralized government, with little to no consultation with indigenous peoples, local communities or civil society organizations."

Corroborating that the R-PIN process had gotten off to a poor start, the World Resources Institute (Davis et al. 2009) noted that many of the fundamental conditions are weak or absent within developing countries that might participate in REDD. Without them, it concluded that it would be difficult, if not infeasible, to reduce rates of deforestation and degradation at the national level and there could be risks of leakage.

Furthermore, WRI (see Davis et al. 2009) found that all six areas it had identified as critical to achieving good governance – law and policy development; land tenure administration and enforcement; forest management; forest monitoring; law enforcement; forest revenue distribution and benefit sharing – were found lacking,[22] putting the FCPF in breach of its own guidelines for the R-PINS in twenty-five countries. This finding is consistent with the perception of the FCPF promulgated by others (Dooley et al. 2011).

The authors went further in stating that the FCPF had not followed its own rules in violation of some of the World Bank's own safeguard policies, that R-PINS were authored by "large international conservation NGOs", or else they exercised considerable informant sway over the process, undermining national ownership of the R-PINs, and that plans for future consultations were vague, and human rights or land tenure issues, aside from the R-PIN in Nepal for human rights, were not the subject of risk analysis. FPIC meanwhile was not even discussed.

The overall conclusion was that

> [t]he poor quality of some R-PINs that have been approved suggests that the Bank's carbon finance unit is keen to get the Facility up and running as quickly as possible, and this accelerated approach has meant that approval of R-PINs has been rushed and corners have been cut. Furthermore, the review finds that the FCPF is not meeting some of the key social commitments it has made.
> (Dooley et al. 2008)

Once off to such a poor start, it is reasonable to wonder if and how the gaps in knowledge as well as process can ever be satisfactorily filled in.

Readiness Plans

Readiness Plans (R-Ps) are the culmination of the R-PIN process. The best analysis of the state of R-Ps was done by World Resources Institute (Davies et al. 2009) after their R-PIN analysis. Their critique was essentially an extension of similar issues already raised in regard to the R-PINs.

Overall the assessment noted that "The R-Plans do not adequately address fundamental governance issues as key drivers of deforestation and forest degradation in their REDD strategies" (ibid.). Furthermore they stated that:

> The purpose of supporting readiness should be to help countries to complete an honest review of the barriers and challenges to reducing emissions from deforestation and degradation, and begin to map out a process for coming to terms with these challenges. There are profound problems that underpin deforestation and degradation. The governance issues that we have highlighted in our analysis of the FCPF R-Plans and the UN REDD Joint Program Documents [JPDs] will not be easy to address. Nevertheless, it will not be possible to achieve the objectives of REDD without addressing underlying challenges around issues such as land tenure, land use planning, law enforcement, or the integrity of systems to manage forest revenues and incentives. In addition, given the complexity of these issues a robust consultation process is absolutely necessary to assure stakeholders that these issues will be dealt with fairly.

That is, the same problems exhibited in the R-PINS were repeated in the R-Ps.

Based on public domain information in late 2012, it does not appear that from the perspective of national stakeholders in many of the countries where R-PINS were developed that issues identified in 2008 and 2009 had been resolved. "Robust consultation" is still lacking. NGOs and civil society groups in the DRC complained about the REDD consultative process (Greenpeace 2010) leading to NGO withdrawal in 2012 (Forest Peoples Programme 2012b). In Guyana REDD activities depend on demonstration of improved monitoring of forest sector activities, a point of contention among civil society experts in the sector (Bulkan 2012). In Panama, the National Coordinating Body of Indigenous Peoples (COONAPIP) contested the UN-REDD driven planning process coordinated with the FCPF overall, and in detail as to implementation (COONAPIP 2012; Lang 2013). In Indonesia civil society reports that the REDD process has been inadequate and has not addressed basic issues (Indradi 2012). For example, all reacted negatively in 2011 or 2012 to national consultation processes under REDD, which relate back to the FCPF as well as UN-REDD.

Weak political capital and poor R-PIN and R-P results

The lack of political capital (Dove 1993) or local democratic decision making (Ribot 2010) has been argued as key constraints in attempts to design activities to combat deforestation that are feasible. In many respects, weak political capital explains the weakness of the R-PIN and R-P processes that NGOs and civil society groups have frequently decried since 2007.

Research has shown that partial devolution of tenure rights will not be sufficient to induce the participation required to achieve the sustainable forest management needed to reduce risks from deforestation (Cronkleton et al. 2012). Yet, in most REDD countries, partial devolution of these rights has in most cases not even been obtained, undermining feasibility from the outset.

Rather than being treated strategically as partners, local peoples remain peripheral players. Despite the language employed to create the appearance of strong collaboration, participation, and FPIC, it is hard to see how local peoples are incorporated as partners, versus treated as threats by leadership driving REDD. The same appears the case at more granular levels where project developers design projects and consult with local populations to fulfill CCBA and VCS validation requirements in pre-compliance, voluntary market initiatives. The conventional wisdom that answers will be secured through the Readiness Phase is therefore widely suspect.

And on the land tenure issues alone, Bruce (2012) has provided a clear overview of the challenges that land and resource tenure pose to policy formulation.

Unfortunately, effective beneficiary identification and design of benefit streams remains inadequately highlighted in the current guidance provided to those preparing proposals for funding for the various REDD programs. Consultation is treated in existing guidance, but the guidance does not address minimum acceptable thresholds. The topic of land tenure and land institutions is mentioned, but has not been well treated either. There has been, in particular, a failure to advise of the need for serious early studies, prior even to local consultation and negotiation, of both the legal framework for REDD and the nature of normative and de facto systems governing resource access and use on the ground.

In short, a tremendous amount of enabling environment work remains to be done before projected local beneficiaries stand a high probability of benefiting from REDD. This work was to have occurred in the R-PIN and R-P Phase of REDD. That it has not raises serious questions about the logic and feasibility of REDD at national levels all the way down to projects.

Norwegian oil and REDD

To frame the divergence in perspectives about the importance of REDD more pointedly, Norway's Environment and International Development Minister, whose country has pledged US$2.8 billion under REDD+, said that REDD+ is "the biggest success story so far in global climate change negotiations", noting that reducing deforestation "can only be done realistically with governments of developing nations in the driving seat, and that payments should be results-based" (Aurora 2011).

In the meantime Norway's support for REDD programming in both Guyana and Indonesia has been a subject of continual scrutiny and critique in some NGO and journalism circles. Corruption and mismanagement, along with an actual tripling of deforestation rates in Guyana in year 1 of the agreement with Norway (Lang 2011b) and continued mining of the Tripa peat swamp in Indonesia (Lang 2012c) continue to be discussed in 2013.

Given that over 60 percent of Norwegian emissions are from the production of exported products, with the extraction of oil and gas accounting for 32 percent of the exported emissions (14 Mt), not including actual emissions from the combustion of the oil and gas, it may be fair for some to question the root of Norway's motivations. For example, if the emissions resulting from the combustion of oil and gas were allocated to Norway, Norway's emissions would increase by over 600 Mt CO_2 making Norway one of the biggest emitters in the world (Peters 2008). Clearly, this may partially explain the rationale for Norway's quest for global leadership in REDD.

Other formulations for addressing the underlying problems

The work of Gregersen et al. (2011) on forest-adding countries, or so-called FACs, does not challenge the underlying logic of REDD itself. REDD is seen as a complementary programming objective to afforestation, reforestation, and restoration of degraded lands. Nonetheless, a legitimate question emerges from this work as well.

As an alternative to REDD, given the many controversies and methodological challenges, might a combination of reforestation together with community driven forestry offer a better alternative for carbon sequestration, GHG mitigation, and greater proven benefits for the rural poor in forest areas than REDD? Or as IUCN's George Akwah suggests, isn't the issue really about "sustainable forest management" anyway (Akwah 2013)?

With this type of alternative policy approach, the additional layer of carbon commodification, its marketing, and its benefit sharing, all integral to operationalizing sustainability and permanence in REDD, would be avoided. This is in line with arguments that REDD may be contributing to a broader trend in globalization and land grabbing (Karsenty and Ongolo 2012; Fairhead et al. 2012).

On the other hand, Costanza (2012) suggests that instead of commodification and privatization which critics are justified of being wary of, valuation of natural capital and ecosystem services should not be jettisoned. The key issue involves creating common property institutions, or "common asset trusts", that empower collectivities of "stewards and trustees with interests in maintaining and increasing the long term value of the assets", instead of private individuals. Clearly, this is an alternative that is worthy of exploring its feasibility, as it builds on the last twenty years of work on communities, common property, and their management.

Finally, academic analysis of the broader PES' markets, of which REDD is one, has generated a diversity of opinion from enthusiasm from the disciplines of forest economics, to caution or pessimism in other social sciences (see Castree 2008, 2011; Farley and Costanza 2010; Sullivan 2011).

The one unifying concern pivotal to REDD's success eludes consensus. This involves favorably answering the question as to whether there is a place for establishing markets for carbon assets over goods that have a predominant common property character. Yet, despite the fact that a critical mass of opposition to carbon commodification appears to be growing, this does not appear to be impeding the roll-out of the green economy in UN fora, judging from programming at Rio+20.

One wonders if Costanza's proposal for common asset trusts may turn out to be the best means of conceptually bridging community groups and activists, with social scientists and political theorists to identify a workable model. Getting there, however, will remain a challenge in itself, as agreement on common asset trusts will depend upon negotiation of a new social contract for approaching environmental governance in lands where rights and ownership claims overlap.

REDD and "green grabbing"

Wunder (2008), Karsenty (2011), and Angelsen (2008) provide cornerstone analyses of forest economics and PES, and how REDD situates within wider trends of PES. Karsenty and Ongolo (2012) add additional context by placing REDD and PES within the increasingly noticed phenomenon involving "land grabs" (Bunting 2011; Global Justice Ecology Project 2012) orchestrated by a range of international actors from Europe, North America, and particularly China and India.

In recent years, the pace of land- or green-grabbing critiques of REDD have amplified. Nonetheless, REDD and REDD+ are proceeding in a manner that may inadvertently facilitate infeasible and inequitable "green-grabbing" (Vidal 2008; Fairhead et al. 2012), rather than feasible, equitable, and thus sustainable development. Biofuels play a role in this (Carrington and Valentino 2011).

As Vidal (2008) notes in his introduction to the phenomenon of "green-grabbing":

> Foreign conservationists have a dreadful record in developing countries. First colonialists took control of countries and communities in order to expropriate their resources, then the conservationists came and did exactly the same thing – this time, in the name of saving the environment. Tens of thousands of people have been evicted in order to establish wildlife parks and other protected areas throughout the developing world. Many people have been forbidden to hunt, cut trees, quarry stone, introduce new plants or in any way threaten the animals or the ecosystem. The land they have lived on for centuries is suddenly recast as an idyllic wildlife sanctuary, with no regard for the realities of the lives of those who live there.

So too as others have noted (Survival International 2009; Fairhead et al. 2012) green-grabbing is now occurring under the guise of in REDD. This means that the implications of REDD programming extend far beyond the technical to embrace moral and ethical considerations, livelihood and social welfare considerations, and global and local environmental considerations too.

The conventional wisdom of carbon trading challenged

More widespread still, critiques have centered on the inherent weaknesses in using market mechanisms focusing on carbon offset trading (The Rainforest Foundation UK 2012; Karsenty and Ongolo 2012; Poynton 2012; Sullivan 2011) as opposed to other finance options. A range of weaknesses inherent to carbon offset trading are forwarded, fairly summed up by The Rainforest Foundation UK (2012):

1. It is *highly questionable* whether a forest carbon market will reduce the cost of tackling climate change or generate billions for forest protection.
2. The proposed forest carbon market is *distorting 'readiness' preparations* for REDD so that they are more focused on creating a tradable asset than outcomes that are beneficial for forests, forest peoples and biodiversity.
3. The *ownership of forest carbon* – the underlying asset of the proposed market – is contested and unclear, and its trade is particularly susceptible to fraud.
4. Potential REDD emissions reductions credits *may not represent genuine reductions* in greenhouse gas emissions, due to inflated baselines and leakage. Trading them in an offset market could lead to increased total global carbon emissions, and prolong existing heavily polluting activities.
5. *Alternative financing options* and approaches exist and are viable.

UN-REDD (2012d) countered that the Rainforest Foundation's conflation of the UNFCCC compliance mechanism standards with those of the VCMs underestimates the rigor and robustness of the UNFCCC standards. Yet, a lack of clarity of land tenure and carbon rights combined with the fact that carbon credits are intangible assets, making trading virtual, has made some VCM projects susceptible to fraud. Under the UNFCCC, this will not be possible, it is argued, because of the stringent MRV requirements for transparency and comparability of results. The contention is that Annex 1 parties to the UNFCCC will not invest in REDD+ unless they are confident that the mechanism guarantees that they actually get what they pay for (UN-REDD 2012d).

This conclusion is, however, speculative and problematic for investors. The UNFCCC may well invest in REDD+ activities at national levels *on the assumption* that the data shows they "actually get what they pay for". When it comes to risk analysis for genuine FPIC on forest lands with multiple property rights claimants, and risk analysis for any intangible carbon derivatives that have been commoditized from such common pool forest resources, UNFCCC REDD+ will be subject to the same risk uncertainties as any VCM carbon transacted. That a nation state may claim authority and successful risk mitigation, while other customary stakeholders may contest rights, does not signify that guarantees provided by the UNFCCC mechanism will deliver what investors pay for.

Alternative models for avoiding deforestation and sequestering carbon

Despite the fact that a decision has been made at international levels to use a market and carbon trading approach to avoid deforestation and sequester carbon, as noted the jury is still out on many aspects of what the nature of the problem is, and the best medicine to solve it.

The reason for equivocation is that the rules under which contributions by developed countries to the global forest fund will be calculated, and the negotiations over the rules defining the baseline calculations (Figuieres et al. 2010) are themselves areas that remain in much contentious debate in the UNFCCC process.

On the side of critics to REDD, evidence and analysis is being made available from various types of participatory forest management (PFM) initiatives that challenges REDD's implicit logic. These include experiences from community based forest management (CBFM) and joint forest management (JFM) initiatives from around the world which suggest that mechanisms that empower local peoples living in communities, and working in alliance with other stakeholders, may in fact be *as or more effective* than hypothetical outcomes from REDD.

This PFM data supports the work of Elinor Ostrom (1990) and colleagues (Ribot et al. 2010), that the principles of subsidiarity and democratic decentralization are crucial enabling conditions for natural resource management and community conservation to work in remote forests of the world. If provided the chance with clear responsibilities and authorities that people buy into, communities can more likely than not tackle much of the job as well, or better, than PA managers if sustainability and cost-effectiveness indicators are employed. This is consistent with Farley and Costanza (2010) who suggest that PES transfers resources between social actors to create incentives for aligning individual and/ or collective land use decisions with the social interest in the management of natural resources. Brown et al. (2008) aslso suggested in considering gorilla conservation in protected areas where communities maintained customary tenure and common property resource use rights, a community driven approach including *both* analytical and decision-making capacity building, with negotiated payments for ecosystem services would work best.

When rights and responsibilities are fully devolved, incentives appear to be sufficient for communities to invest in forest restoration and long-term management (Blomley and Iddi 2009). This would suggest that other market incentives, while not irrelevant, may not be *the* principal incentive that guides participation and community level decision making in natural resources management.

This is consistent with lessons learned from the CARPE which showed that communities are prepared to participate even for years in capacity building activities, if they are convinced this will lead to greater leverage over land use, learning, and generic empowerment skills for analysis, decision making, and advocacy (Brown et al. 2008). To date, it is unclear if REDD policy makers have fully assimilated these types of lessons.

Why alternative theories for how REDD may have evolved are important for adaptively managing REDD

There are many possible theories, all conjectural, as to how and why REDD has evolved in the manner that it has. They are important because they introduce considerations which up to now *have not* been part of critiques of REDD. These critiques have consistently focused on policy, dubious analysis of the role of markets, concerns with participation and FPIC – all important issues, but not providing comprehensive sense of how REDD continues to lack wide traction.

While the "banking of nature" (Sullivan 2011) is increasingly common in academia and civil society circles, there are a host of other possible theories from the social sciences that have yet to be explored.

Psychological explanations

When combined with neoliberal critique of REDD, I believe consideration of other factors, beyond simplistic maximization of profit or utility, provide a far more robust theory for understanding where things stand with REDD. They also provide additional insight into why it will be so difficult to change the prevailing paradigm for REDD, despite the fact that doing so would improve chances of its success.

The strong neoliberal critique of the so-called "banking of nature" (Sullivan 2011) does not provide a full explanation for why REDD is imperilled. The aforementioned work of Trivers (2011), Kahneman (2011), Ehrenreich (2009), Frankfurt (2005) and Macintosh (2006) provides a set of frameworks to approach understanding REDD to complement the neoliberal critique that critics, oddly, do not usually explore. Each provides a complementary perspective for how, respectively, "self deceit", "confirmatory bias", "bright-sidedness", and "bullshit" in everyday life, as well as the intrigues of accounting systems, may be used to better understand how REDD has emerged, evolved, and endured in the manner it has.

For example, while highly theoretical and polemical, Triver's (2011) emphasis on the advantages to individuals, versus groups in natural selection, may ironically best explain why biodiversity conservationists normally wedded to Darwinian explanations for evolution of species, ecosystems, and more complex landscapes appear indifferent to the ineffectiveness of conservation approaches from the standpoint of sustainability. Otherwise, the TMAs advanced by conservationists would logically be putting people first, to enhance efficacy, as it is posited that most threats originate from human sources.

Decades of not generating data on results and impacts

As O'Neill and Muir (2010) and Salafsky (2010) made the point for colleagues in the Conservation Measures Partnership (CMP), verifiable indicators of impact do not match up against the rhetoric of conservation practice. This begs a series of obvious questions that should be of interest to conservation agencies taking a lead role in REDD implementation on the ground, and who also participate as thought-leaders at UNFCCC COP fora. For example:

- Assuming that either local people or others further removed are the cause of biodiversity loss and deforestation, how can conservation or REDD work if most resources are not allocated to addressing people-level issues?
- Why has biodiversity conservation not prioritized the social dimensions of conservation if effectiveness and sustainability depend on people changing social and cultural behaviors?
- If in fact people, their institutions, their problems, beliefs, and actions impact resource use, why are these not assessed to the fullest so that appropriate strategies are devised accordingly?
- Why persist in believing that IEC campaigns and just enough local incentive – a school and clinic here and there – that are then coupled to GIS and good biological data, will generate sustainable behaviors of remote peoples living with wildlife?
- When these peoples suffer externalities from the limitations on resource use that PA systems demand, are conservationists' strategies either equitable or enough?

Answers to each question are of direct relevance to the success of REDD.

Much of the best practice that is leading ACC/AGW mitigation in REDD is based on questionable methods, results, and impacts from the biodiversity conservation and development arenas that have limited proven effectiveness. Conversely, application of current best practice will increase the potential costs and risks posed to livelihoods and local and indigenous ways of life from REDD, suggesting that their local stakeholder perspective is fraught with risk.

Public relation campaigns for REDD launched by intergovernmental organizations and BINGOs, with the backing of Northern governments and strategic private sector partners, have been successful enough to sustain modest funding levels during the Readiness Phase. Projects are being piloted and national consultative multi-stakeholder planning processes have and continue to be undertaken. The urgency of ACC issues, coupled to biodiversity conservation and poverty alleviation gains, provide the principal rationale for moving forward. Unfortunately, the approach for participation and FPIC has hardly been innovative, and REDD's future funding path and actual efficacy, unclear.

USAID, biodiversity conservation, and REDD

One recent USAID request for proposals for its US$200 million Indefinite Quantity Contract entitled "Measuring Impacts (MI)" (SOL-OAA-12- 0000-50) notes that for biodiversity conservation, evidence as part of performance-based monitoring in biodiversity conservation has been sorely lacking (USAID 2012).

Are USAID's biodiversity funded programs and activities effectively conserving biodiversity and natural resources? USAID's programs are based on a number of prominent hypotheses in the field of conservation and development that aim to protect biodiversity and natural resources while improving human well being. Although these hypotheses are used widely by conservation organizations and development agencies, *little information exists about their efficacy in practice* [emphasis mine]. Despite longstanding programs in priority biodiversity regions, there is still a lack of long-term datasets and baselines to analyze the effectiveness of conservation activities.

The implications of the quote above are far-reaching. USAID support for REDD+ programming is part of its conservation and forest sector programming that measuring impacts will focus on. So, if as USAID suggests, we are not able to analyze the effectiveness of conservation activities, and if conservation activities and best practice are actually the principal intellectual and methodological sources for how REDD TMAs are being shaped, *are we not in effect admitting that when it comes to REDD, the partially blind will be leading the blind?*

A possible new analytical angle

The bottleneck for proponents arguing that ACC/GCC *is* urgent, that the means for tackling it *are* at hand at both global and local levels, and that local solutions *are* equitable for what is experienced as a global and regional problem, is three pronged.

The first pertains to technical means for establishing credible baselines and credible methodologies for measuring the historical rate of carbon loss, and current rate of carbon sequestration that is key to accounting and envisioned carbon transacting through REDD. The second relates to finance, and the mechanism used to mobilize the lower end estimated US$5 billion, to US$17–33 billion annual requirement on the upper end (see Eliasch 2008; Stern 2006) as the target required to mitigate deforestation and degradation. The third relates back to proven track records applying the TMAs proposed in REDD.

The strategy that has been employed to date by REDD proponents and thought-leaders replicates a long-standing tradition where wishes and vision are conflated with the ability to design and implement projects on the ground that work. In the broader development context, critique of the efficacy of development policies has proliferated in recent years (Easterly 2006; Collier 2007; Moyo 2009; Cohen and Easterly 2009), with the argument made by Easterly and Moyo that policy and practice is so flawed *overall*, that developing countries would largely be better off on their own. That is, external intervention does not make much positive impact, and really derails development that would be better left to nations working out issues on their own directly with external investors.

Banarjee and Duflo (2011) take things a step further. They eschew sweeping generalizations that they argue are counterproductive, and challenge development planners to base program design on metrics that will enable incremental change over issues that matter, and over which the global community has the ability to intervene. They suggest if formulaic thinking that reduces every problem to a set of general principles is resisted, and practitioners "listen" to the poor instead, planners will be able to construct a toolbox of effective policies and will better understand why poor people live as they do. Tools can be provided to enable the poor to escape from poverty. Rigorous empirical testing is what underpins their approach (see Banerjee and Duflo 2011).

This does not sound much like REDD or the strategy of thought-leaders with the most leverage over REDD project design and implementation. But arguably, it *should* sound a lot more like REDD if REDD is to work.

If anything, REDD policy and programming to date represents the antithesis of the type of "patient understanding" (Banerjee and Duflo 2011) needed to construct a policy toolbox that will help in understanding the poverty related drivers of deforestation. REDD policy continues to be rolled out on the stereotypical assumptions about deforestation drivers and solutions; data as it exists is used to justify conventional wisdom about problems and solutions, versus establishing strategies based on data-driven inductive reasoning. There is little empirical evidence in the public domain in terms of results and impacts, and the methods employed to achieve them in either REDD or its methodological forebears, biodiversity conservation and development, to justify scaled up investment in a roll-out of REDD programs at national, sub-national or project levels as is currently occurring.

Meanwhile, the methodologies developed for the principal REDD standard setters – the VCS on the technical side of methodology development, and the CCBA on the social side – do not demand the kind of rigorous analysis in planning REDD interventions that Banerjee and Duflo (2011) argue are needed for developing successful development programs, such as REDD. As project developers are not required to demonstrate social feasibility, but can substitute formulaic checklists of activities that demonstrate that *some* attention to social issues has been paid, I believe that REDD activities are set up to fail, as weak ethnographic analysis is coupled to rapid participatory consultative processes for coverage of social issues in planning. It is time that more of Taleb's (2009) criterion of injecting local *know how*, instead of external, expert economic driven *know what*, was made.

CIFOR's analysis

The myriad "optimisation" challenges faced in REDD implementation are well covered by Seymour (2012). She notes that there are multiple objectives and dilemmas needing resolution in REDD. So even though the opportunity cost of avoiding deforestation is argued by many to be relatively low, establishing an internationally acceptable system that can achieve and monitor REDD objectives remains a daunting task, even after years of preparedness planning.

CIFOR (2012) released a study at Rio+ 20 entitled *Analysing REDD+: challenges and choices*. The study's central premise is that "REDD+ as an idea is a success, but implementing it is fraught with challenges". While many may debate the claim that the idea of REDD+ is a "success", the main conclusions of this large body of analysis are worth reflecting upon:

The pre-Copenhagen expectations for how REDD+ would play out have not been met. In part, this resulted from the fact that the global community failed at COP15 to reach an overall climate agreement to replace the Kyoto Protocol, and will not now do so before 2015 at the earliest. The prospects for significant REDD+ finance generated by a carbon market under such an agreement have correspondingly declined. While negotiations continue to make incremental progress on global REDD+ architecture, the relative importance of the UNFCCC as a top-down driver of the necessary finance and rules for REDD+ has diminished significantly. As a result, there are now multiple REDD+ policy arenas populated by aid agencies, big international NGOs and various domestic actors. The participants in these arenas often compete for funding, leadership in standard setting and influence over the discourse on how REDD+ should be defined.

Another set of changes arose from the fact that REDD+ emerged just as the world entered a period of economic and financial turmoil. In the mid-2000s, the global economy experienced a commodity price boom, with prices for food, fuel, and metals reaching unprecedented levels. These high prices – and the associated fears about food and energy insecurity – led to a global rush to secure access to land for agriculture and minerals development. Increased competition for forestland will probably lead to increased costs of implementation. These increases may outpace the improvements in land use planning necessary for REDD to be considered as a viable option. Then, the global financial crisis that struck in 2008 distracted attention away from climate change; pressure on national budgets will probably constrain the volume of aid funds available to bridge the REDD+ financing gap caused by the lack of an international climate change agreement (Seymour and Angelsen 2012).

Where REDD critics have also missed the boat

While safeguards are part of the tools employed to prevent harm from occurring, other TMAs are needed for working with people to achieve development objectives. This side of the story in REDD is often muffled by discussion of safeguards, however worthy.

While legitimate concerns about REDD architecture and mechanisms, human rights and equity standards, etc. have been raised, focus on TMAs used for approaching social issues in REDD at the practice level has lagged. Technically, policies pertaining to carbon finance, reference baseline establishment, the specifics of accounting systems as part of national, sub-national, or nested programs (see Terra Global 2010) have been higher priorities than understanding how stakeholder participation in REDD must be shaped for social feasibility criteria to be achieved. Moreover, safeguards have been focused on as the priority in addressing the social dimensions of REDD. This may deflect needed emphasis on key program and project design issues upstream.

Unless TMAs directly become part of policy discussions, notions such as "permanence" of carbon assets accounted for in specific places, or "leakage" of deforestation from place to place, will remain more theoretical than verifiable, as social feasibility will be lacking.

TMAs will need to address land and resource tenure issues, participation and a host of governance issues, and capacity and so-called "social and human capital" issues. These issues are not being comprehensively addressed through the palette of options that industry thought-leaders advance through either the FCPF's Strategic Environmental and Social Assessment (SESA) methods, or social and biodiversity impact assessment (SBIA) for VCM activities (Richards 2011). Even early on in the Readiness Phase, concerns about the viability of SESA were noted (BIC 2010).

By promoting a "theory of change" approach to design, monitoring, and evaluation of REDD activities, SBIA enables the design and financial support of projects that do not have to explicitly address and meet social feasibility criteria. Rather, if a project developer is able to enunciate relationships between an ultimate objective (REDD) with activities and outputs put in terms supported by the ethnographic literature, this appears sufficient to satisfy auditors of the principal REDD voluntary standards that a theory of change is being well employed, and a basis for moving forward is established.

While critics have identified a range of substantive and procedural weaknesses in REDD requiring fixing that pertain to the social dimensions of REDD, the TMA arsenal proposed to address these has not drawn serious focus or attention, save in one research project that principally addresses biodiversity conservation from the perspective of biodiversity conservation organizations (see McShane et al. 2011).

While TMAs now employed in REDD have certainly been marginally improved upon over the past twenty to thirty years in the development and biodiversity conservation arenas, they have not produced strong results and impacts. Lack of informative data on results and impacts in biodiversity conservation has recently been noted and begun to be acted upon (USAID 2012). If however TMAs remain inadequate to address the challenges posed in extremely complex programs like REDD, or even in biodiversity conservation, or most development programming, on what methodological basis then is REDD moving forward?

Most TMA issues are treated as concerns at an implementation level. Yet by addressing TMAs as an issue for technical work groups, instead of addressing weak TMAs as a policy issue, the groundwork is laid for replication of weak methodological standards at the project level across programs.

Moreover, if TMAs are designed and implemented by reputable BINGOs, for all intents and purposes they can be accepted as best practice, as there is no public domain data and analysis to provide credible counter claims, as BINGO thought-leadership is widely accepted and preponderant in intergovernmental fora when it comes to forests and conservation. Credibility and validation has been, almost exclusively, a function of *who promotes* the TMA, and how loudly.

As there are no standards or criteria for judging the fit between methods employed and results and impacts obtained, the REDD industry is by and large flying on one wing of conventional wisdom and another that is largely "seat of the pants". This point has been repeatedly hammered home by activist groups, perhaps most clearly in the Chiapas Declaration in response to the convening of the GFC (Comunidades de la Región Amador Hernández, Chiapas, México et al. 2012).

Participation and REDD

The "illegitimate or unjust use of power" (Cooke and Kothari 2001) to further REDD programming at the expense of politically marginalized, impoverished forest peoples is a topic that has yet to be widely discussed as REDD has evolved. For after all, when multiple claimants to resources believe they have legitimate claims, even if some are state agencies leveraging their power, as appears to be the case in REDD, establishing stakeholder prerogatives in practice that reflect the spirit of a nation's constitution remains challenging in many countries.

As Ribot (2010) points out, if REDD only attempts neutrality it will deepen inequalities. To become a positive force for forest dependent populations, REDD must be guided by clear standards for democratic representation of local populations in REDD decision making, as well as access to benefits. This should include prerogatives over local control over forest resources, and access to REDD marketing activity in which local peoples are enhancing the value of conserved natural resources.

The implications of any illegitimate or unjust use of power will be particularly felt by the poor through manipulation of what are termed "benefit distribution systems" (BDS) that depend upon ensuring local participation in the decision-making process of the sorts of payments to be made in either cash or in-kind benefits. According to SNV Netherlands, to determine what the options for BDS are in practice, decision making also involves empowering the poor with the ability to decide who will receive benefits, when benefits will be distributed, and how the benefits will be made practice (Enright 2012). While seemingly straightforward, these are complex issues that involve social and political factors, along with understanding of who is doing what to sequester carbon assets on the ground, for which payments for sequestering services are being made. Employing PRA will likely not answer these questions.

One interesting question to be asked of REDD planners involves BDS precedents: Where have local people been empowered to set the terms for BDS in practice, are there models for replicating this across REDD participating countries, and what are the preconditions for their successful design?

Another interesting question involves investment decisions: do REDD planners really believe that current investments in "the social dimensions" of REDD will get the job done? In looking at the weight afforded social dimensions in REDD, there appears to be a structural bias in favor of what are considered to be more "technical" aspects. Spending patterns frame what is looked at, and how one understands problems.

Using investment in MRV as opposed to environmental and social impact aspects of projects through public funding used as a guideline, the former is significant in comparison with the latter. For example, the CAR budgeted

US$43,000 for work on environmental and social impacts in its REDD Readiness Preparation Proposal (R-PP) which is approximately 0.6 percent of a total budget of US$6.7 million in comparison to approximately 20 percent budgeted for MRV of carbon (The Rainforest Foundation UK 2012).Where BDS comes in under "social impact" in the projects is unknown.[23]

This spending pattern appears consistent with spending in the biodiversity conservation arena, where expenditure for technical tools – GISs, biodiversity surveys, threat assessment tools, monitoring tools for measuring biodiversity trends, and adaptive management tools for righting the course in the event of misalignment – is, from anecdotal evidence, taking up a large percentage of program resources in comparison to community capacity-building tools,[24] or investments in community development as part of ICDP programming.[25]

Unrealistic calculation of opportunity costs

At the heart of the debate around carbon finance is the issue of opportunity costs or forgone benefits – what it will take in terms of compensation to convince resource users to alter their behavior to keep forests intact.

Some believe that opportunity costs are "usually the single most important category of costs a country would incur if it reduced its rate of forest loss to secure REDD payments" (Pagiola and Bosquet 2009). For this reason, accurate calculation of opportunity costs is essential. But how accurately can these be calculated, and what will happen if overestimates or underestimates for opportunity costs veer too far off?

The assumption that opportunity costs for local stakeholders in forest landscapes are low, and they will therefore readily adopt REDD incentives is perhaps fueled by early enthusiasm for REDD being a highly cost-effective means to mitigate climate change (Stern 2006; Eliasch 2008). Yet, the assumptions underpinning the opportunity costs themselves are highly questionable, and undervalue true costs to communities to adopt substitute behaviors encouraged by REDD. Had an economic anthropological perspective been used to shape REDD policy on incentives, along with standards for validating project development, a much different incentive framework for REDD would in all likelihood have been developed. Such a framework would develop understanding of the full context of community level decision making, placing equal or greater value on the social context within which both individual and group decisions are made.

The indifference to the range of opportunity costs is apparent from reading key documents at UNFCCC, UN-REDD, FCPF, CCBA, and VCS levels, as well as in PDDs developed by BINGOs and private-sector promoters of REDD projects.

In impoverished communities around the world, current opportunity cost calculations lead to values that are abysmally low, despite the fact that the value of the resources which communities may enjoy collective, customary rights over, may be tremendously high.

In situations where BINGOs or for-profit companies beholden to shareholders are able to maximize profits over resources that are recognized as common property, but that governments have managed to nonetheless transact with private

Box 3.1 The opportunity cost conundrum

Relying on opportunity cost analysis to underpin evaluation of PES in REDD presents a conundrum (Karsenty 2009). Among the many problems posed by opportunity cost calculations Karsenty notes are:

> The need to accurately determine causes of deforestation and its impacts to design appropriate solutions; who will actually capture benefits from ecosystem service provision; why opportunity cost calculations may only apply in practice for the poorest, leading to ethical dilemmas, along with potentially leading to unnecessary "hot air" payments as deforestation would have been avoided in any case in such situations; elite capture or inappropriately skewed impacts on resource users where some users are under-compensated; and perhaps most telling, obligating resource users into decades long situations where future development prospects for them may be frozen based on the nature of the agreement signed.
>
> Estimating the magnitude of opportunity costs gives a fair estimate of the pressures for deforestation. Understanding how opportunity costs are distributed across groups within society tells us who would gain and would lose from REDD...
>
> Estimates of the opportunity costs thus provide inputs not only into the costs the country would bear from REDD, but also into the causes and distributional implications of deforestation and, hence, the types of interventions needed to actually reduce deforestation and the potential need for mechanisms to avoid adverse social consequences...
>
> In a carbon market (voluntary or regulated) with a single price per ton of CO_2 resulting from supply and demand, some agents providing an avoided deforestation service will have opportunity costs that are lower than the value of avoided emissions, calculated on the basis of the price per ton of CO_2. This difference between the "production cost" of avoided deforestation and its "purchasing price" creates a surplus. This surplus may be conserved by the agents, but will more likely be captured by carbon market brokers or PES project promoters, who will thereby pay themselves to varying extents.
>
> Moreover, conserving forests in agricultural frontiers in the Amazon instead of cultivating soybean, or in South Asia instead of planting oil palms, generates opportunity costs that are often high since these crops are very lucrative. PES programs will therefore concentrate on forests that are under less threat at the risk of paying actors who have nothing to lose by avoiding deforestation (zero opportunity cost). PES is caught between two stumbling blocks: where the opportunity costs are high, the sums available are often not enough; but where the opportunity cost is low, the risk of paying for environmental services that are not

endangered (lack of additionality) is high. Verifying additionality would require significant means in order to analyse local situations, which would imply higher costs.

A major problem where PES and their social acceptability are concerned is that compensation based on the opportunity cost is inequitable for the poorest populations. Freezing user rights such as clearing, hunting or even the prospect of working in a forestry company *deprives people of opportunities to lift themselves out of poverty* [emphasis mine]. Moreover, within communities, it is often the poorest who depend on natural resources. By giving up certain activities, they lose vital access rights that are not generally offset by the payments, which are based on the average opportunity cost for the whole community. Nor is it unusual for these payments to be monopolized by the "elites". Simply compensating the opportunity cost for very poor farmers therefore raises ethical objections and is enough to justify envisaging another basis for payments.

Finally, adopting the opportunity cost as a basis for compensation does not prepare for the long term. Compensating for the loss of income from giving up certain subsistence activities may free up working time but does not release any new resources to acquire the capital needed to implement new agricultural or agroforestry technologies.

sector interests, red flags may legitimately be raised, particularly if transparency in public domain information has been limited. These are often the opportunity cost contexts in which voluntary, pre-compliance REDD projects are being designed and implemented today.

Obtaining FPIC in contexts of desperation

The notion of "FPIC" as obtained by sophisticated First World interlocutors wishing to transact carbon sourced in developing country communities would be comical, were risks for abusing opportunity cost calculations, along with widespread corruption risks, not so obvious. This makes obtaining FPIC such a challenge.

REDD is premised upon FPIC. FPIC's credibility is underpinned by a cohort of reputable thought-leaders and world bodies like the UNDP, UNEP, and FAO (UN-REDD); renowned NGOs like the Union of Concerned Scientists and the major big conservation NGOs, e.g. TNC, WWF, CI, and WCS; major foundations like the Gordon and Betty Moore Foundation and the Ford Foundation; the principal standard setters like CCBA and VCS, and national governments around the world including Norway, the UK, and the US.

Major governments support FPIC principles too, though here the US is among the few to agree with the World Bank that FPIC involves consultation, versus consent. This is likely due to the abundance and complexity of legal agreements that could be put into question with American Indian tribes in the US if the US were to support the principle of consent as opposed to consultation.

This support for REDD, and the FPIC principle underpinning it, be it where the "C" stands either for consent or for (effective) consultation, leads to an obvious question: Do these thought-leaders and decision makers *really believe* that FPIC, be it consent or even "informed consultation", is being achieved in REDD landscapes?

Do REDD thought-leaders believe it is justified to move forward given the risks peoples around the world in REDD landscapes face if consultation is ineffective, and if consent in fact is not obtained?

Is it a case where thought-leaders believe that REDD is now the only realistic chance to make *the appearance* of addressing the viscerally alarming climate events that have recently been occurring, which one leading climate change scientist (Hansen 2012) now argues are so statistically improbable *not* to correlate with climate change?

Is it a case where thought-leaders see that loose standards for FPIC will be key to at least offset a portion of Northern emissions? This is due to the fact that the politics of emissions reduction (ER) requiring retooling of industries and consumption patterns in the US and the largest developing economies remains intransigent.

Or is it the case where a bit of each is true, but where the global community is going about REDD as it is anyway, akin to what one senior manager at a BINGO confided to the author in a personal email:

> We are all hopeful about REDD+. And of course we all know it will be a disaster. It would be a miracle if some of the $4.5 billion were to drop from the tables of the consultants and managers like you and me[26] to the tables of 'the poor'. History shows it won't.

That the private sector and shareholders to financial investors and carbon project developers and intermediaries are willingly participating in facilitating and obtaining FPIC is understandable. If enabled, why *shouldn't* the private sector seek to generate profits by moving REDD projects through the pipeline in manners that are certifiable as meeting the highest standards of "best practice"? Whether it is through up-and-coming frameworks like California's AB 32, or simply to stock up speculatively on pre-compliance market offset credits in the event a compliance market does eventually come on line, this behavior is understandable from the capitalist perspective upon which REDD currently is premised.

Overview of what a social feasibility framework would accomplish in REDD

A social feasibility framework is pivotal to avoiding deforestation, regardless of the financing mechanism in question. Achieving a framework based on social feasibility will require that new agreements, what I refer to as "social contracts", are negotiated between stakeholders at international, national, and local levels as a first step.

Regardless of the challenges to MRV and establishment of RELs that REDD has made progress on but that still remain contentious, the social dimensions of REDD remain the weakest link in projects and the entire architecture. Given the challenges the various social dimensions pose, the logic for a more focused strategy would appear self evident.

The issue, however, is that what may be assumed as implicit to all REDD planning processes is anything but that. Social feasibility, and its attainment, is currently not a standard or objective employed in either development or biodiversity conservation programming. Rather it is an implicit, cross-cutting assumption in programs – e.g. proponents and project developers appear to believe they are addressing social feasibility through the methods they employ, even while these are in no way explicitly targeting social feasibility. REDD is no different.

Had major issues like land tenure been considered through a feasibility and risk analysis framework upstream, it is debatable whether REDD and REDD+ would have been as enthusiastically kicked off after the UNFCCC in Bali (Bruce 2012). Yet the fact that the program exists, that best practice is guiding it, and that specific projects are being designed and implemented, *despite* the deficiencies Bruce notes for land tenure alone, is indicative of the leverage that wishful thinking has had over solid feasibility assessment for REDD planners and decision makers. This should change.

Despite the existence of best practice for REDD project developers that the VCS and CCBA have approved on the dynamic, voluntary side of REDD, where the best energies at present reside, there is no guarantee that on-the-shelf best practices promulgated for externally designed projects using SBIA guidance (Richards and Panfil 2011) address social feasibility issues at all. The point in the SBIA guidance, far and away the most thoughtful guidance on addressing the social dimensions of REDD, appears *to do enough* so that funding is secured, best practice standards are followed. Emphasis is on reducing institutional and transaction costs to a minimum for project developers, while addressing social issues. This position, arguably defensive or what some may label as realistic, precludes social feasibility from being addressed, as the complexity of social feasibility issues requires more robust methods.

While SBIA is clearly sensible in *ex post* analysis *after* events and impacts have happened as part of an adaptive management program, its use in upstream design is more problematic, as predicting specific impacts are speculative. It is during design when social, and broader feasibility issues, are best addressed.

Recognition of the need for continuous adaptive management as a principle in development and conservation programming suggests that reconsideration of approaches and methods is never too late. By accepting the principle of adaptive management, a virtual best practice mantra in the conservation community (see Conservation Measures Partnership 2010, 2012), a rethinking of assumptions and methods would follow if methods and assumptions were objectively assessed.

Chapter 6 presents a graphic for the adaptive management cycle proposed by the CMP, many of whose members are also part of the CCBA responsible for standard setting for social dimensions in REDD. Whether the CMP would be willing to put this into practice in looking at SBIA or SIA in comparison with a social feasibility framework is worth proposing, as its members are drivers of REDD+ programs.

The fact that some of the big conservation NGOs submitting Readiness Phase REDD projects are those that have supported development of the CCBA standards has not been a subject of analysis in regard to their efficacy for addressing social dimensions in REDD. Oddly, this has not been a focus of opponents of REDD.

Applying SIA is an effective way for enabling projects to be launched, enabling them to gain funding and momentum. The hope is that the best practice guidance recommendation proves good enough to adaptively manage biodiversity conservation (Springer 2009), FPIC in REDD (Richards 2011), carbon permanence, non-leakage, absence of conflict, etc.

Personally, I believe this is delusional. While social soundness and social safeguards both consider what needs to occur for "no harm" to be achieved, social feasibility demands specifications for what *specifically* requires doing for things to *actually* work from the get-go. Although development and conservation planners have never considered it explicitly, it represents a more demanding standard than social soundness, social safeguards, or social impact assessment for program and project design.

4 What do Pygmies circa Mobutu's Zaire have to do with REDD?

Pygmies and other IPs have a lot to do with REDD. Or rather, they should. Yet, it is questionable whether their role is any more central than other remote, poor forest communities equally impacted by REDD. For REDD to work, this must change.

The material context for Pygmies has worsened in many respects as it has for other IPs and many local communities over the past thirty years, even as international human rights treaties supporting IPs have been drafted and signed off by national governments. This creates an extremely challenging context for REDD for achieving FPIC, as there are thousands of dispersed, semi-autonomous Pygmy groups across central Africa who lack effective leadership to facilitate their participation in REDD. This opens the door to simplistic understandings that would form the basis of any "consent" obtained (Brown and Mogba 2010a).

This chapter is a story about a first encounter in 1974 with tropical forests, gorillas, and Pygmies, and how the interface between an anthropologically "exotic" culture, universalist notions of progress and development, absolute dictatorship, and the realities of poverty and remoteness shaped my perspective on REDD today. The issues I encountered in the Ituri forest in 1974, and over the years until today with other Pygmies and neighboring communities in different Congolese forests, provide a fair backdrop for assessing the plausibility of REDD, and its prospects for success or failure.

Beni to Mambasa

In early 1974 I crossed on foot from Rwanda into what then was called Zaire. I was on a university year abroad tour of Africa. My objective was to reach what I imagined to be the most remote group of Pygmies on the planet. Reaching remote Pygmies on foot in 1974 seemed to be a very cool thing to do for a university student from suburban Boston.

I postulated at the time that the Pygmies were the premier "exotic" people in the world, and that they were still living uniquely as *National Geographic* would have readers believe. With academic interest in the rapidly evolving field of international development[1] – improving social welfare of poor, "underdeveloped", "Third World" peoples – the Pygmies represented an ultimate year abroad case study for me to satisfy academic and personal ambitions.

In those days, handheld, electronic information systems (GIS) or cell phones did not exist. My crinkled Michelin maps, the intrepid traveler's tool of choice in Africa, were all I had to guide me. On the maps, the names of well-known ethnic groups had been superposed over color-keyed areas, the latter indicating broad geophysical characteristics. Deserts were brown, forests were various shades of green, bodies of water were blue, etc. These helped point me towards the "g" in Pygmies, nestled comfortably in the greenest portion of one of central Africa's premier tropical forests, the Ituri.

I began my walk specifically towards the Ituri northwest towards Mambasa. I followed a thin white track on the map from the town of Beni. This leg of the journey began at the foothills of the famed Ruwenzoris, which had been dubbed as the "mountains of the moon" by the Greek historian Herodotus. As I had just learned, despite omnipresent March rains, the kaleidoscopic Ruwenzori rainforests could make forest biodiversity conservation champions out of even youngsters who had grown up in and around Boston's suburban malls. I anticipated that the Ituri would be even more amazing than the Ruwenzoris.

On another level it was.

On day 3 of the walk from Beni, I encountered what appeared to be, to use the term coined by then Zairean President Sese Seko Mobutu, an "*authentique*" group of Pygmies – very short, surprisingly strong, gloin-clothed men and grass skirted women without dress tops. They superficially appeared to meet all the pre-conditional stereotypes I had – seemingly joyful, curious, smiling people living in very tiny hamlets characterized by huge *mongongo* leaf roofed huts.

Upon entering one Pygmy hamlet, the dancing and smoking of long pipes with thick wafts of marijuana (*bangi*) melded well with children scampering, and muscular men arriving carrying loads slung across their back with the weight distributed in a strap across their forehead. One super-muscular, body-fat free, small man was seen running out of the village with a spear. From a distance too, laughing and singing echoed through the forest in very unfamiliar patterns. The setting was conducive to the type of neophyte fieldwork I was conducting. *Bangi* was a commodity apparently able to be imbibed at all times of the day, woven seamlessly into the fabric of work and play.

Solitary Pygmies sauntered in and out of the village over the next few afternoon hours, returning from neighboring Bantu villages and foraging and hunting forays in the forest, I presumed. What seemed like a village of ten became twenty, thirty, and then perhaps forty by the time infants were accounted for. This was all just as Colin Turnbull's ethnography, *Wayward Servants* and more popular *The Forest People* suggested would be the case. Lots of smiling people carrying nets around for hunting animals. Other men had long spears hidden away for occasional group elephant-hunting exercises, which with luck could produce a ton or two of meat that could be traded for cassava, or serve as the basis for advancing petty cash loans with neighboring Bantu villagers.

Could there be any more "true", authentic, Zaireans than these Pygmies, the original inhabitants of the rainforest whom I had "discovered" smack on the "g" of the Michelin map? Was this not what Mobutu should have been extolling? I was feeling extremely self-satisfied, as it seemed as if I had really found genuine Pygmies!

I immediately began capitalizing on my rudimentary *Teach Yourself Swahili* primer, and I was able to receive permission for a visitation with this Pygmy village for an unspecified duration. Based on the indifferent head nods of the eldest Pygmy in the village who was attached to his pipe of marijuana, the stay was to be at my discretion.

Over the course of the next ten days, I was struck by how skilled the Pygmies were in their knowledge of the forest environment in which they lived. People of all ages rapidly distinguished trees, plants, insects, and animals of edible and medicinal value while in the forest. They navigated more smoothly forest paths than I ever succeeded doing years later in Manhattan with street signs. This was all as it was supposed to be with indigenous forest peoples.

Their manner of joking – a combination of irony, sarcasm, and intimate camaraderie – was also not so foreign from my experiences in college dormitory hallways. This disabused me of any lingering assumptions that Pygmies were of inferior intellect or, worse, were sub-human in prejudicial ways that still prevail in many central African countries, and that inevitably influence want-to-be anthropologists as well.

There was something that felt so, well, "universal" about them as people – true to what people used to be, should be, but have largely lost through millennia of cultural and institutional layering of complexity along the way. So yes – culturally unique they were for sure; but sub-human?[2] Hardly.

On the other hand, this Pygmy group's illiteracy in geography 101– being unaware of other countries on the African continent, other continents period, even the name of the country they lived in – was an eye-opener, and clearly related to their remoteness and formal (as opposed to functional) illiteracy. When I asked them about the neighboring countries of Uganda or Rwanda, blank stares were elicited. Had they heard of the United States? More blank stares. Zaire – blank stares yet again. True, my Swahili was a grade above rudimentary, but this did not explain the illiteracy.

When I drew a rough map for them of the continents and seas, labeling them as many a fifth grader in Boston would, this generated pure curiosity. This most modest of geography awareness-raising sessions elicited a brief comment from the principal elder in the village when done – *Akili mingi*. I had exhibited "great intelligence" or knowledge. This was apparently the baseline for Pygmy knowledge of geography in 1974.

Clearly, on the one hand the Pygmies were obviously knowledgeable and tremendously skilled in matters pertaining to their forest-based livelihoods. Then on the other hand, they were in abject ignorance about other basic issues in the world.

Unfortunately, for the vast majority of Pygmies in 2013, with the exclusion of what has become a small elite of well educated, urbanized Pygmies, the same would be true if the exercise were repeated in the hinterlands.

How was "development" then to address the chasm of disconnected realities between one of Africa's most charismatic peoples living as they had for millennia, with the universalist cries for development and human rights demanded by many outsiders to this world? Was this imploring for development by President Sese Seko Mobutu through the term *authenticité*, further undermining Pygmy self-determination and enabling Mobutu to define what it meant to be an "authentic" *Zairois*? What did the Pygmies themselves actually want, anyway? And was it even a fair question to ask considering certain basic information deficiencies?

As I lay inside my recently constructed hut looking up at the enormous, fairly freshly cut mongongo leaves, I pondered more specific questions:

- Where in fact would one begin with "Pygmy development"? What would the objectives be?
- Would it be to bargain with the devil and have them learn some things about either agriculture or the city so that they would achieve some hybridized, sedentarized status that would create ambiguity, anxiety, disease, and probably exacerbate material poverty? When I say "have them", who exactly is the one having them do anything – Mobutu? The World Bank or a foreign government?

On the other hand, to "let them" stay as they were in what for some Westerners may have represented a romanticized twentieth-century version of the Neolithic, noble savages to be preserved for their living museum qualities, seemed equally questionable from a moral standpoint as well.

While the Pygmies did "appear" to be living in harmony with the forest, I saw over those few days that they were in clear-cut subservient relationships with the neighboring Bantu farmers who had moved into their midst over the centuries. Watching how the paternalism and joking played out in face-to-face situations between individuals in the two groups over those ten days was fascinating, as each party on the surface apppeared comfortable in their role.

As the Pygmies did not seem to be integrated into government programs, with kids not even going to school, beyond the intention to sedentarize and modernize them, they appeared politically marginalized to the extreme. I learned that their births were not noted; their citizenship was in fact understood, but unofficial, which maybe explained why they didn't really resonate with Zaire – it was out of their frame of reference, as Zaire was not an applicable framework in their everyday lives.

Under the circumstances, who would make decisions about what would be appropriate for the Pygmies, and on what basis would one determine a logical manner of going about it? Ideally Pygmies would do so themselves, but I was still stuck on this troubling point – could this possibly be *fair* given the unequal starting point? And how in the world would defining what would be "appropriate" be decided?

Mbandaka to Lac Tumba

Fast forward thirty years. I am driving on a "road" in Equateur province, DRC in 2005. After the 1997 overthrow of the Mobutu regime, the former Zaire has become the Democratic Republic of Congo. I am in a swamp forest many hundreds of kilometers to the west of my 1974 site visit with the Ituri Pygmies. The country is struggling to emerge from the aftermath of years of civil war; millions of people have died indirectly from the collateral damage of conflict spread across the country. Provision of public administration and social services is abysmal. Corruption is rampant. Parts of the country remain hell on earth, as measured by the estimated several million deaths at the time in the aftermath of Mobutu's overthrow. Things across the DRC could be much better. While CNN and BBC stories abound, the world apparently cares little more in practice than in the time that King Leopold was carving up Africa with European colleagues and then plundering the Congo.

My former professor Chinua Achebe had clearly loathed Joseph Conrad and his renowned *Heart of Darkness*; a gratuitous, racist text that was deforming of Africans and the continent, and played into the prejudices of Conrad's white readership (see Achebe 1977). Yet as I slog through the muddy roads of Equateur province thirty years after Mobutu was determined to have developed the country, where apparently most of the US$4–5 billion destined for that development ended up in his own pockets, there is something about Conrad's confabulation on the heart of darkness that bothered me. Why in the world are the roads across the Congo still so profoundly rudimentary and impassably muddy in the rains, and were the Congolese not in far worse circumstances as they appeared to be in 1974? I don't feel I'm being a naïve, liberal, racist in the mold of a Marlow by asking these questions, as Achebe suggested Marlow certainly was by dehumanizing Africa and Africans.

Certainly, similar questions about corruption and dysfunctionality could be asked in reference to many other African countries. Still, the Congo case just seems so extremely disappointing given the early optimism. How and why did this happen?

* * *

The vehicle I am driving in belongs to the now defunct NGO that I had founded in 1997 – IRM. IRM was working on community based biodiversity conservation issues in central Africa. By 2006 this was primarily in a 30,000 km² area of swamp forest that includes the confluence of the Oubangi and Congo Rivers. Conservationists refer to this area as "the Lac Tumba landscape". It is part of a larger Lac-Tele-Lac Tumba landscape straddling the Republic of Congo and the DRC. IRM was testing methods that could be replicated to mobilize communities to take conservation action at village levels, and to scale those up across conservation landscapes such as Lac Tumba.

IRM's financing was through a sub-grant from the WWF to IRM from a USG program called the CARPE, the overall objective of which was to conserve the Congo Basin's tropical forests. IRM had been working specifically on community mobilization methods to impact biodiversity conservation in Congo Basin landscapes since 1998, focusing on southern Cameroon and then the DRC. Large scale participatory mapping and the definition of a participatory toolkit for communities to analyze costs, benefits, and risks of different land use options was the mission. We were determined to see if a feasible model for engaging communities in conservation could be designed and replicated.

IRM was also working on freshwater ecosystem management issues along the Congo River and major tributaries, transport related corruption along the Congo River and major tributaries, and agricultural rehabilitation issues in the principal food producing and forest production provinces of Bandundu and Equateur. Our perspective held that all these issues were all linked, and that holistic approaches integrating anti-corruption (Brown 2007, 2009), rethinking of the requirements for rural food security, and a host of tenure, benefit sharing, and public participation issues would require addressing if development was to become effective and sustainable. In the section of Lac Tumba landscape that IRM was working in, the villages from Equateur's regional capital, Mbandaka, to Bikoro, the administrative center on the shore of Lac Tumba 125 km away, figured prominently.

Stopping our Land Cruiser midway on the seasonally challenging mud track, in a village where we had been training Pygmy and Bantu villagers together on participatory mapping as a first step in helping communities develop their vision for possible sustainable development options for both the environment and livelihoods, I am reintroduced to a group of Pygmies conferring with several Bantu chiefs. The Pygmies appear deferential to the mud and feather coiffed chiefs, who control the center of the village space on our arrival.

I strike up a conversation with the first of the Bantu chiefs who came to greet us. It became the type of absurd, joking conversation that travel or work in hot, remote, impoverished, post-conflict ridden places occasionally, at times for the better, engenders.

My intention is mainly to provoke *esprit de corps* and some laughs as part of a long-standing working relationship IRM was engaging with people in the region. Laughs and camaraderie aside, the conversation begins to touch on land tenure rights, law, and government policies, none of which the chiefs and local peoples are well versed in.

By the time it is over, I have been unexpectedly taken back to my brief 1974 visit with the Pygmies, when a similar set of issues and reactions on my part were evoked. I find myself again challenging very basic assumptions about the nature of people in the DRC or the US, and what we need to know to function in contemporary society, be one Congolese or American.

As studies on African land tenure have long clarified, tenure is complex across much of the continent. While the state usually claims rights to lands that local peoples inevitably find themselves using and inheriting through customary claims, the inherent overlaps lead to conflict between local peoples and the state. These overlaps often paralyze rational land use, and can lead to a phenomenen of land grabbing, decried by many (Fairhead et al. 2012; Alden Wily 2012; Karsenty and Ongolo 2012).

In the case of the Pygmies, despite centuries or millennia of sustained resource use, the situation is further complicated by the lack of recognition of Pygmy citizenship – no, the oldest and most "authentic" people in Africa still do not enjoy universal citizenship status in their country of residence, be it the DRC, Republic of Congo, or Cameroon.

More often than not, Pygmies have no way to prove when or where they were born, and hence have no basis for obtaining identity papers. Things are further complicated by the Congolese state's recognition at policy levels of customary use rights. Depending on the viewer and her interpretation of applicable codes and legal perspectives, interpretation of these rights might give prerogative to either Pygmies or Bantus. While Pygmy rights are nominally recognized in forest codes, the DRC codes have lacked specific application measures for their implementation in practice since 2002. This leads to ambiguity and impasse, a major constraint to REDD.

* * *

The following conversation with the Bantu chiefs in a small village thirty kilometers from Bikoro, with Pygmies listening in silently on the side, again raised similar questions from the ones I posed in 1974 that had remained only partially answered.

Even if the questions were to be raised in 2013, the answers would likely be no clearer. Yet, I did not *really* anticipate what was to follow in the 2006 conversation.

> IRM president (the author, translated from French): Hello chiefs. We're on our way to Bikoro to check up on our participatory mapping training. How are things in the village with the training here?
>
> The chiefs: [Heads nodding, and smiling] Fine. Our boys are happy to be trained.
>
> IRM president: Listen, I was just curious: would you be interested in selling me your forest? [A really preposterous, politically incorrect question!]
>
> The chiefs: [Justifiable looks of puzzlement.]
>
> IRM president: What I mean is, I may be interested in buying your forest and am curious if you'd be interested. [Again, a ridiculous, politically incorrect request that contradicted our NGO's mandate and immediate mission.]
>
> The chiefs: [Looking back and forth among themselves with conversation presumably in Mongo, the local language.] Well, what do you mean?
>
> IRM president: Well, how much forest could you actually sell me, first of all? How many hours' walk in that direction towards Bikoro [south], that direction towards the dense forest and the bonobos [east], that direction towards Mbandaka [north], and that direction [west] here by the river [west]?
>
> The chiefs: [more or less] 2.5 hours' walk that way [south]; 7 hours' walk that way [east]; 4 hours' walk that way [north]; and 5 minutes' walk that way [west].

IRM president: Oh, OK great, so sounds like you've got a lot of forest to sell.

The chiefs: [Some nods, and looks back and forth amongst themselves that are indecipherable to the IRM president.]

IRM president: OK, so I'll offer you a bottle of whisky, a case of beer, and $50 for all that forest going in all the directions you just told me about, because what are you doing with it anyway? I mean, it's just mostly swamp forest right, though you probably have some pretty big hardwoods out in the east and south-east [to be sure, with individual trees delivered in Kinshasa and then Europe valued at perhaps US$500–1,000 a tree, or even more…].

The chiefs: [Conferring among themselves…] No. We are not interested.

IRM president: [Concerned by the fact the chiefs had actually been conferring…] OK, then how about US$75, two bottles of whiskey and two cases of beer?

The chiefs: [More serious conferring this time, with a voice or two raised.] No.

IRM president: [Becoming alarmed, as they appear to be taking me seriously, regardless of the fact that a transaction could never ensue.] OK, my last offer $150, three cases of beer, two bottles of whiskey, and my invitation to come out to dinner and a night on the town with me in Mbandaka.

The chiefs: [Wholly animated conversation, and a minute later the chief with yellow and red feather in his mud coiffed hair says…] Yes, we accept!

IRM president: [Quickly to himself: What situation do we now have here? Local leaders from at least three principal villages we are committed to doing "capacity building" in, to set the stage for land use planning as a precursor to "sustainable forest management", saying they are prepared to sell me their forest for a pittance (that they absolutely cannot do – at least I think they can't…!). Only the date is now 2006, not the Belgian colonial era. And these are village chiefs from our principal "beneficiary" communities. Do the communities realize they're prepared to sell them and the forest down the river, if they could?] OK, chiefs. I'm not a forester and I don't have my figures straight. But you see that one massive tree over there [tentatively identified, I speculate, as a wengé (Millettia laurentii) that could bring in thousands of dollars delivered abroad if allowed to mature further, but could likely fetch US$500 in Kinshasa anyway…], let's say that wengé tree alone is conservatively worth a few hundred dollars in Kinshasa, and a couple of thousand dollars in Europe. And you chiefs have no idea of the value of that one tree. Even if you could, and I believe that you can't, you'd be ready to sell your whole forest for less than the value of that one tree! Do you really think you could sell me your forest? Who actually owns it? You, the chiefs? The government? The villages? [Silence from the chiefs; IRM president frustrated, uneasy, and patronizing.]

IRM president: [Am I more exasperated because I am failing in my capacity building mission, or because the Sisyphean struggle is so much steeper than we'd imagined, or because of the simple fact that these people are in *such* a precarious position with no tunnel, let alone light in sight, that they'll negotiate *anything* and deal with the consequences?]

With this encounter, our efforts to bolster our NGO's analytical and decision-making capacity building program for Congo Basin peoples redoubled in urgency. From this encounter, my own reckoning about the mission and methods of foreigners working in remote, outside-the-rule-of-law places like DRC's massive Equateur province, re-tripled in urgency.

These guys had been serious enough to negotiate themselves into some sort of fix. Did they actually think they could do a transaction? Did they understand the ambiguous rights to resources they held?

My personal anecdote with the Pygmies is one that could be replicated by many researchers, development practitioners, and tropical foresters across remote forested regions of developing countries. My experience is *not unique*.

While the strangest things can happen in remote places like Equateur province of the DRC, I just do not see how REDD is addressing the real human and social capital deficits on the ground in remote, charismatic, rich forests like those of the Equateur or Bandundu provinces in the DRC. These are not easily bandaged up with either PRA or SIA.

Implications of Congolese anecdotes for REDD

Anecdotes can of course be debunked by virtue of not being scientific and, thus, arguably not representative of anything beyond the storyteller's experience. Nonetheless, they do offer windows into situations where deeper truths may be found.

When we read project development documents (PDDs) from voluntary REDD projects that are validated and verified by standard setters like VCS and the CCBA, the complexity, ambiguity, and socially capital challenged reality of places like the DRC is reduced or eliminated from the documentation. These documents seem overly simplified instead.

PPDs employ all the standard buzzwords that are habitually found in development or conservation documents. They create a veneer of apparent understanding and management control over situations that bears little to do with the complex realities they supposedly describe and control.

This format neutralizes any attempt at serious analysis or discussion, particularly when the standard setters *do not themselves demand it, perhaps believing, perhaps not, that greater detail and analysis into complexity is unfounded when it comes to REDD*. Yet, this messy reality in the Equateur provinces of the world is inescapable in REDD, whether or not the VCS and CCBA, and their accredited auditors with Forest Stewardship Council (FSC) credentials in many cases, validate and then verify the voluntary REDD projects, particularly regarding FPIC among remote forest peoples (Brown and Mogba 2008).

Meanwhile, REDD planning at national levels is supported through the FCPF "Readiness Package" (R-Package) processes consisting of five core elements: (1) a REDD strategy, (2) an implementation framework, (3) an MRV system, (4) a reference emissions level (REL) scenario, and (5) safeguards. Yet how well will the R-Packages support the realities of the Equateur provinces of the DRC? While there is concern that criteria and indicators for the evaluation of the R-Package will not be strong, that there may be no provisions for what happens in cases of non-compliance with standards, and that civil society and IPs' role to participate and validate what their governments say about progress on REDD readiness will be unclear (BIC 2011), how is credible participation in the R-Package process itself to be secured given the myriad constraints posed on the ground?

Whether visible or not, these remote places where REDD is focused are where multiple stakeholder interests converge and then diverge. Those specializing in description of the plight of the commons describe them well (see Alden Wily 2012; Fairhead et al. 2012). As in colonial days, stakeholders in Brussels, Paris, and now Vancouver and San Francisco have sights set on adding value to the forests of Bandundu and neighboring Equateur, albeit "improved social benefits" are central to the program. To this are added Chinese, Americans, Norwegians, and others, along with Rwandans and Ugandans in the subregion, all of whose efforts are facilitated through non-transparent agreements between largely unaccountable administrators in countries with abysmal corruption rankings. In the case of the DRC, for example, the Ministry overseeing environment and conservation has the right to grant rights to private sector operators to develop REDD+ projects in former logging concessions, again with "consent" from local communities, and to then own and sell carbon credits generated from REDD+ activities. Alden Wiley (2012) notes that "World Bank structural adjustment programs aid and abet this, coercing governments to accelerate privatization and sell off untitled lands (including the commons) to foreign investors". While conservation concessions on former logging concessions are different, for local peoples the process or end result may appear, correctly or not, to be the same.

Some refer to multipurpose REDD landscapes as pristine wilderness worthy of setting aside into PAs (as they already often have been where wildlife is abundant). Others from afar see them as the lungs of the Earth, key to the survival of Parisians, Londoners, and New Yorkers. Still others simply refer to these remote places as "home". For the latter category, this is where millions of people are born, where they hunt, maintain sacred sites, collect and grow crops, and have been essentially left to their own devices by governments either unwilling or unable to provide viable alternatives for what essentially still pass as twenty-first century subsistence livelihoods.

Others wish to extract wood and mineral products for global consumption from these spaces. Others promote conservation of forests and biodiversity to save the so-called pristine wilderness, while averting climate change that deforesting spaces would provoke. These conservationists are working more and more in alliances with private sector interests seeking to earn profits from climate finance.

Governments seeking rents from these forests, or from conserving the biodiversity in them, are also part of the emerging alliances. Whether it is through

AD in new conservation concessions or "sustainable" forestry operations in logging operations, public administrators have tremendous leverage over the fate of these forests and the lives of people whose subsistence depend on them. That these administrators often use this leverage to satisfy their own personal needs, perhaps because their salaries have not been paid for eight months or more, or perhaps because they are amassing a war chest through grand corruption in places like the Congo, of course raises questions about the credibility of agreements. Of the many ways REDD is seen to potentially induce corruption, TI (Transparency International 2012) prepared a handbook to assist civil society help bring potential corruption associated with REDD more into the light in countries where forestry sector corruption is business as usual.

For local peoples with little information, limited social capital, and little flexibility for investment of time and resources in risky ventures, it is now imperative *for their own welfare* to sharpen understanding about the value of resources over which they may have legitimate rights. The stakes are now ratcheting up in these intensely contested spaces where local and international interests collide.

Unless local peoples have the ability to analyze and negotiate, and not *only* advance rights-based claims that states and other stakeholders often choose to ignore, the inability to interact with greater sophistication with external stakeholders will leave communities overwhelmingly stuck in vulnerable positions. Rights-based claims alone, in all likelihood, will not be enough for local peoples to engage equitably and effectively with more powerful agenda-setting stakeholders.

With REDD, these remote places have now come directly into the crosshairs of both international and local stakeholders. The space is characterized more by competition than cooperation. Though the quest for sanitized linguistics in REDD is apparent, this type of situation is what van den Berg and Quarles van Ufford (2005) suggest generates "increasingly far-fetched, abstract and artificial images of a world order". This far-fetched, abstract, and artificial world order is striking across the REDD programmatic literature.

Implications when social capital is lacking

Whether it concerns the Ituri Pygmies of Zaire in 1974, or Bantu chiefs in patron–client relationships with Equateur province Pygmies in the DRC in 2006, the limited capacity of local communities to objectively value assets is indicative of a much broader malaise. Having this capacity is key for either capitalizing upon, or defending, local rights. This malaise constrains peoples' ability to participate effectively. It hamstrings achievement of conservation, and ultimately undermines REDD and sustainable development objectives as well, while perpetuating dependency.

Without community capacity to analyze and coherently engage in conversation, any *equitable* negotiation between communities with decentralized governance entities, elected leaders, logging and mining companies, or international NGOs, is limited. Thus, examples like that of the absurd conversation between the author and Bantu chiefs of Lac Tumba is indicative of a much broader challenge facing conservationists and proponents of sustainable climate change mitigation through REDD. This challenge involves determining how human and social capital can be squared with weak official recognition of customary rights of indigenous and local peoples to forest resources. With the critical mass of evidence growing that forest peoples can and should assume a more prominent role as AD stewards under REDD, what policies and practices can help make this happen?

With weak social capital at multiple institutional levels, and scant reference to *systematic capacity building* or the clarification of customary and "modern" tenure rights to resources, can the basic objective of permanence of carbon assets in either fund-based or market-based REDD+ possibly be achieved?

If a REDD project developer suggests that it will be through seasonal jobs monitoring forest resources, with some full-time monitoring jobs in conservation concessions sprinkled in, and that this together with a clinic and schoolhouse will satisfy the community and provide insurance for carbon permanence for rationalizing donor and private investment, than I have some good snake oil to sell that project developer too. All the more so if all benefit flows must pass through the hands of local chiefs, who can no longer be assumed to be representative or working in the best interests of the communities.

5 Science and policy

This chapter presents an overview of the principal science and policy issues related to REDD's success or failure.

Science background

According to the IPCC, the premier consensus-based scientific body on all aspects of natural and anthropogenic climate change, warming from anthropogenic causes of the climate system is widespread. The concentration of atmospheric CO_2 has increased from a pre-industrial level of about 280 ppm[1] to 379 ppm in 2005 (IPCC 2007: 25).[2] The concern is that we have reached a tipping point that could lead to permanent damages and significant, difficult to evaluate, costs.

It is believed to be *"very likely* that anthropogenic greenhouse gas increases caused most of the observed increase in global average temperatures since the mid-20th century" (Solomon et al. 2007). Of this, fossil fuel use, agriculture and land use have been the dominant cause of increases in greenhouse gases over the last 250 years (Solomon et al. 2007).

The costs of this tipping point are becoming clearer: the deaths of nearly 400,000 people a year costing the world more than US$1.2 trillion, and wiping 1.6 percent annually from global gross domestic product (GDP) (DARA and The Climate Vulnerable Forum 2012).

Looking to the future, 50 gigatons of carbon "equivalent" (GtCO2e) are predicted to be emitted between now and 2020, assuming BAU conditions under which no international mitigation pledges are actually enacted. An unmitigated gap of 6 GtCO2e is now foreseen (UNEP 2011).

Previous consensus had it that 2°C represented the upper limit for what could be tolerated prior to irreversible damages to the environment and global economies. This has been "revised upwards, sufficiently so that 2°C now more appropriately represents the threshold between 'dangerous' and 'extremely dangerous' climate change" (Anderson and Bows 2011).

While debate continues over rates of climate change, along with mitigation strategies given costs, evidence is mounting that the pace of GCC is galloping ahead far quicker than previously estimated. Peters et al. (2012) reported that emissions from CO_2 from fossil fuels, the main GHG, jumped 5.9 percent in

2010, the sharpest one-year rise on record. The report also said that carbon emissions cumulatively had risen by 49 percent since 1990, higher than any previous estimate.

Meanwhile, the probability of exceeding 2°C rises to 53–87 percent if global GHG emissions are still more than 25 percent above 2000 levels in 2020 (Meinshausen et al. 2009), a near certainty based on recent analyses (Anderson and Bows 2011).

The IPCC states that models project substantial warming in temperature extremes by the end of the twenty-first century, and that

> it is *virtually certain* that increases in the frequency and magnitude of warm daily temperature extremes and decreases in cold extremes will occur in the 21st century on the global scale. It is *very likely* that the length, frequency and/ or intensity of warm spells, or heat waves, will increase over most land areas.
> (IPCC 2011)

For those who have recently lived through a series of once in a lifetime weather events that are occurring now on an annual or biannual basis, the IPCC pronouncements may even appear anecdotally to the worst hit, to be undershooting the need for alarm.

Policy and stakeholder responses to deforestation

At present the global community does not have a systematic politics of climate change, with some suggesting as a global community we must adapt and develop policies for a sustainable low-carbon economy for our survival (Giddens 2009).

With the inability of the UNFCCC process to develop a mechanism for REDD, this has led to the splintering of independent initiatives where BINGOs, private sector groups, and financial backers are trying to establish the standards and best practice upon which not just voluntary, but eventual compliance markets are based. In this regard, failure at international levels has created space for entrepreneurs to fill gaps and define terms, and a number of groups, with financial backing, are picking up the gauntlet. As of the 2012 Doha UNFCCC COP 18, the mix of ambivalence and wishful thinking among key thought-leaders on carbon markets, along with the prospects for admission of REDD into ADP negotiations for the 2015–2020 period remained striking (see Silva-Chavez 2012).[3]

At the same time, to fuel this dynamic, still volatile situation further, resistance to REDD appears to be hardening, if not expanding. One 200 million strong farmers' movement alone – La Via Campesina – is pronounced in its denunciation of both forest carbon and soil carbon markets (La Via Campesina 2012). Numerous IP coalitions, and NGOs in their websites and listservs are coming out with regular broadsides to REDD on multiple fronts (see greenpeace.org, Friends of the Earth, The Rainforest Foundation UK, REDDeldia Chiapas, Forest Peoples Programme, and REDD-Monitor.org to name a few).

A key technical challenge to REDD which similarly was encountered in the CDM under the World Bank is the principle of "additionality". That is, REDD projects must demonstrate that whatever transpires *would not* have happened in the absence of the project activity. The principle is a very difficult one to prove and work with. This is because additionality is difficult to assess at project levels. Even with clear boundaries of the planned activity and knowledge of historical data related to a given area, deforestation is a result of numerous factors that involve both people and climate, and extends far beyond a single project (Karsenty 2008). Attributing "additionality" directly to projects provides impact evaluators with undeserved sway in their interpretations of data that more likely, is attributable to multiple causes.

The IPCC, which comprised over 2,000 scientists who assess all evidence pertaining to climate change, has estimated that the global forestry sector is responsible for over 17 percent of global GHG emissions. This places it above the transport sector at approximately 13 percent of global emissions.

In fact, the range of estimates for deforestation is wide. The lowest estimates come in at 12 percent and the highest at 25 percent, though as of 2013, an anecdotal consensus *appears* to be at around 17 percent.

REDD first arrived in international negotiation fora at the COP to the UNFCCC in 2005.[4] At the 2005 Montreal COP, Costa Rica and PNG argued that deforestation contributed huge impacts on climate change. They submitted a joint request to the UNFCCC to create a mechanism to reward reductions in GHG emissions due to deforestation and forest degradation. Thus was born the notion of providing incentives for countries with high deforestation rates to conserve future forest cover. With the political traction gained from countries that heretofore had been excluded from the CDM, linked as well to the interests of forest industry lobbies and conservation advocacy groups, the impetus for REDD was launched.

At the Bali Climate Conference in 2007, the transition from REDD to REDD+ was made, with the "+" covering the preservation or enhancement of carbon stocks. Since the Bali conference, numerous public and private initiatives have been undertaken to support REDD+ actions at both local and national levels. Unlike REDD, a REDD+ mechanism has never been formalized by the UNFCCC. Now eight years on after first initiating discussions on the need for REDD+ at the UNFCCC COP in Montreal, REDD+ remains awkwardly waiting at the altar for the long-discussed event to be finalized.

What happened at the subsequent Bali UNFCCC COP was intriguing. Countries with extensive forest cover and actual low rates of deforestation gave birth to a new lobby for virtuous, non-deforesting nations, was born. Led by the Congo Basin countries of the DRC, Republic of Congo, Gabon, and Cameroon, they lobbied for the REDD mechanism to cover not just the "worst offenders" but also the so-called "more virtuous nations" who up to that point had largely succeeded in preserving their forests. In their case, this was due to conflict, remoteness, or simply economic policies not conducive to logging operations. In the DRC, the locus of the Congo Basin's principal forest carbon sink, protracted underdevelopment, conflict, and the hefty transaction costs of entry into the logging sector were all at play.

As the DRC was moving to democratic elections from protracted civil strife, the specter of potential deforestation was potentially a risk there. This provided a logic for creating incentives to avoid what had *yet to happen*, but which clearly could. Combating this incentive to deforest became the basis of the DRC's negotiation for recognition of its counterfactual premise underpinning REDD (and other countries with historically low deforestation rates such as Guyana, supported by the Norwegian government with a US$30 million planning grant and an additional US$250 million through 2015 if all goes well).[5] That premise is the following: if not provided incentive through PES (presumably under REDD) to *not* deforest now, forests in countries like Guyana and the DRC will likely be converted. The growing demand for wood products, minerals, and cattle is likely to increase pressures in the next few decades that will be unavoidable, absent externally provided incentives. Thus emerged what many have come to consider a form of developing country blackmail.

The Cancún COP (UNFCCC 16) in 2010 went further and established eligibility criteria for REDD+. Eligible actions for support were identified as those aiming to enhance forest carbon stocks or mitigate their reduction. The proviso was that it be demonstrated that any such actions constitute demonstrable and measurable steps forward from a BAU scenario, and that a system for MRV of carbon stocks be implemented. So too, safeguards would need to be put in place to ensure that the full participation of indigenous peoples, preservation of biodiversity, and conservation of natural forests (Samyn et al. 2012) be part of the process.

Holes in the logic

Holmgren and Marklund (2007) outlined the perceived MRV challenges to REDD from a UN-REDD perspective. Many experts continue to believe that "viable REDD measurement and validation systems can be implemented" (Wertz-Kanounnikoff and Verchot et al. 2009). While it is straightforward to establish the framework for an MRV system for REDD, whether it is effective is still being determined. Technically, progress from year to year in MRV has been evidenced.

On the other hand, "full participation" of IPs and local communities is far more problematic. The combination of complex logistics to reach the tens of thousands of remote IPs' hamlets scattered across REDD-eligible countries, combined with illiteracy, functional illiteracy, and the inability of female IPs to participate as equals with men, makes nonsense of the aspirational notion of full participation. This is true for UN-REDD consultations for national level planning, as well as for individual PDDs. The conclusion: *Full participation*, as planners and rights based groups would like for it to mean, is impossible to attain in 2013, or in the foreseeable future. This has direct implications for FPIC.

What is key is to determine in every REDD country setting, given the mix of logistical and social capital challenges, what is necessary and feasible to seek for the quality of participation among IPs and local communities. While this could be set as a verifiable process level indicator that auditors could validate, for the time being, emphasis appears primarily to be on checklisting stakeholder meetings.

Considerable debate remains as to what will constitute "catastrophic climate change", when this change can reasonably be expected, and what it will take to mitigate it in terms of actions and resources. The commonly accepted threshold for when CCC will occur is measured by an average temperature rise of more than the 2°C globally against the pre-industrial revolution average. There is a growing body of scientific analysis that the probability for crossing this threshold is rapidly accelerating (Hansen 2012), and that we are, in fact, on a pace to surpass the threshold far sooner than envisioned even five to ten years ago.

Some argue as to whether the 2°C maximum temperature rise should remain the focus at this point, as opposed to "large-scale discontinuities" being the biggest cause for climate concern (Lenton 2011). Radiative forcing, which involves changes in energy fluxes, could in addition to global warming, contribute to other watershed events such as irreversible loss of major ice sheets, reorganizations of oceanic or atmospheric circulation patterns and abrupt shifts in critical ecosystems (Lenton 2011).[6]

Meanwhile, policies to enact mitigation measures to put even a dent in the accelerating pace of anthropogenic GHG emissions have met stiff resistance from a range of stakeholders, most notably from the US, China, and India. With each passing UNFCCC event, a mix of cynicism, skepticism, and moral outrage confronts bureaucratic platitudes about progress achieved.

If concern were to be accurately measured in terms of critical journalism, it is fair to suggest that serious concern, combined with widespread confusion about the climate mitigation process, now far exceeds rosy optimism at multiple levels. As time passes, a mishmash of data points confounds generalization:

- clear indicators for indisputable Arctic melting become more visible (National Snow and Ice Data Center 2013);
- "Climategate" scandals tarnish the objectivity of climate science and some of the principal scientists involved even though claims for supposed "clandestine data manipulation" have been laid to rest as bogus (Pearce 2010);
- freak weather patterns and superstorms become more normal;
- the global community appears ever more divided and unsure about whether to care at all and do anything about it.

Meanwhile, what seemed a mere five years ago to be a surefire bonanza in future REDD official development assistance (ODA), combined with a potential tidal wave of on-the-sidelines finance waiting for the next great investment opportunity, has shriveled in the aftermath of the 2008 global recession. The situation remains more than a bit muddled in 2013 as to the future of GCC mitigation.

REDD as a mitigation strategy

REDD, as one mitigation strategy among many that will be needed, is predicted to address the 1.3 to 4.2 GtCO2e of deforestation emissions per year through a reduction in deforestation, and changes in forest management that increase above and below ground carbon stocks (UNEP 2011). However, given that "it is not now possible to ensure with high likelihood that a temperature rise of more than 2°C is avoided" and accepting the view that reductions in emissions in excess of 3–4 percent per year are not compatible with economic growth, the premises underpinning REDD appear wobbly, e.g., can it, and will it, make a crucial difference?

In the UK, the Committee on Climate Change is, in effect, conceding that avoiding dangerous (and even extremely dangerous) climate change is no longer compatible with economic prosperity (Anderson and Bows 2011). It simply is perceived as costing *too much* during a period of economic downturn.

How this will play into UK participation in the global mobilization of US$5–33 billion needed annually to address deforestation remains to be seen.

REDD in the context of major policy initiatives: the Paris Declaration and MDGs

The Paris Declaration on aid effectiveness (OECD 2005) offers a twelve point plan for what the objective of all development practice should be about. When referred back to REDD, serious questions emerge.

According to the Paris Declaration, development is the "arrangement for the planning, management and deployment of aid that is efficient, reduces transaction costs and is targeted towards development outcomes including poverty reduction". So, too, it also contributes to satisfaction of sustainability objectives per the MDGs.

It is difficult to see how REDD meet the criteria of the Paris Declaration definition in terms of efficiency, reduced transaction costs, and poverty reduction as of 2013. While purposefully by choice reducing transaction costs, poverty reduction objectives in REDD will be a function of the untested theory that communities will benefit from setting a market value on environmental services and then reorganizing customary resource use accordingly. In fact, poverty may be exacerbated rather than reduced through this approach.

The MDGs "provide a framework for the entire UN system to work coherently together toward a common end" under the UNDP in 177 countries and territories (UNDP 2012). Lewis and Mosse (2006) note that despite the bold MDG declarations, few of the MDGs are likely to be met in practice.

The gap between vision and reality in poverty reduction is because the latter "is more prosaic and, in many zones of the world, tragic" (Gough and Wood 2004), with "rapid increases in global inequality, the continuing spread of HIV/AIDS, famine and conflict and the fact that in many areas where capitalist economic growth has taken place it has been associated with increased instability, insecurity and vulnerability" (Lewis and Mosse 2006). So MDGs are espoused on the one hand, and disappointing development indicators still prevail on the other. This failure of the MDGs is relevant to REDD, as REDD in turn hopes its success will further MDG outcomes.

Learning from Joint Implementation and the CDM experience

Joint Implementation (JI) under the Kyoto Protocol involves joint projects between "Annex 1 countries"[7] that save on CO_2 emissions. The Kyoto Protocol produced JI as an instrument between Northern countries and the CDM, which enables investments in Southern countries to be credited to Northern ones (Newell and Paterson 2010).

On the surface, it would seem as if the CDM experience and REDD share limited similarity. In fact, many aspects of the experience are germane to REDD, and foreshadow problems REDD may eventually face.

In both the CDM and REDD, emissions credits go to countries that reduce GHG emissions *overseas*. In the CDM, this includes credits from wind, solar, energy efficiency, and other technologies, while at the same time promoting employment and helping conserve biodiversity (Newell and Paterson 2010). In REDD, Northern investors can purchase offset credits through voluntary markets. By 2020 it is assumed that compliance mechanisms will be in place which are to create incentives leading to higher prices and greater market predictability for avoiding deforestation.

The structure and mechanism for developing projects is very similar for CDM and REDD projects, with project development documents, agencies which validate methodologies proposed and employed, and auditors who certify that business is conducted per the terms of the proposal and the standards against which the project developer wishes to be evaluated. As the High Level Panel on the CDM Dialogue (2012) suggests, REDD is a potential program area that is seen as offering advantages for progressive CDM involvement. Past CDM failures may, ironically, be mitigated through success in REDD.

While far from clear, the CDM could end up ultimately housing a REDD mechanism as encouraged by the High Level Panel (2012). Figure 5.1 provides an idea of what may reasonably be expected of the administrative architecture for a REDD framework housed by the CDM, should things evolve in that direction (UNFCCC 2012a).

In it, a project sponsor hires a Designated Operational Entity (DOE) to validate claims developed in a Project Development Document (PDD). Once validated, host country government approval is sought through a Designated National Authority, after which it arrives at the CDM Executive Board for approval. If it is using an already approved methodology for accounting, it then passes directly to the CDM Executive Board. In the event a new methodology is proposed, another consulting firm (the DOE) must validate the methodology. Implementation follows with monitoring responsibility placed on the project developer, and it is again certified by another auditor. This DOE sends a report to the CDM. The Executive Board issues credits, or CER units, which are the fundamental currency in the global compliance carbon market, and are traded by investors as assets (Newell and Paterson 2010).

Figure 5.1 CDM project cycle: a possible basis for a UNFCCC REDD mechanism

Source: UNFCC http://cdm.unfccc.int/Projects/diagram.html

The three phases of REDD

Under the current REDD+ orientation being promoted through UNFCCC negotiations, there are three phases of REDD+ envisioned: (a) the *Preparation and Readiness Phase* where "national REDD strategies should be built in a participatory way, including and recognizing the rights and roles of indigenous peoples, local communities and other vulnerable groups such as women", (b) the *policy and measures phase* where "national policy frameworks and reforms in the forest sector will be developed and the building of links with other related sectors such as agriculture, energy and development", and (c) the *performance-based payments* phase, where "mechanisms such as the carbon market and fund-based mechanisms should *deliver performance-based payments for emissions reductions* or carbon stock enhancements. National and local REDD-plus projects should demonstrate results in this phase" (IUCN 2009).

Appreciation from various stakeholder perspectives of the implementation of the three phases of REDD+, particularly in the current Preparation and Readiness Phase, are highly divided between proponents who see REDD+ activities going through predictable pains in a "preparation and readiness phase", and critics who see the Preparation and Readiness Phase as not doing nearly enough to prepare and get REDD+ ready for actual implementation.

The Readiness Preparation Proposal Process

The R-PP process is the principal motor behind REDD programming at country levels. It is driven by the World Bank's FCPF and UN-REDD, the latter comprising UNDP, UNEP, and the FAO.

That said, the R-PP process contains deficiencies, particularly the apparent inability to meet the FCPF's own standards in many R-PP reviews for consultation. In addition, because the standards do not require it, there is no regard to feasibility principles, and social feasibility in particular. It is unlikely that the template employed (FCPF and UN-REDD 2012) is designed to sufficiently capture information needed to design feasible projects.

Taking one example, the first Technical Advisory Panel (TAP) Review of the PNG R-PP on standard 1b for "information sharing and early dialogue with key stakeholder groups" conducted on September 12, 2012 (Sathaye and Cobb 2012) notes that the R-PP provides no sense of the effectiveness of the national consultation process in terms of who participated, what the participants noted about their willingness to engage in REDD+ activities, funding needs, etc. So long as this part of the REDD planning process is weak, there is no way to achieve social feasibility or long-term sustainability in REDD.

Nepal presents an example of how one country is approaching Readiness Phase planning (Government of Nepal 2010). From document review, planning appears to be top-down, and many key elements of what constitutes the social dimensions in REDD are left to be determined, presumed to be resolvable in due time, or else do not require attention. A reading provides little indication that anything resembling feasibility analysis is being attended to. Rather, the approach is more procedural and abstract in terms of what is being done, when, and why. This does not inspire confidence.

The following is adapted from Nepal's R-PP (Government of Nepal 2010):

> Nepal has demonstrated that community involvement in forest management can significantly contribute to reducing deforestation and forest degradation especially in the Mid-Hills and that this significantly contributes to forest conservation and enhancement of carbon stocks.
>
> Management rights in most of the community management models are transferred to communities whereas land tenure rests with the government. A key principle is that carbon rights should be tied to land and forest tenure rights.[8] In the next few years carbon ownership for all types of forest *need to be resolved* [emphasis mine] as a priority during RPP implementation. Similarly, clear and legally defined benefit sharing mechanisms that can deliver benefits to grassroots level communities, will be an important factor for REDD success.
>
> To clarify institutional arrangements, further studies and consultations with stakeholders from local level to national levels will be carried out. A hybrid approach to REDD implementation at both national and sub-national levels is proposed *although details of these arrangements still need to be finalized through pilots and further consultation and studies* [emphasis mine].

In addition, Nepal is proposing to use a trust fund model for financial transactions from which payments are made on the basis of a public carbon registry maintained by the REDD cell. Although there are a couple of pilot projects that are already being implemented in Nepal, their findings will not be sufficient to develop the implementing framework. Therefore the process of finalizing the institutional arrangements includes conducting and synthesizing studies, policy development, and designing and piloting the institutional framework.

(Government of Nepal 2010)

As an indicator of just how difficult the task of "*finalizing the institutional arrangements*" is proving to be, the policy challenges that Kanninen et al. (2007) outlined remain *still* to be credibly addressed six years on into the first REDD programming cycle: e.g. policies need crafting to address diverse local situations; regulatory and governance reforms are needed; perverse subsidies that provide incentives for clearing forests need removal; reform of forest industry policies allowing unsustainable extraction needs implementing; and devolution of resource rights and management responsibilities to local forest users, and recognition of forest-based environmental services (in addition to carbon storage) must occur.

TMAs and the impact of deploying theory of change in REDD

The root of the challenge confronting all stakeholders to REDD, proponents as well as critics, is determining what the terms "readiness" and "preparedness" *should* mean, and currently do mean. This is clearly the case as well with TMAs employed in REDD.

While it has become politically correct in development to be non-prescriptive in terms of TMAs, the lack of clarity in content and thresholds for what actually works, as opposed to what approaches have been tested and practitioners believe with expert judgment are worthy, leads to very weak standards and practice.

It is an unspoken, conventional wisdom to believe that non-prescriptiveness represents a strength, as it enables users to choose what is most appropriate for their circumstances. However, the lack of clear directions as in a "roadmap" constrains effective action, though, as Taleb (2009) correctly advised, no map at all is better than an incorrect map to begin with.

Assuming a roadmap can be accurately designed to help REDD practitioners negotiate through the complex steps of designing and implementing activities that work at national or project levels could save practitioners considerable time and uncertainty. Presently, practitioners are left with considerable leeway regarding multiple methods to employ in REDD, whose efficacy is left to users to evaluate.

In SBIA guidance for the expanding field of voluntary REDD projects, where entrepreneurialism and dynamism is best exhibited, the palette of options provided to approach design and implementation perfectly meets the needs of fulfilling "theory of change" requirements (Richards and Panfil 2011).

Box 5.1 Early lessons from FCPF R-PPs and UN-REDD National Programmes (adapted from FCPF and UN-REDD 2012)

Assessment of early R-PPs by the Participants Committee and by the UN-REDD Programme suggests the following early lessons for countries preparing an R-PP, as well as preparing and implementing National Programmes:

1   *Develop some form of cross-sectoral REDD+ working group:* The working group composition and national REDD+ management processes need to be cross-sectoral and engage relevant sectors and stakeholders. Some working groups described to date in R-PPs were dominated by a single agency and did not include other key agencies (e.g., agriculture, mining, transportation) and interested forest agency. In some countries, forestry and environment agencies have to learn to work more closely together and cooperatively with civil society and indigenous peoples, as all have competencies related to REDD+.

2   *Create platforms for meaningful and effective stakeholder participation:* Participation and engagement is critical to developing viable REDD+ strategies and implementation frameworks, and should begin as early as possible when a country begins considering participation in REDD+. REDD+ requires extensive information-sharing with and consultation among interested stakeholders including multi-sectoral government agencies, civil society, private sector, indigenous peoples, and development partners. Stakeholder consultation processes not only ensure wide-ranging acceptance and interest in REDD+, but also build the trust of stakeholders and support their capacity to participate in REDD+ in a meaningful and effective way as an ongoing process. The readiness process needs to establish both formal and informal mechanisms for engagement and feedback to ensure adequate consultation among all these stakeholders.

3   For countries participating in the UN-REDD Programme or that otherwise follow a policy of Free, Prior and Informed Consent (FPIC), the engagement of indigenous and traditional forest-dependent peoples is an on-going process, rather than a single event. Adequate time needs to be allowed for the careful management of awareness raising and engagement with local authorities communities as well as with national indigenous organizations and relevant intermediary groups. Provision of enhanced local capacity for effective awareness raising and discussion of issues is important, as is carefully structuring an FPIC process and documenting its decisions.

The rationale for theory of change underpinning REDD design and impact analysis is explained as follows (Richards and Panfil 2011):

> Having reviewed potential SBIA approaches, it was concluded that the *theory of change* approach was the most cost-effective and appropriate methodology for assessing the social and biodiversity outcomes and impacts of land-based carbon projects. While projects applying to the CCB Standards are of course at liberty to use a different approach, a *theory of change* methodology seems most appropriate for the following reasons:
> - It uses a very similar logic and sequence to the CCB Standards;
> - It should contribute to a strategic project design (at least of the social and biodiversity objectives);
> - It involves a high level of participation by stakeholders and can contribute to project ownership and project–stakeholder relationships;
> - The external advisory costs are significant but not excessive – other impact assessment approaches such as the quasi-experimental method are much more demanding.

The *theory of change* approach is also consistent with the principle of "appropriate imprecision" (as opposed to "inappropriate precision") promoted in rural development participatory learning approaches (Chambers 1983). Constructing a robust project *theory of change* and backing it up by carefully chosen indicators and participatory impact assessment, as suggested by Catley et al. (2008), is more important than striving for precision through a more sophisticated or quantitative approach. It is also much more likely to be understood by the project stakeholders.

A theory of change approach thus does not require proof of feasibility but, rather, creates a flexible framework. This flexibility allows for subjectivity to enter into the design and evaluation process.

From an unstated, readily apparent tactical standpoint, emphasizing theory of change enables project developers to buy time to work out complex issues. This ostensibly is sensible, save that it opens the door to mediocre and substandard design in the process. Nor does it account for how powerful organizations affect local institutions when they come into a new place in undertaking REDD (see Ribot and Larson 2012). The guidance for utilizing theory of change places loose demands on REDD project developers to receive CCBA approval, and largely avoids consideration of structural issues pertaining to power relations and the impact of REDD on social stratification. While the theory of change approach to all appearances kills several birds with one stone, it ultimately is undermining feasibility by not creating mandatory consideration of key social and institutional issues.

Where policies promoting theory of change in REDD break down

The problem with the theory of change is apparent from analysis of one CCBA verified and validated project from the Mai Ndombe that employs it.[9] The DRC and specifically Bandundu province and the Mai Ndombe district is an area the author knows well, as the organization he ran (IRM) conducted US$14 million in projects across multiple sectors from 2000–2008 in the country, with several involving the Mai Ndombe. Yet, the project is CCBA validated and verified (ERA and Wildlife Works 2012), and it appears to fulfill all the guidance provided by Richards and Panfil (2011). Moreover, it does so well.

Rather than demonstrating insight into how local institutions do or do not function effectively for REDD purposes, why, and what to do about it to raise probabilities for success, REDD project developers are enabled under current CCBA standards to present a theory of change that is able to avoid addressing these complex issues, while basing reasoning on what inarguably are important, albeit simplistic development deliverables. It will only be through enlightened donors and the standard setters themselves that this situation changes.

As the CCBA policy of prioritizing the theory of change approach to design and monitoring of projects is, in fact, an excellent facilitator of projects that are self-billed as "appropriately imprecise", decision makers and donors have accepted this premise. This has likely been acheived for reasons of expediency; e.g. social issues are addressed through what to all appearances is a logical theoretical framework; costs have been reduced as a result of the guidance; and, best of all, local stakeholders do participate.

In many respects, the less information a project developer needs to present, the better, as the situations are so complex, and the understanding about how to work through the complexity so basic, that REDD projects likely would not be funded or invested in if this complexity was to be made transparent and broadly understood.

Obvious gaps in the theory of change in Mai Ndombe

The theory of change approach bypasses regard for feasibility assessment. Basic rural development needs are packaged into theory. Few would disagree with the project developers that rural communities in the Mai Ndombe, for example, do not place a priority on (a) access to quality health and education, (b) access to potable water, and (c) food security and economic alternatives – this actually is true across virtually the entire DRC.

What would be more instructive in a theory of change for the Mai Ndombe is reference to how the project developers are going to achieve buy-in for project activities considering the outstanding tenure ambiguities that communities may perceive over land transferred to ERA from a prior logging concession (SOFORMA) by the Government of the Democratic Republic of Congo (GDRC) in which the Ministry of Environment (MECNT) granted ERA the rights to develop its REDD+ project in its concession area *with consent from local communities* and to own and sell carbon credits generated from REDD+ activities. While ERA has a contract with the GDRC, it is less clear if this will facilitate ERA establishing a social contract with the Mai Ndombe communities they may fully buy into.

Box 5.2 How a theory of change approach can fail to address social feasibility

The example presented here is from the CCBA validated Mai Ndombe REDD Project in the DRC. It covers the project monitoring plan from the March 2011–October 2012 period. The Plan has been informed by CCBA guidance from Richards and Panfil (2011) and Richards (2011). It relies on a theory of change approach to create a conceptual or causal model for monitoring and evaluation.

The case is drawn from ERA and Wildlife Works (2012a, 2012b).

- The project proponent is ERA-WWC Joint Venture, a joint venture between ERA Ecosystem Restoration Associates Inc. and Wildlife Works Carbon LLC. In March 2013 ERA Ecosystem Restoration Associates Inc. changed its name to ERA Carbon Offsets Ltd., after aquisition of two companies to form Offsetters Climate Solutions Inc.
- Per Richards and Panfil (2011), verification audits against the CCB Standards that assess whether a project has actually generated net-positive social and biodiversity benefits must be carried out within five years of validation, which may be too short a period for some changes to be measurable.
- Through the ongoing assessment of the extent to which anticipated outcomes and impacts are being achieved, assumptions about how project activities will result in expected changes (the theory of change) will be tested.
- Community and biodiversity monitoring for the Mai Ndombe REDD project at this early stage of implementation (1.5 years into a 30-year project crediting period) is focused toward project outputs.
- Quality education and health care are the main community level priorities.
- Since the early 1990s there has been no significant state, NGO and/or private sector investment in infrastructure for communities in the project area, and existing infrastructure is in a state of steady decline. There is no improvement on the horizon with regard to the reform of political and administrative institutions at all levels, and corruption is common in spite of many exemplary efforts on the part of individuals.
- Through participatory rural appraisal (PRA), three focal issues emerged as themes "that are currently most important to local communities with respect to their current conditions and well-being": (a) access to quality health and education, (b) access to potable water, (c) food security and economic alternatives.
- "Among the assumptions made in this theory of change analysis is that the formation and training of committees to manage community decision making will lead to increased community capacity, ability and desire/ motivation to collectively respond to local issues, and an ability to plan locally for and manage development initiatives aimed at improvements to local well-being. These cause and effect relationships are often employed within contemporary, empowerment focused community development work".

In this theory of change, there is no indication from public domain information that ERA is validating that the Mai Ndombe communities do not continue to perceive residual rights to land. Yet, given the contentious nature of land tenure and rights-based claims that are known historically to underpin resource use in Bandundu province, this theory of change does not account for this. Instead, emphasis is on external deliverables, the assumption perhaps underpinning the theory of change may be, not unreasonably, the communities will accept the conservation concession and ERA's ownership and rights to sell carbon credits from the concession, if SRCs actually deliver the goods.

In former logging concessions, what is known in French as the *cahier des charges*, the documentation noting the responsibilities of the contractor to communities, is key. How the theory of change builds on the difficult legacy of overwhelming corporate non-compliance with *cahiers des charges* responsibilities, leading to community disillusionment, will be key to successful negotiation of acceptable SRCs in the Mai Ndombe.

So too, addressing how conflict is to be addressed in the theory of change is also pivotal to assessing feasibility. In October 2012, reports of conflict in the Basengele sector of the conservation concession area circulated through email traffic among agencies working on REDD in the DRC.[10] The customary chief in Bongo was apparently shot at, and his house reportedly burned for having been perceived as signing away customary lands to foreigners without full community consent. Politicians from Kinshasa supposedly had "instrumentalized" a youth uprising against leaders who some saw as "selling out" the community, and its lands, to foreigners. In short, there was a lot of *apparent* confusion, finger pointing, and inappropriate activities occurring in remote parts of Mai Ndombe. How much any local disaffection over the status of the social responsibility contract (*cahier des charges*) applying the "conservation concession" in question was really a contributing factor is anyone's guess. This is not a place where clarity in information customarily abounds.

The manner in which the theory of change accounts for conflict events and their potential triggers, provides little sense for how the project will respond if communities demand greater transparency as part of the FPIC process, and do not receive it. For example, what if greater transparency on agreements between the project developers and the DRC government that awarded ERA a conservation concession were required? Where FPIC is provided, one would imagine this being a reasonable precondition on the community's part. Understanding the nature of any agreements that implicate lands and derivative carbon rights over which communities believe they hold customary claims would seem fundamental for approaching a theory of change in the Mai Ndombe.

For example, how will communities react if they feel that revenue sharing arrangements are inequitable based on their understanding of the price that carbon from the concession is being marketed at? This is one type of information the theory of change would ideally be accounting for. In the absence of transparency, communities in Mai Ndombe and elsewhere could conceivably challenge the basis of FPIC. While clearly different from the US public's challenge of its "consent" for the bank bailout under the TARP program which coincided with loss of 3.5 million homes and US$7 trillion of homeowner wealth (Barofsky 2012), communities 'notions of equity, fairness, and notions of consent can change over time, be they in the US or the DRC.

Who is to say that the Mai Ndombe communities will not perceive the project developers in a similar way – as beneficiaries of a privileged relationship with a corrupt government that has enabled them to potentially profit at the expense of communities who may over time believe they have accepted triflingly little in exchange. Ideally, a theory of change would be monitoring for such contingencies.

As Box 5.2 alluded, conflict over FPIC may have already begun to materialize in the Mai Ndombe. The events of September 2012, when village youth were reported to have attacked a customary chief along with a government appointed chief suggests that some may question, legitimately or not, the validated CCBA FPIC obtained. It is also consistent with a critical posting about ERA's process for obtaining FPIC that appeared on Forest Peoples Programme's website (Forest Peoples Programme 2012a). It notes that the Mai Ndombe consultation process was highly flawed to begin with. The shootings and house burning of one important customary chief of Bongo in Basengele sector of the Mai Ndombe, if in fact FPIC related, would confirm this.

While carbon credits had been mentioned during some of the preliminary meetings with ERA staff, according to FPP (2012a) none of the communities had been informed about what the carbon market actually is, or how it works. According to FPP, they were told simply that the project would generate carbon credits and that they would benefit from this. The communities were not told that ERA had licensed the rights over the entire 300,000 hectare concession and were therefore charged with the management of the entire area. They had also not been told what would happen to the management of their customary areas covering both forest and farmlands across the concession, even though FPP estimated that at least one-third to one-half of the ERA concession overlaps local communities' customary territories. Communities were not informed of the scale of profit that ERA might be able to achieve by taking over the communities' forest, nor were they told what proportion of this profit would be shared with them. To FPP this is not in line with the principle of FPIC.

The description provided by ERA and Wildlife Works (2012b) below of their approach to "initial consultations" with communities does little to shed greater light on the credibility of the theory of change they employ, nor why the CCBA and its auditors verified and validated the Mai Ndombe project. One wonders if "at least two days" is sufficient time for communities to analyze and reach momentous decisions.

Information sessions were held in each of the 26 major villages in the project area. In these meetings ERA Congo was introduced, its history, and the concept of REDD including a discussion of global climate change. Meetings typically lasted between two and six hours and were attended by between 20 and 100 adults. Villagers were encouraged to ask questions. Before leaving a community, ERA would introduce a consent form and encourage stakeholders to debate the merits of consenting to project development without the company present. After at least two days had elapsed, ERA returned to each village to answer further questions, continue with information sharing, and collect consent forms if they had been signed. Consent forms were signed in one hundred per cent of communities where they were proposed.

That this process passes muster for FPIC, a viable theory of change for a REDD project, and CCBA validation, is surprising for anyone familiar with the complexities of the Mai Ndombe and Bandundu province in the DRC. It also raises serious questions about the rigor of CCBA verification and validation processes overall, for this situation in the Mai Ndombe is *not* unique in terms of the quality of information provided leading to project verification and validation.

How the project developers, and the CCBA and its auditors, will address agriculture within the conservation concession will be of interest, as the precedent was well established in logging concessions that agriculture was permitted. If agriculture is to be prohibited in the conservation concession, as may be a requirement, it will be key for the theory of change to address such modifications to land use. Yet from available information, this central element to communities impacted by the concession, who have apparently provided FPIC, appears lacking. "Normally", the theory of change might assess how well communities are adapting to demands in change of land use patterns, with the case of agriculture in the concession area an issue with a long-standing reference level. Should there in fact be community disaffection in the Mai Ndombe, it will be central to the theory of change to account for this, along with the adaptive management plan for community disaffection and any conflict.

Presumably, if a social feasibility framework had guided the Mai Ndombe REDD project developers from the outset, the probability of community-level misunderstandings, or any conflict, could have been reduced. Unless theories of change, and project design itself, consider the the range of dynamics impacting project feasibility, their usefulness becomes questionable.

The transaction costs bugaboo

Transaction costs for addressing social issues have not been incorporated into full cost calculations for the REDD mechanisms during the Readiness Phase. Nor have they been incorporated into opportunity cost calculations in assessing what will be required at local levels to induce community engagement in REDD on a sustainable basis.

What has been done is to instill fear of high transaction costs as a disincentive to investors. However, this is a way to dumb down what is needed to design feasible REDD projects. The rationale for doing so is "appropriate imprecision", which is simply a euphemism for avoiding complexity and feasibility requirements in design.

It is doubtful Chambers (1983) would have encouraged appropriate imprecision. For projects where private sector operators are able to negotiate agreements with governments for hundreds of thousands of hectares of forest concessions, now deemed "conservation concessions", and where tenure rights involve multiple claimants, irrespective of any attribution of rights to a foreign entity by the state, appropriate imprecision may not be good enough to promote conflict-free sustainability.

And it is doubtful that appropriate imprecision would be used for projects where, depending on assumptions, ROI from REDD to private sector interests could be in the tens of millions of dollars (or more) to investors with publicly traded companies. If all goes well, and where the theory of change underpinning a CCBA validated project holds that schools and clinics and jobs will be good enough to satisfy the basic needs of communities in the conservation concession, and that local development committees will be the key local facilitators of the process, it is hard to see how this would even meet the appropriate imprecision definitional standard, if risks are objectively considered.

It is an understandable strategy for REDD proponents intent in generating optimism, funding, and momentum to reduce complexity and minimize short-term costs. Yet, by not attending to the full set of capacity building needs that fall under enabling conditions for REDD from a *sustainability perspective*, not a project development or certification perspective, this simply creates the illusion of feasibility where in fact it may well not exist.

These capacity building needs are likely part of key assumptions that were left out of Stern's (2009) *Global Deal*, in which the CBA that was used to make the case for climate action was not to be "taken too seriously". To Stern, the numbers he employed in his CBA are "very sensitive to assumptions", "leave out conflict", and are "weak on risk and biodiversity".

As avoiding conflict and risks to IPs and other forest peoples is clearly part of the REDD compact, brushing these under the carpet while making the case for a global deal on REDD appears to be both suspect and perilous indeed. This is showing signs of beginning to play out on the ground.

Constructing illusions of order: PES and REDD

PES payments must be "*additional*" to BAU practice (Wunder 2007). From the equity point of view, "many consider that those who conserve their forests, and therefore deliver an environmental service, should be paid regardless of their opportunity cost to conserve this forest" (Karsenty et al. 2012a). There is thus a risk that PES based on opportunity cost rewards potential forest destroyers.

It is probably too ambitious, and somehow dangerous, to expect economic instruments designed to address environmental problems to also be levers for social justice and poverty alleviation, especially when those instruments are market-based (Karsenty et al. 2012a). The question then becomes one of whether social justice, "rights based" policy levers are logical to mix in with market based initiatives. This is the exact pathway that REDD planners have chosen.

Market involvement in PES and REDD is seen to some as "market masquerades" (Milne and Adams 2012) which neither benefit people or the forest.

These masquerades are arguably the result of a disregard for feasibility analysis, in lieu of bright-sidedness (Ehrenreich 2009), self-deceit (Trivers 2012), or, worse, bullshit (Frankfurt 2005).

Illusions and masquerades involve the manner in which project developers engage in planning practices which unduly simplify complexity within communities to better enable market transactions to occur. As noted, this tactic works well with approaches that employ theory of change as the basic strategy for framing a project design process that leads to indicators to measure success against over time.

While expedient and conforming with the industry standards for project design, it is not clear that these standards will generate successful projects. Transaction costs key to undertaking preparatory activities may be covered by multilateral or bilateral REDD funds, and market transactions in the short term may occur. This does not mean that the projects will actually achieve objectives or prove sustainable, nor that the array of projected co-benefits will be realized.

The full gamut of transaction costs that actually require coverage for REDD to become sustainable will inevitably be higher *if* principles of additionality and the objective of biodiversity conservation and poverty alleviation are considered. Given prevailing human and social capital in most REDD landscapes, lengthy consultation and capacity building, analysis and decision making, and negotiation would inevitably be needed if "genuine" FPIC is to be obtained. This will increase costs. This, however, does not appear to be part of early project developer strategy.

On the anthropological side, REDD is another case where "illusions of order" are constructed that "obscure a far more complex and messy reality" (Lewis and Mosse 2006). The point was also made in Chapter 2 regarding assumptions about rational choice and maximization of economic utility.

Assumptions about the effectiveness of managerial order and "instrumentalized knowledge products such as participatory rural appraisal (PRA), indigenous knowledge, and social capital" (Lewis and Mosse 2006) are also pivotal in REDD. When coupled with the highly technical requirements of REDD MRV that REDD is prioritizing, a strong intellectual facade for the undertaking is established. On the other hand, the fact that consensus and strong stakeholder backing prevails for current methodological approaches to REDD in no way substantiates the conclusion that the espoused orthodoxy will work in practice.

Practitioners of neoliberal critique to conservation offer a broadside of challenging the "illusion of order" constructed for PES as applied to REDD.

Six features of neoliberalism that include privatization, deregulation, market-friendly re-regulation, the use of market proxies by governments, the encouragement and facilitation of civil society support, and the construction of individualized, market-compliant individuals and communities have been identified in a framework (Castree 2011).

Transnational backed, neoliberal approaches to biodiversity and conservation natural resources management have been undermining rural peasantries (Corson 2011). Through projects aimed at capturing carbon credits and expanding protected areas, Corson believes that conservation advocates are staking claims to peasant-utilized commons in the name of "the global commons".

In the neoliberal era of downsized national governments, these conservation claims are often initiated, shaped, implemented, and even enforced by private and non-profit actors operating in collaboration with, and often under the auspices of, the state (Corson 2011). The implications of this form of "rule of experts" (Mitchell 2002) are serious, with alliances forming in corridors during international meetings and in foreign offices or domestic capitals – arenas that are largely inaccessible to rural peasants. Yet, parties within these alliances negotiate not only rural peasants' rights and access to land and resources, but also the authority to legitimize these rights. That is, external agents plan for, and presumably act in the interests of, the rural poor.

The implications of "the rule of experts" is what led Chambers and Belshaw (1973) to introduce their pathbreaking and paradigm shifting PRA approach as an antidote to top-down planning. What appears unfortunate is that this pathbreaking methodology has been called upon to tackle methodological challenges that go far beyond its pretensions or capabilities when the basis for genuine FPIC is considered in complex REDD contexts.

Critics of neoliberalism also note that the nexus of relations between corporations and big conservation NGOs who have come to enjoy predominance in thought leadership in conservation, and now REDD, must be appreciated in order to craft appropriate strategies for dealing with the emerging green economy (Brockington et al. 2008; Sullivan 2011). This reinforces earlier "insider revelations" (MacDonald 2008) of the burgeoning phenomenon of environmentalist–corporate alliances, with the sub-field of anthropology and political economy that focuses on the "banking" or commodification of nature flourishing in its own right.

The activist position of the 20 million member strong La Via Campesina, and emergent civil society coalitions like those opposing the Governor's Climate and Forests Task Force (GCF) in Latin America and Asia (see Comunidades de la Región Amador Hernández et al. 2012), complements academic contributions on the illusory, misleading nature of the green economy which critics attribute to the "structural crisis of capitalism" (La Via Campesina 2012b). The contributions of activist groups are discussed in greater detail in Chapter 10.

Prevailing best practice in REDD

"Best practice" ("BP") approaches for social and institutional issues in REDD such as participation, FPIC, and tenure rights, are by and large treated under the umbrella of "Social and Biodiversity Impact Assessment" (Richards 2011), itself a subset of social impact assessment for conservation initiatives (see Schreckenberg et al. 2010). For example, WWF and its partners conducted a workshop called "Building the Treasure Map: REDD+ Experiences in Latin America" in July of 2012. Code REDD is "like a trade guild or industry association" that will promote best practices internally and is funded by one organization, Wildlife Works. Code REDD projects must be validated by VCS and CCBA, as well as by Dutch energy provider Eneco in accordance with guidance from WWF, the International Finance Corporation (IFC), and CI

(Forest Carbon Portal 2012). On the other end of the spectrum, Alcorn (2010) has focused on best practices in terms of safeguards that protect IPs' rights and enhance livelihoods.

Tools to promote best practice approaches for social and institutional issues in REDD such as participation, FPIC, and tenure rights are treated comprehensively under "Social and Biodiversity Impact Assessment" (Richards 2011).

SBIA has many useful tools that are broadly recognized. These reflect on current best practice supported by many well known organizations in the non-profit arena. They also appear to be looping into newer initiatives like Code REDD, which builds upon NGO best practice from CCBA, VCS, CI, and WWF.

Meanwhile, social feasibility is *not* among the explicit objectives of SBIA. This can perhaps be explained by three factors:

1. A focus on safeguards.
2. A principled determination to keep transaction costs down.
3. Reliance on theory of change as the framework for validation of REDD projects.

The three factors are tied to what may be a strategic decision to optimize expediency with responsibility – i.e. if REDD projects can minimize costs to local peoples, they will in fact be able to be evaluated favorably in comparison with many standard development and conservation projects. If this is labeled as best practice by most thought-leaders within the industry, who will oppose it as such?

By keeping transaction costs down, this will offer a more favorable investment prospectus to generate momentum on that end, all the while arguing that best practice standards are being adhered to in individual project design. Furthermore, by using the theory of change framework, a tautological framework for verifying and ideally validating and funding projects can be readily established. This, however, may have little bearing on the feasibility of what is proposed, while nonetheless conforming to guidance for a credible theory-of-change driven REDD project design and implementation.

That this approach does not emphasize feasibility or sustainability is an issue that neither SBIA standard setters or project developers employing methods drawn from SBIA explicitly address, since a magnanimous evaluation time frame of twenty-five to thirty years is built into the guidance by virtue of relying on its theory of change guidance; pressure is minimal to demonstrate measurable results even on a biennial or triennial basis.

In addition, safeguards are used to bolster best practice in project design. Safeguards are perhaps the most critical element in obtaining and sustaining verified and validated projects.

At the same time, safeguards also introduce considerable subjectivity as regards efficacy. By emphasizing safeguards, project developers can *appear* to address social issues responsibly while, in fact, they may simply be enabling a series of detrimental policies and actions to take place while highlighting the presence of safeguards. Most importantly, an overemphasis on safeguards diverts attention from long-term development aspirations that may not be part of either safeguards or the core proposal supported by industry best practice. Rather than contribute to development, the collective package of theory of change and safeguards, coupled to FPIC if it is obtained, may worsen peoples' *overall* social welfare over the long term.

In this regard, SBIA completely avoids what rights-based groups and a number of CBNRM and conservation analysts believe will represent a key issue impacting REDD success and failure at site levels: local stakeholder ownership of REDD processes and decision making. Rather, it creates a logical framework for perpetuation of externally designed, managed, and driven projects that are largely beholden to shareholders and donors as opposed to people with customary rights over the resources transacted in REDD.

Based on the empirical record from earlier conservation and CBNRM efforts, it is reasonable to hypothesize that local ownership and capacity to participate effectively in decision making and management is a principal factor in establishing a basis for achieving "carbon permanence". This can be accomplished through transformation of inappropriate agricultural or forest use practices into "avoided deforestation". SBIA, however, does not offer an explicit methodology to identify the *type and extent* of capacity building that is required at a minimum to enable local ownership of REDD. Nor does it discuss what constitutes, and how project developers measure, credible local decision making as part of FPIC in REDD. There is no plan through SBIA for how this fundamental issue is to emerge. Rather, general principles and guidance is provided.

This approach appears to mask what is needed for REDD to work and become sustainable. In all likelihood, based on the empirical record of development and conservation from which SBIA best practice guidance is drawn and offered, there is no basis for believing this will work.[11]

Indigenous institutions: problem or solution?

Nepstad et al. (2006), Bray (2008), Ellis and Porter-Bolland (2008), and Porter-Bolland et al. (2011) suggest that forests managed by local or indigenous communities can be as or more effective than forests managed by national agencies. Hence, communities are seen as a solution to forest management in the medium to long term.

While governments and internationals NGOs are often hesitant to empower communities with full management responsibilities, from a cost-effectiveness standpoint as Hockley and Andriamarovololona (2007) suggest from their analysis of CBNRM outcomes in Madagascar, they may well wish to reconsider this position. Even more radically, should more stringent demand for "informed" FPIC become a more serious objective of funding and implementing agencies, this could upset the current best practice applecart. With sufficient training, communities have proven capable of assessing costs, benefits, and risks in land use planning and then come up with plans that balance ecological, economic, and social variables (Brown et al. 2008).

While outcomes data from forest management, biodiversity conservation, CBNRM, and PAs management are not commensurate, there are enough indications that communities, if given the chance to take on roles in REDD that historically have been reserved for external agencies or government, may be able to produce results equal to or better than external agencies. They could also do so at far reduced costs. Clearly if indigenous institutions were seen to be a more integral part of the solution, as opposed to rhetorical or instrumentalized players as they now primarily are under REDD, this would have major political and economic implications.

This equivocation on whether indigenous institutions and communities should be drivers of programming, as opposed to consulted beneficiaries, is due to myriad issues of policy assumptions, problem framing, prevailing best practice, donor culture, as well as organizational and technical capacity issues that do preclude most indigenous institutions from fully capitalizing on their comparative advantages and long-term potential.

Scholarship on community resource management suggests that communities are capable of assuming significant responsibilities in REDD programming (Agrawal and Angelsen 2009), just as they have under activities labeled as CBNRM (Hockley and Andriamarovololona 2007). The experience from biodiversity conservation, its successes as well as failures, lends further credence to the idea that progressive empowerment of communities in REDD will be critical to success, and does not pose great risks if compared to standard best practice approaches that industry leaders from BINGOs or the private sector offer.

Practically speaking, this does not mean that communities are well positioned to assume management responsibilities *tomorrow* in REDD. In the absence of well tailored institutional and technical capacity building, capitalizing on their potential is doubtful if taken to scale.

Most parks of the world, especially in developing countries with weak nation states, have been poorly protected. They have not been able to adapt to the challenges of internal ecosystem dynamics or socioeconomic drivers such as demands for land and migration of people (Duffy 2006a and 2006b). This led some to call for innovative systems of protected area governance (Borrini-Feyerabend 2004). Many conservationists realize that local people establish and sustain self-organizing institutions with positive outcomes for the protection of biodiversity (Berkes 2004; Gibson et al. 2000), challenging the prevailing assumption oftentimes guiding program design that local stakeholders are the principal threat to forest and biodiversity conservation. This ambivalence between framing people as threats, versus treating people as solutions to problems, remains a source of contention between conservation planners, social scientists, and CSOs.

Box 5.3 presents the vision of two of the premier thought-leaders in REDD, the World Bank and CI, for why and how REDD is approached in one region of Madagascar. The Ankeniheny-Zahamena Corridor (CAZ) in Madagascar offers a good example of the vision that the World Bank and CI have, together with the Malagasy government, for one of Madagascar's most precious sites for biodiversity, with over 2,043 plant species, 85 percent of which are endemic.[12]

While numerous studies document the content, functioning, and role of community-level institutions in conservation and social–ecological resilience (Fairhead and Leach 1996; Berkes et al. 2000), there is still a lack of understanding of community dynamics, and their full potential to be leaders in community forest

Box 5.3 An appealing narrative: the Ankeniheny-Zahamena
Corridor

The Ankeniheny-Zahamena Corridor (CAZ) is marketed as a win–win–win
for biodiversity, climate, and local livelihoods in Madagascar. It is an AD
REDD project that suggests it could represent an innovative break from BAU
approaches to promoting community-driven conservation.

Madagascar is a country that has received enormous ODA support and
targeted funding for biodiversity conservation over the past twenty years.
The outcomes of all this effort are dubious on a biophysical and human
welfare level, as biodiversity loss, erosion rates, and poverty indicators have
not improved. Moreover, the country continues to face enormous governance
and credibility challenges for several years in the aftermath of a *coup d'etat*.

Whether the project, the CAZ, which covers approximately 425,000
hectares and is home to hundreds of local Malagasy communities, can
deliver results as the following promotional material from the World Bank
contends it will, will be interesting to follow.

> The project description: In the humid rainforest of Madagascar, 14
> threatened species of lemurs, found only on the island, will soon be able
> to breathe a little easier. A corridor of forest that stretches along the
> eastern coast of the island called the Ankeniheny-Zahamena Corridor
> (CAZ) covers approximately 425,000 hectares. CAZ is a region of
> rich biological diversity and home to hundreds of local Malagasy
> communities. The diverse forests provide essential ecosystem services
> upon which local people rely for their daily subsistence, but they are
> under pressure from slash-and-burn agriculture, illegal logging, and
> other unsustainable practices. By protecting these forests and by focusing
> on alternative livelihoods, it has been possible to see the link between
> healthy ecosystems and human well-being. With the fall in deforestation,
> there has been a significant reduction in carbon emissions. A project is
> now being developed that will bring monetary benefits to the community
> for protecting these forests.

management (Agrawal and Angelsen 2009). This constrains the ability to formally
provide an alternative model for projects like the World Bank/CI CAZ. While on
the surface the CAZ appears plausible, there is no basis for assessing why it should
be feasible given the dismal legacy of conservation in Madagascar and how this
will be either built upon, or broken from, in the CAZ's own theory of change.

As data continues to build, illustrating that the efficacy of community level
management relative to other forest management modalities is potentially equal to
or superior to that of more traditional approaches, and more cost-effective at that
(Hockley and Andriamarovololona 2007), at some point the policy logic for CBNRM

or community forest management approaches may be looked at in a more level-headed way. Until then, the shift in favor of grand landscape level approaches where policy and international finance now take precedence, before one has demonstrated the ability to get projects to work at smaller scales, is the priority.

What could be wrong with the CAZ strategy and policy?

On the surface the CAZ illustrates exactly what REDD seeks – establishment of a "triple win" for biodiversity, climate, and people. Whether the CAZ can achieve and sustain this triple win, a number of issues will need to be addressed in the coming years which the public relation materials alluded to do not provide answers for. For example:

- Was FPIC obtained among the 100 association members, and, if it was, on the basis of what criteria?
- Are the associations in the CAZ operating under the same principles of *transfert de gestion* reported on by Hockley and Andriamarovololona (2007), and how will the shortcomings faced by COBAs previously in Madagascar be more effectively mitigated under this project activity?
- Hockley and Andriamarovololona (2007) noted that the viability of COBAs was threatened by "a naive view of forest conservation as a pure win–win scenario for communities and external agencies". Is the CAZ a break from this naïveté? Or are the same mistakes about to be repeated, where the presumed benefit sharing from future VER sales, coupled to any capacity building to strengthen associations, will not be enough to foster conservation?

While numerous questions about the specifics of capacity building, rights over forest resources, decision making over resource use, etc., could be asked to comprehensively assess social feasibility, these are the most basic social feasibility questions that this new model for achieving the "triple win" in Madagascar will face.

Standard setters in REDD+: UN-REDD and FCPF

REDD is led by the UNFCCC in the preparatory Readiness Phase. UN-REDD (UNDP, UNEP, and the FAO) and the FCPF of the World Bank are collaborating on all activities that provide the context for standard setting.

The two main multilateral readiness platforms for REDD, the UN-REDD Programme and the FCPF housed at the World Bank, are well aware of the challenges REDD countries face in successfully preparing for and implementing REDD. As a result, the two initiatives are actively coordinating their efforts. The FCPF and the UN-REDD Programme work together both at the international level, harmonizing normative frameworks and organizing joint events, and at the national level, where joint missions and sharing of information are producing coordinated support interventions.

The UN-REDD Programme and FCPF agreed early on to coordinate their global analytical work in a manner that builds on and leverages their comparative advantages. For example, the UN-REDD Programme has taken the lead on providing its technical expertise to furthering methods and approaches on how to best meet country needs for MRV, while the FCPF leads in the area of economic analysis for REDD strategies (UN-REDD Programme 2012c).

The revised R-PP template jointly developed by UN-REDD and FCPF is the best window to understand the standard they mutually have established for facilitating national level REDD planning. Chapter 5 provides selections of how the R-PP template addresses social issues in project development contexts.

The VCS and CCBA PDD product

The VCS has become the principal standard for most voluntary market activity (see Diaz et al. 2011). Often VCS validation and subsequent verification is "twinned" with CCBA validation to enhance the credibility and value of voluntary emission reductions (VERs) and CERs.

More and more, voluntary market innovators in REDD are pursuing VCS and CCBA validation simultaneously (Peters-Stanley et al. 2011). This section presents information drawn from public domain sources that illustrates how REDD project developers preparing projects in the Readiness Phase are going about their work to meet social standards specifically developed by the CCBA. These projects may be for piloting purposes or for getting these projects compliance-eligible once guidance on compliance markets mechanisms are clarified. They also may simply be designed for voluntary markets under current guidance.

The section further explores the plausibility of how the social dimensions of REDD are treated in a sample of VCS/CCBA-validated projects. The information in these validated projects was not explicitly prepared to address *the feasibility* of the projects. Nonetheless, the PDDs themselves often read implausibly when the challenges posed by actual situations to address the social dimensions of REDD are fully considered.

The exercise here is a quick "snapshot" of what could be undertaken in a more comprehensive analytical review. The point is to confirm that (a) the standards do not appear demanding enough for addressing social dimensions, and (b) the information provided by auditors who validate the CCBA standard is highly subjective and unconvincing based on a small sample of validated initiatives submitted to the CCBA.

The implications of this are important, as social issues that the CCBA standard treats are ultimately subject in the real world to an array of complex challenges that determine success or failure.

CCBA background

The CCBA is considered to be the leading standard as of 2013 for verification and validation of social and biodiversity co-benefits from REDD+ projects.

One notable aspect of the CCBA is that it builds "on the use of existing certifiers authorized under Kyoto[13] or by the Forestry Stewardship Council (a private forestry products certification scheme) for example, as third party evaluators of whether a project deserves to be certified" (Newell and Paterson 2010).

Based on information from its website, the CCBA standards evaluate land-based carbon mitigation projects in the early stages of development,[14] fostering the integration of best practice and multiple-benefit approaches into project design and evolution. The Standards

- identify projects that simultaneously address climate change, support local communities, and conserve biodiversity;
- promote excellence and innovation in project design;
- mitigate risk for investors and increase funding opportunities for project developers.

In a concurrent validation and verification, the auditor must assess whether the Project Development Document (PDD), and actual project, has been implemented in a way that conforms to the requirements of the CCB standards. It must show that it has delivered net positive climate, community and biodiversity benefits. The verification is an assessment of the implementation that has already occurred and there must be adequate monitoring records to demonstrate delivery of these net benefits.

(CCBA 2012)

This section considers the types of information that REDD project developers prepare and submit for review for validation under the CCBA. Social issues from one case study are focused on, since these continue to represent the weakest apparent link in the overall REDD process.

Case study: April Salumei sustainable Forest Management Project in Papua New Guinea[15]

April Salumei is a REDD+ project with considerable controversy.[16] Validation was approved under the CCB Standards Second Edition Gold Level on June 13, 2011 (see Scientific Certification Systems (SCS) 2011).

No information was provided in the PDD on how consent among 163 landholding groups was obtained. All the information on social and cultural issues appears to have been drawn from older ethnographic overviews. These provide little sense on current *social dynamics* regarding stakeholder analysis, social capital requirements, resource tenure nuances, and local conflict issues. Information from the PDD, public comments from WWF (below), and information from the final auditor's Validation Report is available in the public domain.

Babon et al. (2012) raised the issue of how public information on REDD in PNG is limited and biased in favor of optimistic and supportive government positions in the media. The voice of local stakeholders, they note, is particularly weak. In this context, and where corruption is known to be prevalent in the forest sector (ibid.; *The Ecologist* 2010), obtaining FPIC is either a notable achievement or else to be sincerely questioned as to the manner in which it was obtained.

Apparent implausibility

This case illustrates a disconnect between the PDD, WWF's valid concerns raised,[17] and the auditor's final Validation Report. Questions surface regarding the quality of the standards, the quality of the standard-setters' oversight, and the ability of validators to be objective in their interpretation and application of the standards.

Readers can make their own judgment as to whether the Validation Report, and the underlying standards, are sufficient to address the diverse social complexities raised by WWF's Matte Leggett in 2010 public comments to CCBA on the PDD. Reading the materials indicates otherwise.

As both the standard and its interpretation is open to subjective appraisal in all respects as regards quality of content and threshold indicators, the fact that checklist requirements were met in the PDD tells little about what really is going on in April Salumei. This simply confirms Leggett's initial reservations:

1.   Community testimony and research findings indicate that the level of community consultation and understanding of the project in the region is insufficient to guarantee the project has ensured free, prior and informed consent of landowners.
2.   The proposal does not adequately recognize or account for existing disputes over land tenure and landowner company representation in the region.
3.   According to recent statements by the Office for Climate Change and Development the development of voluntary carbon projects is not currently supported by the Government of PNG.
4.   The standard for FPIC and consultation remains a concern.

The conclusion here is that the project confuses the presumed need for a REDD+ project to eliminate the possibility for a poorly regulated logging activity, with the actual feasibility of undertaking the activity in practice. The information provided does not satisfy any minimal threshold of core social feasibility requirements.

In its validation report, SCS (2011) highlighted that there was "some animosity" between stakeholders in the project area that represents a project risk "with regard to future conflicts as funds begin to be realised by the Project", and that the status of a gazetted Wildlife Management Area outside the carbon areas that has been under dispute since 1997 remains unclear. Yet, SCS (2011) judged that the project implementor has tried to manage these potential conflicts through incorporating design elements into the project such as holding the money in trust and encouraging transparent elections on the board of the project developer, the Hunstein Range Holdings.

SCS found "a high level of understanding of the benefit sharing, enterprise development and services and infrastructure detailed in the PDD which supported the Project Implementor's claims of recent stakeholder engagement trips to the area". Yet, as in the Mai Ndombe, no threshold criteria are provided. Thus, it is impossible to know what would constitute a "high level of understanding". From an informational and feasibility standpoint, the information is thoroughly unconvincing, and does not appear to answer any of WWF's Legget's public comments.

Similar to the situation in Mai Ndombe, the information provided, along with its validation and conformance with CCBA standards, reflects acceptance of a high degree of generic pablum – information that is vague, general, indicative of some degree of truth, potentially misleading, and incomplete. As it neither conveys a full story nor what the real story is for a REDD project to work in the area in question, its utility is questionable. The inconsistency and weakness of the standard was the subject of a review by the Swedish Society for Nature Conservation (2013) which stated:

> The application of the CCB principles and requirements appears to be inconsistent and weak, with an inclination of certifiers to approve projects at the expense of a resolute consideration of the community and biodiversity interests. The definitions and procedural guidance on the application of FPIC and benefit sharing arrangements in particular are weak, and the system lacks a mechanism for challenging certification assessments made by the auditors. *CCB certification can thus not be seen as assurance that communities benefit from the projects, tenure rights are respected, or that FPIC has been ensured* [emphasis mine]. CCB requirements on biodiversity are also of little relevance for REDD type projects.

The implications of this are that "[a]cceptance of dubious methodologies and practices on the voluntary market may thus have more widespread and serious repercussions", as voluntary markets are "setting precedents for emerging and possible future compliance forest carbon markets" (Swedish Society for Nature Conservation 2013).

As a final demonstration in April Salumei of conformance to CCBA standards for community engagement, the validation report noted (SCS 2011) that during community consultation meetings, SCS spoke to a range of people in both formal and informal settings. "These people included ILG chairmen, village elders, youth and women. Some offered answers willingly when asked general questions of the group." The validator found "a high level of understanding of the benefit sharing, enterprise development and services and infrastructure detailed in the PDD which supported the Project Implementor's claims of recent stakeholder engagement trips to the area."

This checklist approach to consultation is not informative beyond indicating that consultative meetings occurred. Were lives and livelihoods not at stake, the quality of the information provided would be amusing. Such is not the case.

What was discussed and learned, not discussed, and not understood, as regards the feasibility of April Salumei can only be inferred from the validated project materials made available by CCBA and the validator, SCS.

This would appear to represent a weak basis for validating the consultative process and REDD project in a country "notorious for corruption" according to *The Economist*,[18] where abuse and manipulation of customary landowners in the logging sector is well documented (Greenpeace 2012), and which ranked 154 out of 182 on TI's 2011 Corruption Perceptions Index.

What should a PDD actually tell a reader?

A PDD should provide enough information and analysis to suggest *sufficient grounds* for justifying validation. Information should:

1. assess risks of proposed activities to livelihood security of to-be-impacted peoples;
2. assess if local people were able to make an "informed" decision that was "consensual" – i.e. representative of diverse ethnic groups, gender balanced, minorities and IPs, etc.;
3. assess if resource tenure and rights is factored into programming;
4. assess if social capital at community levels has been factored into planning to enable credible FPIC, or if capacity building is in fact needed;
5. assess the manner in which the above points are integrated to provide a judgment on minimum feasibility thresholds.

Box 5.4 Development or safeguards – equal priorities?

> Women have no voice. Pygmies have no voice. In the forest areas of the DRC, pygmy communities – widely considered as being 'backward' – are numerically significant. Although some progress has been made at the discourse level, this has had practically no impact on the realities of women and indigenous peoples. These exclusions are serious handicaps to the Congo's broader development needs. … Logging companies reinforce these forms of exclusion.
>
> Theodore Trefon 2006 (quoted in Greenpeace 2007)

Given this state of affairs – the "reference scenario" for the launching of any central African REDD initiative – how projects obtain FPIC and undertake demand-driven development when social and human capital are so constrained in the subregion faces all agencies wanting to work there.

Is the prevailing industry "best practice" that relies on PRA standards going to be good enough to enable Pygmies to escape their seemingly entrenched lot, or will more robust tools be needed?

What does it really mean when CCBA validators validate FPIC in projects where the sense of FPIC cannot possibly extend beyond would-be beneficiaries attending a meeting, to have awareness raised of the project developer's plan?

Employing this analytical framework, the April Salumei PDD would not pass. Projects such as the Rimba Raya project[19] in Indonesia,[20] financed by Shell and Gazprom and designed by Winrock International with backing from the Clinton Foundation, and certified by VCS, would face similar hurdles on social feasibility grounds. This may explain why it was strongly objected to in 2010 by Friends of the Earth, the Indigenous Environmental Network, and the 300 million member La Via Campesina network.

Safeguards in REDD: necessary, but insufficient

Safeguards have become the principal focus of attention to attenuate risk to local communities and IPs in REDD. While safeguards are a crucial component to a holistic plan to address the social dimensions of REDD, they have unfortunately come to assume a default position for almost all things social. This unwittingly undermines the legitimate development aspirations of local peoples, that safeguards are in fairness not designed to address. It also dilutes the value of safeguards.

Article 20 of the UNDRIP (2007) speaks clearly to the rights of IPs to development, not *only* to safeguards: "Indigenous peoples have the right to maintain and develop their political, economic and social systems or institutions, to be secure in the enjoyment of their own means of subsistence and development, and to engage freely in all their traditional and other economic activities."

According to IUCN (2009): "Safeguards should be developed to *ensure* rather than merely *promote* specific principles or actions. Safeguards will provide the necessary building blocks for a future climate deal."

Clearly, safeguards are fundamental in REDD. Yet, in the same manner that threat analysis begins from a dubious premise about local people negatively impacting biodiversity and thereby justifying external systems of conservation management (Brown 2010b), safeguards establish a similar problematic starting point for people in REDD.

By focusing policy on doing no harm, as opposed to creating an enabling environment conducive to bottom-up development, stagnation and continued dependency is perpetuated. For central African Pygmies who already have no recognized status, for example, prioritizing making life better, versus safeguarding the dismal declining quality of life most Pygmies now face, could well take precedence for most Pygmies.

Should safeguards then be concerned with development as much or more as to guarantee no harm is done to existing circumstances? Does safeguard policy unwittingly contribute to a regressive policy for local peoples' future legitimate development aspirations? This is one policy arena that has not been sufficiently reflected upon, even by NGOs specializing in IPs and human rights issues.

As UNDRIP Articles 20 and 21 clarify, IPs have legitimate rights to participate in activities that will improve their economic welfare, to which they have freely consented to. The issue is *not* simply to help them slip back even further from their current position. There is no indication that safeguards promote the development perspective of IPs and other local communities.

Swan (2012) suggests that an indirect risk in REDD+ "is an increase in social inequities caused by disenfranchisement, exclusion or tenure reform reversals", and that any social inequity would negatively impact forest-dependent IPs and local communities. As these very communities often conserve biodiversity effectively through decentralized forest management and governance, this would clearly be a loss for both forests and peoples should REDD+ efforts not succeed.

Yet the safeguards to protect forest peoples from harm are too often notional and aspirational, rather than mandatory and binding. The "Cancun safeguards" present a base of aspirational guidance, though not specific direction, for addressing risks and opportunities of REDD+, and were not substantively improved upon in the UNFCCC COP 17 in Durban. While many see the development of safeguards as a positive factor in REDD, their effectiveness depends on their definition, their actual implementation, as well as the extent of support given through other policy measures (Swan 2012). Placing primary focus on safeguards also provides an easy out for PDD developers and policy makers; if safeguards can somehow be gotten "right" this output will generate impacts that can be reasoned as "good enough". Philosophically and morally, this reasoning is suspect.

If safeguard failure occurs, the global community supporting REDD must also account for the opportunity costs of avoided or failed development for front line communities absorbing the greatest risk in REDD. Yet, there has been *no critical analysis* regarding opportunity costs to front line communities from failed REDD programming.

Lack of critical analysis of the opportunity costs of REDD failure occurring is a noticeable gap. For example, the main "message" from the 2012 International Institute for Sustainable Development (IISD) report on REDD+ (Boyle 2012) states:

> Measurement, reporting and verification (MRV), safeguards and financing are considered by many as the three key elements or "pillars" of REDD+ at present. There is a need to link the international negotiating process with all levels of research in order to understand and identify where discrepancies or information gaps exist. These pillars of REDD+ are where research and policy development should be focused.

The social dimensions of REDD do not even merit a mention in the IISD report, embedded as they are under the notion of safeguards. REDD planners and project developers are thus well positioned to repeat the same errors that have plagued development and biodiversity conservation for decades, as social dimensions and their feasibility remain buried.

Nesting

MRV issues related to baseline establishment and carbon monitoring that are essential to carbon transacting have been of greatest concern to date given the perceived, pressing need for extraordinary sums of carbon finance. To facilitate MRV and create win–win contexts, nested REDD systems are proposed.

The rationale for "nested" REDD approaches[21] to REDD is that nesting "is critical to the successful scaling of REDD since sector initiatives and methodologies to date have mostly developed at a project-level, while compliance systems signal that REDD accounting will be performed at national or regional levels" (Terra Global Capital LLC 2012).

While nested approaches to stakeholder participation and decision-making offer promise,[22] unless local stakeholders have the capacities to participate "well", this normative principle will not lead to efficacy or sustainability in REDD.[23] There is no particular advantage to nesting from a social feasibility standpoint. This is because neither the prospects for FPIC, participation, stakeholder analysis, or what it will take for R-PPs or PDDs to be feasibly designed is holistically taken into consideration.

The genius of nesting is that it will facilitate the embedding of carbon accounting at local jurisdictional levels into national frameworks such that accounting for REDD and transacting of REDD carbon offset credits can be facilitated. This creates a win–win from an accounting perspective. Under AD programs where land use planning was addressed through some form of international or national funds, the appeal of nested REDD architectures would be lessened.

Yet before getting to "nesting", many first order REDD issues remain outstanding. These pertain to whether or not *it is even feasible* to seriously consider PES/REDD types of transactions in socially complex places, like the Cardamoms, Mai Ndombe, Chiapas, or remote parts of Madagascar, where corruption, complex governance arrangements, protest, and conflict exist, if experience in the Chiapas and Mai Ndombe in late 2012 is an accurate indicator.

In placing the "market cart" before the "social feasibility horse", REDD projects are likely sowing the seeds of their own failure. "Nesting" in and of itself, similarly to MRV or GIS systems, are not in and of themselves the answer.

Sub-national REDD carbon offsetting as a transitional gambit

States such as California in the US have directly linked with sub-national jurisdictions in a number of developing countries under the Governor's Climate and Forests Task Force (GCF). The GCF is facilitating sub-national REDD initiatives to link Northern emissions offsetters with carbon sink jurisdictions in Brazil's Amazonia, Mexico's Chiapas and Nigeria's Cross River states, with emitters such as the #3 ranked Fortune 500 company, Chevron, based in Richmond, California. The premise is that sub-national inter-jurisdictional offsetting can provide the necessary finance to create a foundation for mitigating deforestation.[24]

Greenpeace argues the premise for carbon offsets is nothing other than "hot air" (2012), whether it is through sub-national nested systems or not. A 2012 conversation between Greenpeace and a leading proponent of REDD in Amazonia, Dan Nepstadt, is instructive on the very different perspectives of international NGO stakeholders to REDD.

Greenpeace uses the GCF's promotion of sub-national REDD+ offsets in California and other carbon markets as an example of how this action risks exacerbating the climate crisis by allowing industries to continue to pollute, while not providing real emission reductions in exchange. Nepstad (2012) counters that Greenpeace's arguments will exacerbate deforestation: "Greenpeace's new report, *Outsourcing Hot Air*, could help to slow—or reverse—the progress of tropical states and provinces around the world in reducing emissions from deforestation and forest degradation (REDD)", rationalizing that "Greenpeace is criticizing 'an emerging emission offset system that is in the early stages of development and years from implementation'".

Greenpeace replies (Czebiniak 2012) that in a best case scenario, Nepstad "fails to point out that the subnational REDD offsets the GCF advocates would by definition, result in no additional climate mitigation since they would merely allow industrial emitters to continue polluting in California", and that "discussions about 'proof of concept' and 'learning by doing' are fine if a program is still in a research phase, but promoting offsets that would allow companies in California to continue to impact the health and environment of local communities in the absence of significant demonstrated success seems problematic" (see mongabay. com 2013).

For stakeholders who prioritize maintaining forest values over potential for generating "hot air" or externalities to local communities, assuming the risks inherent to REDD projects may be worth it. For stakeholders who either prioritize net global emissions reductions, or minimal risks to communities and livelihoods, the current risk exposures under REDD may be too problematic.

6 Stakeholders and REDD

This chapter assesses different stakeholder positions on REDD. It also projects what different groups could do in the future to enhance REDD prospects.

Stakeholder positions

Appreciating the evolution of REDD requires understanding the range of stakeholders and stakes involved. While these stakeholders from global to local levels can be identified, the potential benefits from REDD as currently framed by policy will accrue, potentially predominantly, to those far removed from the front lines where REDD will be implemented.

On the other hand, the potential costs of failed REDD projects, meanwhile, stand to be inordinately borne by community level groups on the front lines of policy and project implementation. These stakeholders will suffer most directly from negative impacts and opportunity costs of changed resource use patterns fundamental to livelihoods should REDD fail. That said, should private sector investment in pre-compliance, voluntary projects continue to ascend as the data over recent years indicates (Peters-Stanley et al. 2011), private sector investors could suffer losses too.

Proponents of REDD+ as an approach to contributing to mitigating AGW argue that REDD will lead to mitigation of the roughly 17–20 percent of GHG emissions attributable to deforestation that contribute to GCC, while contributing to poverty alleviation as well. Proponents include Northern and Southern governments, investment banks, technocrats from select major GHG emitting businesses, BINGOs[1], and intermediaries transacting carbon offset credits.

With financial and policy backing from the G8,[2] support for REDD is led by the UNFCCC and its principal operational program, UN-REDD, who works in partnership with the World Bank's FCPF. UN-REDD is a program which "builds on the convening power and expertise of the Food and Agriculture Organization of the United Nations (FAO), the United Nations Development Programme (UNDP) and the United Nations Environment Programme (UNEP)".[3] Numerous governments including Norway, the UK, Australia, Canada and the US support REDD objectives. So too do some of the largest NGOs supporting the conservation of nature, including Environmental Defense Fund, CI, WWF, TNC, and WCS.

Opponents are drawn from farmer alliances, NGOs, academics, and journalists.

How values affect stakeholder analysis

It is a fair question to ask if the dominance of global values and the perceived urgency to do *something, somewhere* to mitigate increasingly alarming GCC indicators has led to the depreciation of local values and perspectives in decision making on the same issues in the evolution of REDD. Mounting evidence suggests this to be the case.

To assess how values shape REDD, stakeholder analysis of the drivers of REDD programming can consider the following:

- the importance of GHG ERs for stakeholders in the global community;
- the theoretical value of REDD markets and the correlative financial implications for REDD investors, intermediaries, and BINGOs who stand to financially benefit from donations and projects for both climate change and biodiversity conservation objectives;
- the implications for nation states who stand to benefit from new revenue streams that would not otherwise exist;
- the implications for UN agencies at the forefront of thought leadership and coordination of compliance-based REDD mechanisms through the UNFCCC process;[4]
- the implications for impacted peoples in Northern countries where offsets will enable BAU polluting practices to continue, and in developing countries where credits will shape new resource use patterns that will impact on livelihoods;
- the impact for poor local peoples who stand to bear the greatest opportunity costs and livelihood risks in the face of speculative, potentially modest gains even under successful REDD scenarios.

Anti-REDD advocates

Many civil society groups, led by large peasant farmer organizations like La Via Campesina together with rights-based NGO partner organizations in the North and South, are gaining traction in analysis and public dialogue over the present state and future of REDD. The dialogue is becoming increasingly politicized and contentious.

For example, as part of the run-up to the Rio+20 conference in June 2012, La Via Campesina attacked the advance of the capitalist system that is producing crises of "unprecedented dimensions", which in turn, are enabling corporations to "take advantage of the crisis to extend their domination over territories that have not yet been conquered" (La Via Campesina 2012a).

Rights and Resources Initiative (Pearce 2012) notes increasing activism in the forest sector:

> [T]here is hope – derived largely from local communities and progressive private actors. The local custodians of the world's remaining natural resources are becoming difficult to ignore. The recognition in 2011 of the importance

of forest communities in maintaining vital forest carbon sinks is only one example. A rise of popular politics asserting more control over local resources is challenging business-as-usual and leading to political changes at the national level, which, in turn, is exerting an influence internationally.

Capacities to participate, analyze, and to make the best decisions in consideration of the facts which balance environment, social equity, and economic development concerns require strengthening at all institutional levels, inclusive of government and international NGOs. The issue is not only of concern to local communities, though they clearly remain marginalized so long as their ability to react to challenges and opportunities, or act proactively to take the lead in planning, remains questionable. The issue concerns all interested in the success of REDD.

Who are the stakeholders in REDD?

Identifying stakeholders in REDD is fairly straightforward. Identifying the logical role for stakeholders based on what is credibly required to meet feasibility criteria is, arguably, not nearly so clear.

Among the most obvious stakeholder categories, and prominent individuals in the evolution of REDD, are the following:

- intergovernmental bodies like the UNFCCC, the IPCC;
- the CBD and the CDM under the Kyoto Protocol which may stand to gain through REDD;
- national governments participating in UNFCCC negotiations driving the REDD agenda from forest-rich countries with low historical deforestation rates (Guyana, DRC, PNG, and others);
- sub-national jurisdictions in REDD-participating countries that qualify for participation in compliance mechanisms;
- local peoples at the interface with forest resources eligible for REDD programming;
- consumers or intermediaries who depend upon forest products for energy, habitat or food sources;
- NGOs who receive direct financing for REDD projects and/or who have the right to transact and benefit from carbon credit marketing including CI, TNC, IUCN, WWF, WCS, and BCI to name but a few;
- the Climate Market & Investors Association (CMIA), "an international trade association representing firms that finance, invest in, and provide enabling support to activities that reduce emissions across five continents" (Argus Media 2013);
- for-profit organizations who may or may not be members of the CMIA and who have signed agreements with governments such as ERA Carbon Offset Ltd. and their partner in the DRC Mai-Ndombe, Wildlife Works, (who earlier had obtained the first VCS, approved REDD methodology in Kenya and is the founding member of Code REDD);

- NGOs, academicians, and development professionals who provide technical assistance and information support for or against REDD such as Forest Trends/EcosystemMarketplace, Rights and Resources Initiative, Global Witness, Greenpeace, etc., on the NGO side, along with numerous academics who write either pro or con on PES and REDD;
- intermediaries who facilitate transactions for voluntary emissions reduction credits or CER credits;
- technical agencies who establish baselines and measure carbon emissions, and others who validate, verify or register carbon transactions, report on them, etc.;
- Information support services such as REDD-Monitor and mongabay.com who provide a broad range of information on REDD from official as well as critical sources;
- investors in REDD projects whether for philanthropic, investment, portfolio diversification, or risk-mitigation purposes;
- regulatory agencies such as the Financial Services Authority in the UK who play a watchdog role on carbon trader activity (for insight into regulatory issues and scam activity among carbon traders in the UK see Financial Services Authority 2012);
- and last but not least, the intergovernmental, multilateral, and bilateral institutions promoting REDD including UNDP, UNEP, and the FAO in the UN system, the World Bank, USAID, Department for International Development (DFID), the Cooperation Francaise, Deutsche Gesellschaft für Internationale Zusammenarbeit (GIZ), and others.

What is less clear are the implications from success or failure that each stakeholder group would benefit or suffer from different REDD outcomes.

For example, while local stakeholders stand to benefit marginally from successful REDD schemes, pro forma analyses of what individuals and communities stand to benefit from REDD under different scenarios are not available in the public domain. Depending on assumptions employed, benefit projections are difficult to make given uncertainties over carbon pricing and the availability of both discretionary and investor finance.

As rights based analysts have pointed out (Dove 1993; Ribot 2010), profits from forests and the forests themselves are captured by powerful groups when their value mounts, and when there is a premium, suggesting that the major beneficiaries from any REDD carbon trading schemes will likely turn out to be traders and intermediaries (including government entities allowing trading to proceed within jurisdictions) as opposed to local participating peoples. For companies with private shareholders, who have signed agreements with corrupt governments for conservation concessions and have full rights to carbon and its trading, why would this also not be so?

Governments will garner rents, though Karsenty et al. (2012a) challenge the extent of these. Local people under best case scenarios may receive 50 percent of net revenues from REDD activities as in Makira, Madagascar. Presumably after all expenses and transaction costs are accounted for by the relevant NGO or for-profit venture, these could turn out to be paltry to non-existent at the end of the day. Much depends on the accounting procedures used, which may well not have been part of a negotiation process with local peoples. This is the type of important detail that would ideally be attended to in a negotiated social contract linking stakeholders to REDD in a national or sub-national context.

BINGOs

BINGOs have prominent profiles in Northern countries as well as the global South. Many European airports have prominent donation kiosks for conservationists that thousands of travelers pass by daily. In the US, annual revenues for the biggest biodiversity conservation organizations range from US$200 million to nearly US$1 billion, with total asset values of the organizations in the hundreds of millions of dollars or more.

While forty years ago these may have been perceived as small organizations with green agendas, this is no longer the case, as the biggest conservation NGOs are now unabashedly seeking out and establishing partnerships with corporations that in the early days of the environmental movement would have been unthinkable (see MacDonald 2008; Burkart 2010). This fact clearly influences BINGOs' stakes in REDD, as it shapes financing sources (see Lalande 2012), possible greenwashing of corporate agendas, and the standards promulgated by BINGOs to get REDD projects through available funding pipelines. The vigor of neoliberal critique of REDD along with assorted community and NGO activist positions against BINGOs is partly a function of the prominence of these alliances.

Common ground between BINGOs and the private sector has been established to move toward the "triple win". Yet clearly, as the conclusions of Robinson's (2012) report suggest, this may be occurring because investors in REDD who are accountable to shareholders and boards of directors must demonstrate that their ROIs are comparable or higher than elsewhere.

For example, investment banks like Goldman Sachs may be interested in exploring the potential of land investments (Kaufman 2011) along with carbon markets. They seek to generate returns by participating in public–private partnerships with organizations like WCS, where pilot activities like those in Tierra del Fuego, Chile, involve preserving "what is now a 750,000-acre wilderness area" (Wolstencroft 2009) and is linked to the ability of REDD finance in support of conservation activities over the long term. With the WCS as a founding member of the CCBA, this conceivably may have helped in establishing this type of public–private partnership with Goldman Sachs.

Whether activities of this sort will help further the emergence of an eventual compliance market for REDD credits as well, in which Goldman Sachs could participate in taking REDD to scale, will be determined over time. In this regard, a new *"quadruple* win" could be in the offing, where any "wins" that carbon transactions lead to for investors and intermediaries is part of the package.

Leveraging past success in REDD

BINGOs have succeeded over the past two decades in raising what heretofore would have been considered astronomical sums of funding, from both private as well as public sources. This success helps in leveraging partnerships and investment in REDD. Some suggest that BINGOs may well share considerably more in common with large corporations and financial institutions than they do with small idealistic not-for-profit organizations struggling to eke out a living (see Igoe and Brockington 2007; Brockington et al. 2008; Sullivan 2011).

For example, according to IRS 990 forms posted to their respective websites, TNC reported US$997 million in total revenues in 2011, CI reported US$140.7 million in total revenues in 2010, WWF/US reported US$182 million in total revenues in 2010, and the WCS reported US$206.1 million in total revenue in its 2010 submission.

This growth, from what were small operations in most cases, to the juggernauts some have become, has been accomplished on the basis of effective public relations campaigns. The "twinning" of people and environmental concerns, particularly those linked to charismatic wildlife species such as elephants, lions, whales, and the habitat they live in, has led to the generation of enormous revenues. And while environmental organizations like the Environmental Defense Fund who support REDD did see serious impacts on their bottom line subsequent to the 2008 recession,[5] the largest BINGOs appear to be doing well despite the enduring economic recession. At the same time, this growth cannot be due to verifiable effectiveness. Rather, it is a result as much of successful public relations than empirically verifiable results.

Data that demonstrates results and impacts from the conservation mission remain sorely lacking for the past thirty years (see O'Neill and Muir 2010; Salafsky 2010). For impacts on people via what are referred to as co-benefits of conservation, data is even weaker to demonstrate that biodiversity conservation leads to promotion of social and economic co-benefits for communities and local peoples on the front lines of conservation (Brown 2010a). On the contrary, social indicators from protected area programs suggest the opposite; local peoples have long been suffering externalities from biodiversity conservation programming impacts (Dowie 2009; Brockington et al. 2008; Duffy 2010).

BINGOs have been rewarded despite having little empirical evidence to demonstrate effectiveness in mission, or ROI. This opens the door to an obvious question: *Given this legacy, on what grounds can biodiversity conservation be used as best practice to build upon in REDD?* The conceptual linkage of biodiversity and poverty alleviation "co-benefits" to reducing GHG emissions is what BINGOs are working for – the "plus" in REDD.

The relevance of earlier critiques of BINGOs to their role in REDD

Appreciation of the range and longevity of critique of BINGOs is useful for understanding the evolution of REDD and better understanding the stakes involved. A plethora of critiques of BINGOs in environmental affairs has emerged since Chapin's (2004) watershed piece for World Watch Institute, or Dowie's (2009) analysis of BINGOs' role in creating millions of conservation refugees worldwide through establishment of PAs.

MacDonald (2008) took this furthest in her highly critical case study review of CI, and the manner in which BINGO interests and Fortune 500 companies have come to converge over time. This has led to accusations of corporate greenwashing, as opposed to legitimate corporate greening (Burkart 2010).

Duffy (2010) mapped the growth of these connections between corporations and the biggest conservation NGOs, whose reported assets even several years ago totaled in the hundreds of millions of dollars for the more successful organizations, to nearly US$5 billion in the case of TNC based on their 2011 IRS 990 tax submission to the US Internal Revenue Service (IRS).

In addition to suggestions that conservation organizations may be "selling out" through convenient relationships between themselves and corporations, BINGOs are now fielding questions *from within* as to the palpable benefits they bring to communities from prevailing approaches (Kareiva et. al. 2012) and the need for an overhaul of the "fortress conservation" paradigm.

While the recognition may have set in that local people must be part of the solution, that displacement of millions of people in the name of conservation is unjust and unsustainable, and that the preoccupation with mythologizing the perpetuation of pristine empty spaces filled with biodiversity as infeasible in the face of large-scale poverty and the pressures this brings to bear. Yet, convincing BINGO strategies for overhauling their failing approaches remain distant. Warning indicators are already flashing that lessons learned have not led to significant change in approach when it comes to REDD.

For example, WWF/Tanzania has been accused of being "bamboozled by climate change" (Booker 2012). A combination of farmer evictions associated with a REDD project activity that, together with apparent malfeasance in the WWF/Tanzania office, led to staff firings and negative publicity.

WWF/Tanzania and its programming in the Rufiji Delta of Tanzania was also placed in negative limelight (Beymer-Farris and Bassett 2012) where a transition from CBNRM in the Delta landscape was transitioning to fortress conservation and, presumably, REDD. While WWF has publicly denied that their activities in the Rufiji Delta constitute a REDD+ activity (Lang 2011d), the point is that the situation is complicated, contentious, and potentially a misreading of the African landscape and how actors shape it.

BINGOs, improved transparency, actual results

The approach of BINGOs to both biodiversity conservation and REDD inevitably can best be understood through ongoing open access work through the industry's Open Standards project, available at https://miradi.org/openstandards, a spin-off of the CMP. This is a collective effort among leading conservationists to improve standards and conservation performance.

While admirable, the CMP is premised on the belief that incentives exist for BINGOs to internally generate improved transparency in reporting without external, third party facilitation. This premise was challenged in a proposal to the Betty and Gordon Moore Foundation (Brown 2010a). Given BINGO success at fund mobilization, it is unclear that greater transparency in results and impact reporting will help if the data shows that results and impacts are poor to begin with, or the data does not exist, because the results and impacts are poor.

Box 6.1 BINGOs and REDD

Vidal (2008) provides an overview of the stakes, as well as the skepticism, with which BINGOs working on REDD issues are viewed:

> Conservation could now be about to get even bigger still, exerting more control over local communities than traditional colonialists ever did. Because forests lock up nearly one eighth of all the world's carbon, US hedge funds, financiers, governments, the world bank, private companies and many conservation charities see the chance to make potentially enormous amounts of money by stopping trees being felled.
>
> The big new climate change idea, now snowballing around the world, is for rich countries to pay poor ones not to cut down trees in return for carbon credits. One plan is to give communities or countries cash; another is for a global system of carbon trading where poor countries sell the carbon locked up in their trees to allow rich countries to continue polluting as usual.
>
> It sounds good for the climate and the communities, but the reality on the ground is that it could be disastrous. "Once you get big carbon money going to the world's forests you get questions about who actually owns the trees," says Dr. Tom Griffiths, who works with Forest Peoples Programme. "Is it the people who give the money to save them, or the communities?" The carbon rush, he says, could turn conservation back to the bad old days of fences and guns and guards, with increasing control by governments and big international conservation groups over vast areas of land.
>
> He and others foresee over-zealous officials evicting people to protect lucrative forest carbon reservoirs, more corruption, speculation, land grabbing and conflicts. "All these new schemes being devised have important implications for how forests are managed and what may or may not be allowed to happen in them. Carbon companies are already approaching communities, offering to strike deals so they can obtain carbon credits. We are very concerned," he says.
>
> Observers say a legal nightmare could ensue. A hypothetical case might be a conservation group signing up a community to protect a large swath of forest in return for cash. What happens if the chief agrees without people's knowledge? Is there any guarantee that the money will be paid? What happens if a logging company has the rights to fell the trees? Will it take the money leaving the community nothing? Who does the carbon stored in the trees actually belong to?"

Emerging carbon coalitions

Insofar as coalitions have the ability to leverage political decision making, be it in REDD or elsewhere, emerging carbon coalitions can be considered as a stakeholder to the REDD process.

US$144 billion in carbon trading took place in 2009 (Meckling 2011). REDD was a tiny fraction of this total. Political opportunities that arise during policy crises, combined with available funding and institutional legitimacy, are seen to generate advocacy and the engagement of nation states in promoting carbon trading as a solution to GCC. The Kyoto Protocol from 1989–2000, the EU ETS, and the on and off again interest in carbon trading under US cap-and-trade legislation have influenced the formation of carbon coalitions. The very failure of US cap and trade has played favorably into shaping the logic of more innovative public–private coalitions. These include the GCF linking California and its evolving AB 32 legislation, with numerous state level jurisdictions in Brazil, Mexico, and elsewhere, along with voluntary REDD market ventures. The alliances are between financial institutions, private sector intermediaries, and BINGOs.

The CDM is a model for the first and most important carbon coalition, and its compromised status in 2013 is instructive. Despite one billion CERs being issued under the CDM, this has not led to developed country ER targets. The High Level Panel on the CDM Dialogue (2012) noted the equivocal position that the CDM currently finds itself in. When the question was posed to UNFCCC Executive Secretary Christiane Figueres on evidence that the CDM can encourage developed countries to increase their ER targets, no response was provided (Lang 2012e).

The "Unity Agreement" linking big conservation and environment NGOs with major corporations advocates USG support for REDD programming. It is described in the *Tropical forest-climate unity agreement: Consensus principles on international forests for US climate legislation* (American Electric Power et al. 2009). While the Unity Agreement does stipulate the important role for social safeguards, these pertain more to doing no harm than to promoting local development visions. The latter, however, is as important to local people as avoiding harm per se.

In PNG, local NGOs working with research institutes have formed "a vocal and effective coalition advocating for more inclusive and transparent REDD+ policy processes, which they perceive to be taking too long and with insufficient participation of key stakeholders" (Babon et al. 2012). If participation, equity issues, and credible revenue sharing mechanisms can be gotten right, REDD in all likelihood could resonate quite well in PNG even among clan owners who still own 97 percent of land and forests in the country. This is a grand "if".

Standard setters

More than twelve sets of standards exist for assessing carbon projects in voluntary markets, and each has a slightly different focus and different evaluation criteria (Welch-Devine 2009). According to Peters-Stanley et al. (2011), two have resoundingly emerged as leaders – VCS and CCBA. As of November 26, 2012, VCS and CCBA "announced key new provisions that will make it easier to develop and register projects using both standards simultaneously. Projects that successfully complete this streamlined process are termed *VCS+CCB projects.*" They *must* meet the requirements of VCS and CCBA standards by demonstrating benefits "not only for the climate but also for local communities and habitats".[6] As CCBA provides leadership on social dimensions in the standards, and VCS more so on other technical issues pertaining to MRV, the focus is on CCBA here.

CCBA

Under the CCBA, projects that mitigate or adapt to climate change – forestry projects, or renewable energy plants – are scored for their contribution to GHG emissions, biodiversity conservation, and benefits to local communities. Projects that score at least 50 percent in each of these three components are eligible for certification.

As of 2012, the alliance comprised a core group of NGOs (CARE, Center for Environmental Leadership in Business at CI, TNC, Rainforest Alliance, and the WCS). Institutional advisors included Centro Agronomico Tropical de Investigación y Enseñanza (CATIE), The World Agroforestry Center (aka ICRAF), and CIFOR. Eight "standards sponsors" included The Blue Moon Fund, The Kraft Fund, BP, Hyundai, Intel, SC Johnson, Sustainable Forestry Management, and Weyerhaeuser.

REDD+ Social and Environmental Standards have been developed and tested by CCBA (CCBA and CARE 2010). These standards, which have become the principal reference point for all things social in the emergent REDD+ industry, consist of principles, criteria, and indicators, and provide generic guidance to countries as they develop and implement national and/or sub-national REDD+ strategies to "ensure that a range of social and environmental issues are taken into account" (Harvey and Dickson 2010). They also suggest a process for monitoring, reporting, and verification on social and environmental aspects of government-led REDD+ programs.

The problem with the standard

EcoSecurities (2010) suggests that additional certification by a recognized standard that evaluates socioeconomic and ecological impacts is highly important for the financial viability of REDD projects. In the voluntary, de facto pre-compliance market, this is presumably what the CCBA and its approach to validation of PDDs seeks to accomplish.

The preparation of a Project Development document (PDD) is central to the certification process of a REDD+ pilot project. The PDDs contain detailed project descriptions, including the pre-project state of biodiversity, the anticipated ecological effects of the project activities, and a monitoring plan.

With no framework for determining social feasibility, current standards enable the accumulation of background information that is arbitrarily structured under the guise of a theory of change. This information as illustrated in the cases of the Mai Ndombe Project in the DRC and April Salumei in PNG is uninformative about key variables impacting whether REDD can actually work in local social settings where projects operate. A leap of faith is required based on the information PDD developers need to provide to achieve project verification and validation, that FPIC, benefit sharing allocation, and conflict mitigation is credibly addressed.

While the PDD template employed to gather information is sufficient to meet CCBA standards, along perhaps with providing enough information for project investors, it arguably does little to de-risk investments over the long term, as social feasibility issues are not considered. Were they considered, transaction costs would rise. Conversely, risks to impermanence and unsustainability would be lowered.

Information now required by CCBA as a standard setter has proven enough for its supporters as well as market participants using the standard. In that respect it is clearly successful. That, however, does not confer appropriateness or feasibility of the approach.

Two questions can be asked about the logic of the overall process:

1. Is an alliance like the CCBA, comprised of NGOs with a mission for biodiversity conservation, and a pressing need for funding, along with technical advisers specializing in forest management, together with businesses with interests in the forest sector (and/or the possibility for realizing carbon offsets), predisposed to be biased to REDD+ and proposal developers?
2. Is there greater proclivity for this type of consortium to establish standards to expediently move projects through the pipeline and gain momentum for carbon finance flow?

Pivotal, potentially conflictual social issues that invite, if not demand, deep analysis are not being asked through CCBA-certified methodologies. The information provided through the CCBA scorecard and checklisting template CCBA auditors utilize would appear to be of questionable analytical and practical value for investors. If investors are hoping that risks will be minimized by virtue of employing CCBA scorecards, they may wish to rethink some of their basic premises.

Yet, investors do not appear to be focusing on this issue that is so fundamental to the nuts and bolts success of REDD, or to the quality of any investments they may make. This fault may come back to haunt the CCBA and those backing it, along with those projects that have been certified and validated to date as best practice. Only time will tell.

UN-REDD

Together with the World Bank's FCPF, UN-REDD is the principal operational driver of REDD programming at national and international levels. Through

2011, UN-REDD had allocated US$108.1 million to UN-REDD programmes, of which US$59.3 million had been distributed to fourteen National Programmes and US$48.8 million to the Global Programme Support to Country Actions (UN-REDD 2012a).

UN-REDD has seven principle outcomes of concern covering MRV, governance, transparency and equity, community impacts, the nature of benefits, shift to a green economy, and knowledge management (UN-REDD 2012a).

With its leverage over Readiness Phase funding, and its role in driving pilot activities and knowledge management, UN-REDD remains in a strategic operational and knowledge management position to shape the future direction of REDD, all controversies with indigenous peoples' groups in central America considered.

Commercial stakes in forestlands and land grabbing

The transformation of large-scale agriculture and forestry sector activities as part of modern day land grabbing in developing countries is a growing, troubling phenomenon (Karsenty and Ongolo 2012; Alden Wily 2012). Much of the land being sold or leased to entrepreneurs in the global South is on "commons' that often exclude permanent farms and settlements (Alden Wily 2012). This fact enables governments to "supply the thousands of hectares large-scale investors want" as stakeholder interest is diffused across hundreds or thousands of people.

Governments achieve this by not recognizing customary, common property rights. Commons are often deemed "vacant and available". This notion of emptiness led Hardin to conclude that commons led inevitably to mismanagement, which in turn spawned a generation of disastrous land use policies in the pastoral sector of Africa particularly in the 1970s and 1980s. It also contributed to the mythologizing of wild Africa which has led to inappropriate conservation policies that disempower Africans at the expense of conservation scientists armed with GIS tools and conservation biology theory (Adams and McShane 1992).

While conservation concessions are not usually spoken of in the same breath as "land grabs", is there a fundamental similarity between the two. If opportunity costs exceed benefits for local people that are generated in the conservation concession, the net social impact may be little different than that of a private sector land grab.

If people may be required to abandon farming within conservation concessions, as may be the case in the Mai Ndombe REDD conservation concession in the DRC, would most community-level people provide free consent to do so if conjectural REDD benefits and perhaps a chance at a seasonal job is the major incentive provided? Will access to a school and clinic (that may or may not have teachers, materials, etc., if it depends on the GDRC at all) tip the decision in favor of the conservation concessionaires' theory of change and proposition to communities, as FPIC verification from CCBA implies?

If the process is free and informed, it is hard to fully conceive why communities would provide consent. For example, in the past, *at least* riverine communities to the former SOFORMA logging concession had reportedly been granted farming rights in the concession. If subsistence security is imperiled for communities in the remote Mai Ndombe by virtue of a REDD concession, and if people have agreed to the concession based on information provided that may not have fully clarified that access to agricultural land use in the concession could be curtailed, would this be perceived as a land grab by rural Mai Ndombe Congolese?

Much analysis of land grabbing comes from Cambodia, a country where the practice is considered rampant. It has been termed a "country for sale". According to the human rights group Licadho, "22 percent of Cambodia's surface area is now controlled by private firms, mainly through Economic Land Concessions (ELCs) held by agro-industrial companies" (Milne 2012a; see Vrieze and Naren 2012). This is despite the fact that the firms are operating on already-occupied lands.

Concession holders can reportedly obtain windfall profits from timber sales by cutting down forests before industrial cropping even begins. The government's role in this is "the issuing of land concessions without warning or consultation", and "has come at the expense of ordinary Cambodians". Meanwhile, tens of thousands of farmers have been evicted or forcibly displaced by ELCs in recent years (Schneider 2011).

REDD projects in Cambodia, as elsewhere, may well find themselves inadvertently associated, either directly or indirectly, with the complex array of land grabbing issues that are transforming rural forest and agricultural landscapes in many of the world's poorest countries.

Rights-based advocates and peasant farmer movements

Members of IUCN's Commission on Environmental, Economic, and Social Policy (Campese et al. 2007) note that "it is now abundantly clear that conservation has too often undermined human rights, most clearly through protected area-related displacement and oppressive enforcement measures." This has recently been confirmed by the Forest Peoples Program in Cameroon and elsewhere (Colchester 2011).

Rights and resource based advocacy is often aligned with IPs' movements and their critique of REDD. The eponymous Rights and Resources Initiative (http://www.rightsandresources.org/) is the most reliable source of rights-based positions on forest sector policy and REDD. La Via Campesina (http://viacampesina.org/en/), the largest peasant farmer movement at some 200 million members, is a vocal opponent of REDD whose analysis and critique often reach international websites.

An example of the concerns of activist organizations is a July 12, 2012, letter addressed to the Governor of California on the subject of AB 32 carbon offset legislation (Activist San Diego et al. 2012). The activists argue that REDD will increase rather than decrease emissions, will lead to human rights abuses, and is a bad deal overall for Californians.

In citing the Munden Report (2011), Activist San Diego et al. (2012) state that "developing a REDD+ offsets program risks wasting finite resources on a policy mechanism that will be inefficient, ineffective and possibly harm the lives and livelihoods of indigenous peoples and local communities. A comprehensive assessment of REDD+ conducted by experts in derivatives trading also found that "using carbon markets to finance REDD... is likely to be a drain of resources, both in terms of money and time, away from the very serious problems REDD seeks to address." To other critics, attempting to regulate carbon offset markets is the ultimate fantasy – "the carbon offset market is an example of such an unregulatable market, and that attempts to regulate it will only entrench its status as a locus of international corruption and exploitation" (Lohmann 2009).

The take home message from the letter is clear: at a minimum, REDD is a controversial issue that is progressively pulling together a multi-stakeholder, increasingly concerted opposition. This opposition has yet to reach critical mass. As information on human rights violations or unacceptable externalities on livelihoods comes in, along with any evidence of FPIC violations, REDD proponents may be finding themselves in the spotlight for all the wrong reasons.

Communities – both local and indigenous

The ultimate success or failure of REDD is about people. Local people can buy-in to REDD and facilitate its objectives. Or they can amass in opposition to REDD, or else increasingly turn to elected representatives in countries where democratic institutions offer people choices in representation.

Due to their presence on the front lines where REDD is implemented, forest peoples therefore have leverage over REDD's success or failure. Reading the PDD literature, the richness of the human potential to either accept or reject REDD, depending on the perceived feasibility of the offering, is completely absent. Making sense of where to go with REDD therefore requires depiction of a clearer *human face* than PDDs currently account for.

Much activity in REDD inevitably will occur in remote areas of developing countries where ethnic peoples – so called "tribal", and IPs – predominate in tropical forested landscapes.[7] While the size of the land area held under customary law without official tenure protection under statutory law and titling is unknown, it may be as high as 90 percent in Africa, and covers significant areas in Asia, Pacific, and Latin America (to a lesser extent) (Alcorn 2011). With an estimated population of 370 million people occupying 20 percent of the world's territory, IPs comprise one-third of the world's poor and live an average of 20 years less than the non-indigenous population (United Nations 2009a).

Clearly, many of this global IP population could be subject to REDD policy and practice. This gives REDD program and project developers enormous leverage over much of the globe's surface, as well as its peoples.

Yet, one of the persistent dilemmas in the literature on development, biodiversity conservation, and now REDD, is its level of abstraction from real world people. Technocratic solutions predominate; strategies to work with, and for the benefit of, people, continue to elude. REDD is saddled with a legacy of anything but best practice when conservation impacts on people are highlighted.

Instead of *driving* the process in defining how REDD should work, local peoples living in communities continue to be treated as instruments of policies and incentive structures drawn up primarily by expatriate experts.

In practice, this has led REDD and REDD+ to recycle ineffective practices, albeit with improved participatory rhetoric over time. In the meantime, where the rubber meets the road, the lion's share of attention and funding appears to be going to "technical issues" involving MRV issues that will enable carbon trading in markets to proceed. Thus instead of being an epiphenomenon of local people driving participatory, collaborative planning, and decision-making processes, carbon baselines, subsequent MRV, and project finance predominate over the social dimensions of REDD. This imbalance represents a surefire recipe for failure.

Rationale for empowering local managers

Analysis of satellite data shows that indigenous lands occupy one-fifth of the Brazilian Amazon (Gorenflo et al. 2012). This is five times the area under protection in Brazilian parks, and currently represents the most important barrier to Amazon deforestation (Nepstad et al. 2006).

Biodiversity is equal to if not higher in areas with more indigenous presence than areas with less, though these are areas where population densities are generally low. Critically, "the inhibitory effect of indigenous lands on deforestation was strong after centuries of contact with the national society, and was not correlated with indigenous population density" (Nepstad et al. 2006). This suggests that IPs' practices have significant sustainability implications in AD and, therefore, REDD.

This recognition among biodiversity conservationists of the potential of IPs and other local peoples to conservation is welcome. It complements other social science research (see Anderson and Grove 1987; Homewood and Rodgers 1991; Pretty et al. 1995; Fairhead and Leach 1996).

Social scientists have recognized that people are embedded in ecosystems, have co-evolved with ecosystems, and have helped shape much of the renowned biodiverse landscapes around the world. People are thus, in large part, responsible as agents for biodiversity richness and ecosystem function in some of the world's greatest landscapes. Far from being wilderness, so-called "wildlands" with great biodiversity values are often indicative of landscapes that have been shaped by the hands of peoples – hunter-gatherers and pastoralists in particular (Adams and McShane 1992). These may not always be officially recognized as IPs.

This perspective of seeing people as assets is in contrast to the conventional wisdom that people pose problems, often need to be removed from biodiversity-rich landscapes, or else need to have their resource management systems modified if forests and biodiversity are to be saved. Based on early indicators, REDD appears to be perpetuating these beliefs in practice, all the while espousing respect for IPs' traditional and indigenous knowledge (TIK).

By creating methodologies that enable the measurement, reporting, and verification of carbon based mainly on assumptions about proximate deforestation drivers, there is little assurance that the fundamental social issues that historically have plagued conservation will be better tackled under REDD. Attribution of deforestation proximate causes – usually unsustainable farming practices, charcoal making, and timber extraction, etc. – are often deemed the problem without broader contextual analysis of their relation to other systemic drivers. For example, the alternative argument can be made that underlying drivers such as lack of viable agricultural technologies, lack of alternative energy sources, and generalized poverty are what really merit programming focus.

Yet clearly, these drivers are far more difficult for implementors to tackle than more manageable issues which national policing and forest management services can support. For example, measuring success on the basis of peoples' approval of monitoring jobs and a school or clinic that validates the project's theory of change while banal and simplistic, can nonetheless be done in sites where customary land use may need to be modified to accomodate a conservation concession management plan. This is more manageable than trying to tackle far more complex deforestation drivers where multiple factors and stakeholders may be involved.

Research shows that it is not just poor people who drive habitat and biodiversity loss (MacKinnon 2011), as the global trade in wildlife for example is such big business that it is often run by organized crime syndicates that also specialize in drug running and human trafficking. With the potential role of the Mafia already red flagged in REDD (Global Witness 2011b; Transparency International 2012), and when coupled to the acknowledged role that the Mafia play in "conflict resources" such as trade in timber of minerals for arms (Global Witness 2011a), it is clear that REDD strategies and policy must clearly transcend the focus on local resource users as *the* drivers of deforestation if it is to succeed.

Yet, REDD projects now being designed do not appear to address these underlying drivers. Nor is it clear how the interface between policy and practice in combating *and* avoiding corruption that REDD may induce is being dealt with.

Simplistic assumptions about stakeholder participation

REDD planning suffers from the same types of flaws in assumptions, policy, and practice, that have plagued development for over forty years. That the same mistakes derived from top-down hubris grounded in the "tyrannical rhetoric of participatory development" (Cooke and Kothari 2001) is fascinating, disappointing, but not particularly surprising.

All indications are that REDD has not learned much from this experience.

Development has proven elusive to generate benefits for rural peoples in much of the developing world, not including South Korea, China, or India, that many parts financed their own respective ascents that have become visible in the 1990s and 2000s.[8] Many of the programming assumptions embedded in REDD boil down to the capacity of rural peoples in REDD-eligible countries to participate in consultation processes such that FPIC can be achieved in REDD. As is proving to be the case, FPIC has been anything but easy to obtain in REDD practice to date (see Colchester 2010), with the question "consent for what?" (Di Gregori et al. 2012) remaining outstanding.

Consultation, participation, and FPIC are as difficult to facilitate or obtain in 2013 as would have been the case in 1974 or 2004. Under current circumstances, BAU-consultation is possible, but is often poor quality due to the demands for expediency that proposal submission processes create (Pretty et al. 1995). FPIC, if it is to be genuine, while philosophically laudable and crucial to achieving sustainability in REDD, will be improbable to obtain in the vast majority of cases REDD developers face.

This paradox – FPIC can be sought and obtained, but in most cases it will not be genuine and hence good enough – will compromise sustainability of REDD programming over the long term unless new strategic approaches are undertaken. This crucial topic is not, however, one that is broadly discussed.

There is a tendency to conflate TIK with the increasing politicization on the part of a number of IPs around the world. IPs are demanding legitimate rights. Clearly these should be respected and granted. The problem comes up in regards to how well IPs and other groups can participate in programs originating in the North. Most IPs and local communities are ill-prepared to interact with equal tactical sophistication as Northern groups can, placing them at a disadvantage so long as analysis and negotiation of terms in agreements are an issue.

Ascertaining whether local peoples have the range of technical capacities required to analyze situations and make optimal decisions to deal on a comparable technical plane to REDD PDD developers is important. This applies to land use planning or benefit sharing terms, for example, and will be prejudiced if IPs have not mastered the terms and procedures that REDD project developers themselves employ.

In REDD, these types of capacities are as important as any TIK that local peoples can bring to bear in arguing for particular land and resource use arrangements. By *not* having these capacities, this simply leads to the paternalization of IPs and communities by Northerners driving REDD programming. For example, PRA preference ranking is employed as part of the consultation process to generate information that underpins a project developer's theory of change – i.e. communities may desire potable water, a school, and a clinic. Future success will be measured over years as to how villagers rate these inputs as key to their newly improved land use practices adopted and attributable to the project.

TIK, coupled with the rights provided through UNDRIP, national constitutions, and other UN charters, while normatively significant, have yet to be translated into effective policy and legislative frameworks for local peoples and IPs in practice. Nor do they necessarily catalyze projects that work in local peoples' interests, as project developers can readily demonstrate how they have checked off consultation requirements pivotal to UNDRIP, etc., without fundamentally integrating communities and TIK significantly in planning and decision making.

Planners and decision makers remain resistant to delegation of decision-making powers to village-level peoples for numerous reasons. Those pertain to biases, assumptions about the magnitude of transactions costs, and the justification to maintain control over programming decisions and resources.

Perhaps most perniciously, the calls for capacity building to enable effective participation have become largely hollow. It has been well over twenty-five years since the global community has understood that local capacities are weak, and that capacity building is crucial. Yet, commitment to comprehensive capacity building seems forever deferred.

So while capacity building at all levels of multilateral, bilateral, and NGO programming has persistently been recognized as needed for decades now, comprehensive capacity building strategies and programming have *never* become fully prioritized at any level of donor programming involving development and environment. As a result, rural peoples have largely remained stuck, overly dependent on external sources to come and resolve their problems. And while there are, fortunately, a number of counterpoints to this narrative, they are far and few between on the ground. In the vast majority of cases, rural communities are at a distinct capacity disadvantage when tackling land use management issues in which they must interface with BINGOs, private sector companies, and national government.

The role of communities in REDD to date is thus not materially different than under prior CBNRM, biodiversity conservation, or development programming. Communities are sought out in UN-REDD convenings. Best practice supported by the CCBA and technical agencies (see Richards and Panfil 2011; Richards 2011) resoundingly underscores the importance of principles long acknowledged and applied from the early days of PRA (Chambers 1986) – i.e. that it is best to involve communities as far as possible. Participation is resoundingly noted as key.

Yet, discretion is *totally* left up to REDD project developers under this guidance to define participation content and boundaries. How much particiaption, and to what level of excellence, is never specified. With no clear threshold standards pertaining to participation in any aspect of REDD established by the guidance, little clarity is added beyond what is already known.

This shortcoming repeats those of PRA as a methodology that is proven strong on description, strong on participant ranking, but weak on identifying the determinants of social feasibility (Brown 2010a). Hence, its value as part of VCS, CCBA, FCPF, and UN-REDD "best practice" methodologies is of questionable value, beyond moving weakly designed projects through funding pipelines.

So long as communities maintain an ambiguous strategic position in REDD – acknowledged for their knowledge, insights, and customary rights, but rarely trusted or empowered to take the lead in design or management processes – REDD will founder.

Common property and communities

Since Ostrom's (1990) major work on the importance of institutions for collective action in governing the commons, there has been an uptick in research concerning the effectiveness of community managed forests as compared with formalized protected area (PA) management. For Ostrom, common pool resources such as African rangelands or Maine lobster beds can be successfully managed when those who stand to benefit are both closest to the resource and most responsible for its management.

This insight, fundamental in a social feasibility approach, leads to the basis for the hypothesis that communities that are reliant upon common pool natural resources can be, or already are, capable of working out strategies together to avoid over-exploitation of common resources (see Ojha et al. 2009; Almeida et al. 2012). Clearly there is no reason why this should be any different with avoiding deforestation and REDD.

To work, peoples' perspectives on fairness must couple with resource sustainability. If empowered, and provided sufficient technical support, communities have proven capable of cost-effectively developing management rules. This normative principle should, in an ideal world, increasingly become the kernel of any social contract negotiated for avoiding deforestation and sequestering carbon. Of course, the world, still, is far from ideal.

When community-managed forests are compared with strict protected forests, based on a statistical analysis of annual deforestation rates for seventy-three case studies throughout the tropics, deforestation is shown to be significantly lower (Porter-Bolland et al. 2011). When provided with clear rights and authority and managed by indigenous people, indigenous areas and multi-use PAs can work to accomplish ambitious conservation goals (Nelson and Chomitz 2011).

Many forests and biodiversity-rich savannah landscapes have been shaped in co-evolution with human activities (see Homewood and Rodgers 1991 on the Kenyan Masai, for example). Separating people from ecosystems they are a part of is artificial and arbitrary. It alson underestimates the positive feedbacks between people and ecological systems from an ecosystem function and sustainability perspective (Berkes and Folke 1998). Contrary to assumptions based on theories of the tragedy of the commons (Hardin 1968) that still underpin much conventional wisdom in development and conservation programming, arguably inappropriate people *do* have the capacity to create rules, norms, and institutions that regulate human use of ecosystems, and have been shown in many places to have sustainably managed resources over long periods of time (Feeny et al. 1999; Ostrom 1990; Dietz et al. 2003). Local institutions that use what Ostrom refers to as 'rules-in-use' (Ostrom 1990), evolve dynamically together with local ecosystems, and build resilience into the social–ecological system (Berkes et al. 2000). The potential to mobilize communities when provided the opportunity to create rules-in-use has long been clear (Brown 2001a, 2001b).

For example, *if certain conditions* can be met, COBAs in Madagascar have demonstrated why they can be invested in as principal institutions in forest and biodiversity conservation (Hockley and Andriamarovololona 2007), and potentially REDD. For US$4.70/hectare, Malagasy COBAs were shown to be able to protect key conservation corridors. This in turn generates a net benefit of about US$54/hectare, once costs are subtracted (Hockley and Andriamarovololona 2007). Whether the ratio is 10:1, or more modest as an ROI, the ROI is clearly worthy of exploring in a program like REDD. When considering that USAID environmental programming alone totaled about US$342 million in Madagascar for the 1984–

Box 6.2 Community empowerment lessons from Kaa-Iyaa, Bolivia

Empowering IPs and local communities, while rhetorically acknowledged in biodiversity conservation for decades under the importance of traditional knowledge and customary management systems, and now under REDD, in practice this rarely occurs. Is the odd case such as the Kaa Iyaa Del Gran Chaco National Park in Bolivia one to look to in REDD for inspiration? Facilitated by WCS, IPs – the Upper and Lower Izozog Authority (Capitanía del Alto y Bajo Izozog), an indigenous Izoceño-Guaraní organization – actually assumed the lead in management planning and implementation of the Kaa Iyaa Del Gran Chaco National Park in Bolivia. The park remains the only national PA in the Americas created as the result of an initiative by an indigenous organization (Painter 2004), albeit strongly supported by USAID and WCS.

If in fact successful, why is this so? Was it because the IPs were adequately empowered? Is the institutional situation at heart different from COBAs in Madagascar (Hockley and Andriamarovololona 2007)? Can REDD planners learn from both? In Latin America "a large proportion of the 1,949 national protected areas in Latin America and Caribbean covering 211 million ha. and 29 million ha. of terrestrial and marine systems, respectively is devoted to ethnic territories", though the effectiveness of these territories is unknown (Toni 2012). Can the Kaa Iyaa model thus be replicated?

While communities are hardly a panacea for the limitations of weak local government entities, their obvious stake and interest in the forest and its sustainability from a livelihood security standpoint, suggest that they become a principal focus for support and investment in avoiding deforestation.

2009 period (Freudenberger 2010),[9] had USAID invested US$34 million in COBAs progressively and systematically over that period, what results would have been yielded? The same question should be posed by planners now in REDD.

Dedicated investment in local peoples' management of natural resources can make a major difference in outcomes and impacts. The basic issues faced in Malagasy CBNRM are similar to those faced in REDD. But one key lesson from Madagascar remains to be learned by REDD planners more widely:

> [B]y providing insufficient support to COBAs, and pretending indifference to the wider benefits of their management, external stakeholders have tried to extract a "free lunch" from communities; securing forest conservation at minimum cost... If favourable conditions are created, COBAs offer an efficient and equitable mechanism for achieving forest conservation in Madagascar. In areas where communities identify strongly with the forest, some form of community management may be the only viable option.
>
> (Hockley and Andriamarovololona 2007)

Proposals to target one component of a system, without analysis of the impact on the broader system, can have deleterious consequences (see Sharpe 1952). The assumption in REDD is that by incentivizing carbon, a suite of sustainable outcomes will follow. There is no empirical basis for this premise that underpins the REDD theory of change. REDD decision makers and PDD developers stand a much better chance of success by bulding on communities and people, than by becoming seduced by atomic number 6 on the Periodic Table.

Unfavorable tenure policies

As REDD has evolved, increasing and deserved attention has been paid to tenure as a principal constraint in REDD (Sommerville 2011; Rights and Resources Initiative 2012). Tenure policies are often unfavorable to communities. Even when implementation regulations exist,

> the act of putting a community tenure regime in place is often mired in bureaucratic requirements such as costly land delimitation processes, the undue requirements for communities to acquire legal status, the complex legalese of applications, the need to provide evidence of the traditional use of forest land, and the short timeframe during which communities must comply with the complicated procedures established by law.
>
> (Almeida et al. 2012)

The net result of this is that state or public ownership of forest land and resources still dominates many of the world's forested countries. Often, these public lands degenerate to "open access" (as distinct from common property), as the state does not have the ability, nor customary owners the means or authority, to manage resources coherently. REDD will be hard-pressed to produce tangible results so long as this remains the case, as tremendous disincentives for local community management and compliance with REDD regimes will prevail. Yet it is unclear how the PDD process through VCM initiatives, or national policy setting through the R-PP process, is mitigating tenure constraints beyond the obligatory acknowledgement of urgency.

Moreover, an additional layer of challenges in central Africa particularly, involves how the legacy of problematic fulfillment of SRCs will impact REDD implementation. Decades of disappointment over unfulfilled SRCs has been well described by Greenpeace for the DRC (Greenpeace 2007). The implications of unresolved SRCs on REDD are not to be underestimated for all of central Africa.[10]

IPs and REDD

Much has been written of IPs from an anthropological and human rights perspective. The anthropological literature indicates that IPs have evolved highly sustainable resource management systems (e.g. The Ba'Aka Pygmies (Bahuchet 1994), the Bambuti Pygmies (Turnbull 1965), and the Kalahari Bushmen (Hitchcock 1996)), to name several notable cases. This is not surprising, as IPs are simply a subset of the many communities that have highly developed common property resource management systems as discussed in the prior section.

The suggestion that hunter-gatherer IPs maintain a conservation ethic has, however, been challenged by some critics who advance that any sustainable management practices among IPs may relate more to practices developed due to low population densities, coupled to disinterest in capital accumulation, than in being strong conservationists per se. Some conservation biologists suggest that the goals of wildlife conservation and the goals of IPs are not always compatible. This leads some to conclude that the notion that IPs have de facto resource management skills in all cases, when data exists that proves otherwise, simply perpetuates the myth of the "noble savage" (Redford and Robinson 1985).

Whichever is the case, IPs do have significant conservation bona fides in comparison with most peoples in the world. This clearly is an area to be built upon in REDD, and has in fact been one point upon which BINGOs have supported IPs' presence in and around PAs, yet rarely to the point where they could be envisioned to be empowered to manage large-scale initiatives. One notable exception to this is the Kaa-Iyaa del Gran Chaco (see Box 6.2).

Much focus has been on the impacts of REDD+ on IPs. The perspective from the International Work Group for Indigenous Affairs (IWGIA) on REDD+ is instructive,[11] as they see climate change impacts already evidenced on IPs across diverse ecosystems, and consider REDD+ as a possible solution.

Examples of IPs' resistance to REDD

Central America has been one hotbed of opposition to evolving REDD programming. If there is one region where a new social contract in REDD is desperately needed, it is Central America.

One indigenous analysis of the potential costs of REDD+ programming to both livelihood security and cultural lifestyle comes from resistance to REDD from the Chiapas State in Mexico. It considers diverse stakeholder positions including the State Government of Chiapas, Starbucks, and CI. The analysis challenges whether PES' markets, and REDD are a logical development pathway for the people of Chiapas (see RED por la Paz Chiapas 2012).

A profound chasm of perception and understanding currently divides local communities, UN-REDD, and government in implementing REDD Programmes in the subregion. The continuing saga of the Panamanian R-PP process involving IPs through their umbrella organization, COONAPIP,[12] and Panamanian authorities and UN-REDD, illustrates the disconnect: "If we are having such problems in a process *that is just beginning* [emphasis mine] and the agencies involved behave in ways that are fundamentally inconsistent with the principles that are supposed to apply to REDD, what can we expect when the REDD strategy actually begins to be implemented?"

This is not the only case of "incongruences and inconsistencies" between local peoples, UN-REDD, and government in Panama. The situation is replicated in other countries, with similar letters expressing similar messages of discontent.

Not all IPs are critical of REDD at the outset

While many IPs are skeptical of REDD, not all indigenous and local peoples are opposed to REDD from the get-go because of carbon markets. To suggest otherwise, is an oversimplification.

It is true that in Latin America there are many IP and NGO federations and alliances that oppose REDD. Some are highly politicized and capable like AIDESEP (2010) or FENAMAD (2010; also see Llanos and Feather 2011).[13] Yet not everyone appears to be in opposition.

Taska Yawanawá's October 27, 2012, letter to REDD-Monitor.org posted on November 2, 2012[14] on PES in Acre, Brazil, whether it is a good thing or a bad thing, for IPs, shows that in fact there is diversity of opinion in the IP community in Brazil and elsewhere. In assuming a position of neutrality – "I want to better understand before making any hasty decision that could jeopardize the future of my people or the planet" – Chief Task Yawanawá has done something that neither REDD proponents nor critics have exhibited: taken a step back and argued for analysis *prior* to taking a position. Far more of this sort of refreshing, "neutral" position in accepting or rejecting REDD is needed.

Yawanawá's reaction does raise a fair question: Are IPs and communities *technically* informed enough, and internally organized enough, to either consent freely to *any* REDD initiative? Yawanawá suggests this may be possible. If so, what is needed to get there?

The case of the 1,000-member Suruí in the Brazilian southwest Amazon is interesting in this regard. To capitalize upon the potential that Article 231 of the Brazilian Constitution potentially could offer IPs in REDD, Forest Trends engaged a DC law firm, Baker & McKenzie, "to not only focus on the specific rights of the Suruí, but to increase the scope of their analysis to reflect the rights of indigenous peoples throughout Brazil". If they could succeed in securing carbon rights for the Suruí, they felt this would establish a precedent for establishing the carbon ownership rights of other IPs throughout the country, potentially impacting 235 ethnic groups, with over 750,000 individuals (Katoomba Group 2009).

Some suggest that the example of how REDD+ has been evolving among the Suruí of Brazil provides hope that REDD+ could, if approached with the necessary capacity building and multipurpose objectives in mind, be a source of strengthening of IPs' rights, enhancing incomes and livelihoods, in addition to combating deforestation (Prizibisczki 2010). Others, as alluded to, outrightly reject the commodification of nature and creation of a category of rights that has no inherent sense from legal or operational grounds (Karsenty et al. 2012a).

From my own personal experience with IPs since 1973 in Africa particularly, outright rejection of REDD for philosophical or political reasons makes little sense in contexts where communities have limited real options. The affinity of Pygmies to *marketing*, whether it be through hunting or collection, poached elephant tusk or legally obtained honey, is an indicator that they could be open to the notion of being paid for environmental services. There would be a willingness to give REDD a try *if* appropriate mechanisms could be created for Pygmies to engage coherently in the process.

The problem, I believe, has more to do with the lack of a social contract to serve as a framework for coherently engaging Pygmies, or other indigenous groups, than it does with the inherent antipathy to markets or the notion of creating a new commodity called carbon as such.

Urbanized Pygmy intellectuals I have spoken with take a pragmatic approach to the dilemma facing Pygmies today in central Africa. The same is true for clanspeople in PNG and elsewhere: the logic of the overall offering will incentivize participation.

The carbon market element of REDD is arguably less politicized for IPs than for international NGOs or academics. Though the latter groups might suggest that that simply reflects a variant of Marxian false consciousness on the part of IPs, or simply reflects an unfortunate apologism among any external agents to IPs'

Box 6.3 Tortuous policy rhetoric and social feasibility in REDD: Indonesia's IPs

Safeguarding IPs' rights is fundamental to REDD. Indonesia is one of the premier countries in the REDD Readiness Phase. UN-REDD notes on its website: "The Indonesia UN-REDD Programme has been launched to assist the Government of Indonesia in its REDD+ readiness efforts, in order to establish and organize a fair, equitable and transparent REDD+ architecture in the country."

On October 1, 2012, Survival International reported, meanwhile, that "Indonesia denies it has any indigenous peoples".[15] This was based on Indonesia's response to the UN four-year periodic review of human rights[16] where recommendations were made to Indonesia to adopt new policies and practices in line with UN principles. Indonesia responded as follows to the report: "The Government of Indonesia supports the promotion and protection of indigenous people worldwide. Given its demographic composition, Indonesia, however, does not recognize the application of the indigenous people concept as defined in the UN Declaration on the Rights of Indigenous Peoples in the country."

On September 24, 2012, WRI, WWF, and TNC announced they were awarding President Yudhuyono with "the first ever Valuing Nature Award"[17] for his leadership in the coral Triangle Initiative which "offers a model for connecting marine conservation to the health and security of local communities".

Survival International noted that the "denial of the very existence of indigenous peoples in Indonesia is symptomatic of the government's total disregard for their rights".

Question: *How will social feasibility for Indonesian IPs who are targeted and "safeguarded" by REDD be achieved?*

society that may suggest the REDD could be beneficial, local peoples in many places are open to the possibilities REDD could offer.

Perhaps it *is* false consciousness that conservation organizations, along with donor proponents, hope to capitalize upon in REDD. Selling a bit of banana beer, thread, used clothing, etc., is enough to make the difference between household livelihood security and impoverishment on a daily basis in many countries. If REDD could offer credible marginal benefits over the current state of affairs, many IPs could readily accept this, in contrast with other groups that may be more politicized.

Maximizing information transparency, and providing IPs the capacity building needed to objectively assess whether they wish to embark on REDD and provide consent, or not, is fundamental to REDD's success. If this can be accomplished, perhaps a good number of IPs at the end of the day may actually opt for REDD as a best option for the present and future.

7 Social feasibility and its components

Development and conservation practitioners have yet to explain why results and impacts historically in either ODA or privately funded development and biodiversity conservation activities have fallen short of expectations. Various explanations looking retrospectively at the data as to why this has occurred were covered in Chapters 1 and 3. Debate still abounds. Its implications shape the future success or failure of REDD.

Banerjee and Duflo's (2011) argument is appealing: by resisting "the kind of lazy, formulaic thinking that reduces every problem to the same set of general principles", and by listening to poor people, establishing "rigorous empirical testing to establish a toolbox of effective policies", the global community will be better placed to alleviate poverty. While the definition for what constitutes "listening" requires precision, along with its measurement, the point is key for public–private programs like REDD. Yet for REDD to work, one important caveat beyond Banerejee and Duflo's proposition on the social side of issues remains if REDD is to succeed.

Poor forest peoples whose FPIC is sought to move forward with REDD projects must not simply express what *they need* in village level meetings so that a box can be checked off, and project developers can move forward in implementation as they are now doing. Poor forest peoples must be able to assess REDD with the same neutral, non-polemical, patience that Chief Tashka Yawanawá in Brazil is imploring IPs in Brazil to display. Chief Tashka leads the Yawanawá, an IP who live in Acre, in the far western part of Brazil. He has worked with Aveda who helped the Yawanawá double the size of their land rights to 178,000 hectares (687 square miles) and has met with Prince Charles and the UN.[1] As such, it is fair to say that Chief Yawanawá is probably not an "average" indigenous person in the Brazilian Amazon.

For IPs to objectively assess REDD, to accept it or reject it in an informed manner, information and appropriate analytical tools are needed. While Chief Tashka also studied in Rio de Janeiro and four years in the US (Ashoka 2008), 99.99 percent of IPs do not have similar experience to capitalize upon in assessing REDD. They have not had the benefit of being Ashoka Fellows. The know-how and analytical tools that most all IPs and other forest communities need to neutrally assess REDD are not currently found in village toolkits. The skills demanded extend beyond what indigenous knowledge systems and traditional values can provide.

We can all think like scientists

The important clarification that Africans, and by extension so-called "primitive" peoples globally, are as capable of producing sophisticated thought processes and analysis as Westerners, and that their "thought-systems" well demonstrate this (Horton 1993) has enormous implications for programs like REDD. Basically, it undermines the paternalistic pretense that Westerners (aka "Northerners") must be those responsible for defining how local-level deforestation and forest degradation is to be avoided, and carbon to be sequestered. If provided the tools, and the capacity building to capitalize upon them and adapt them to their circumstances, remote forest peoples can reason and negotiate with the North.

At present, however, this possibility is precluded.

Development and conservation programs do not even explore the implications of what negotiated, collaborative planning processes with local peoples key to programming success in REDD would practically mean. Despite extolling the virtues of participation and TIK, TIK is treated as a separate category of knowledge distant from scientific knowledge. Social scientists, conservation biologists, and REDD planners of course take TIK into consideration when deciding what is to be implemented in REDD projects as part of best practice guidance. Yet IPs and communities are never provided an equal seat at the REDD planning table when projects are designed.

Since neither the tools nor the capacity building is dispensed to remote forest communities under REDD, in the belief that needs assessments to resolve basic community needs will be "enough", the feasibility of REDD programs is undermined before they even begin. For both moral and practical reasons, this assumption is false. If not corrected, it will undermine any hope of REDD succeeding.

The rationale for a social feasibility framework in the Yawanawá and IPs' contexts

What is it that would enable Yawanawá people, and IPs' leaders from around the world, "to better understand before making any hasty decision that could jeopardize the future of my people of [sic] our planet" (Chief Tashka Yawanawá 2012)? What capacities need strengthening, and to what levels?

For REDD to work, planners must get past the simplistic logic of needs assessments and treat communities with the same respect they would demand for their own community's present and future. The acceptance of this premise, the social contract to work towards it, and the toolkit for actually doing so, have yet to be perceived as key objective, let alone agreed upon.

What then specifically can break through our inability to enable more effective upstream, community level participation in planning to actually enter programs like REDD? This is where a social feasibility framework, as opposed to a PRA checklist, comes in.

Those same poor people that Banerjee and Duflo would have us listen to so as to understand the logic of their choices need to expand their capabilities to better assess whether programs like REDD have anything to offer them. Chief Yawanawá's dilemma is this – is REDD poison for him and his people (as most vocal IPs' and NGO activist groups on the topic suggest), or could it be a pathway out of poverty for the Yawanawá while protecting the environment they cherish? Village-level meetings where PRA checklists are used to rank needs are insufficient to deal with the trade-offs that the Yawanawá, or any IPs, face in REDD.

Background to establishing feasibility

Deciding on what to do in REDD *should* involve the assessment of information about climate change trends, the costs and benefits of mitigating climate change from deforestation, and *the values* that influence different stakeholders who inform and/or make decisions about climate change mitigation. Between trends, cost–benefit issues, and values, the latter is inarguably the most difficult to reach any kind of consensus over at global scales.

Values influence the extent to which costs are willingly able to be imposed by certain groups of stakeholders over others in the name of collective good. If imposed, as has been argued has already occurred in the conservation arena (Brockington et al. 2008; Duffy 2011), conflict is likely to ensue.

Sticking just to the most basic parameter to REDD success – participation – the risk is very high that techniques are being employed that tend to co-opt, manipulate, and mask social complexities as much or more than they promote any genuine participation, consistent with what has been called "the tyranny of participation" (Cooke and Kothari 2001). While participation *is* clearly necessary for REDD to succeed, if the approaches are not gotten "right", superficial participation indicators can be used to obtain funding, while masking serious REDD design flaws.

Based on critique to date, it appears as if REDD is succumbing as one more example of how participation is being poorly approached.

Social feasibility as step 1 in overall feasibility analysis

If the Yawanawá and other IPs and communities confronted with REDD were provided the framework and means to first assess what their minimum requirements are for REDD to *work for them*, to be socially feasible, this would establish a reasonable basis for design of programs and projects that are aligned with other technical elements in REDD.

Yet to do so, most IPs will need skill building to assess the terms being offered them in REDD. Costs, benefits, and risks of the REDD offering will need to be neutrally assessed. If the offering falls short, IPs and communities should be able to offer informed alternatives to potential REDD partners.

Tools to accomplish this analysis and counterproposal offering will need to be in place. Presently, these tools, and thus capacities, are lacking in most instances.

What then is social feasibility?

Social feasibility is a favorable end state. It relates to the adequacy by which programs and projects address social dimensions in their design. To be demonstrated, a number of minimum thresholds as judged by communities impacted by projects must be met. These relate to several domains including: the adequacy of participation and governance arrangements in planning, the adequacy of safeguards for livelihood security, the adequacy of proposed benefit sharing arrangements, and the adequacy of resource allocation dedicated to safeguarding communities' present and future prospects (Brown 2010a).

By referring to "adequacy", this immediately introduces the need for judgment as to what is adequate. Judgment involves objective and subjective criteria. In community contexts, this demands that multiple issues be evaluated against some measure that leads to pronouncement of consent or rejection of a REDD offering. However defined, IPs and other communities will need to conduct their own version of multivariate analysis. If not, *informed* consent will remain elusive.

While not explicit in UN-REDD, FCPF, CCBA, or any other policies or standards, any community provision of FPIC should *normally* presuppose that social feasibility would be considered and the above noted criteria demonstrated. Under current standards and reporting requirements, however, these indicators are not explicitly addressed.

Yet to even achieve FPIC, there are preliminary considerations to be addressed at community levels (Anderson 2011). These include:

- assessment of capacity building needs required such that FPIC *could be* obtained;
- definition of measurable thresholds for ascertaining current community capacities to achieve FPIC;
- compliance in implementing capacity building and measuring key capacities.

Key indicators for ascertaining if a credible FPIC process has been implemented in REDD include the following:

- *Equitable and full community participation* – i.e. are women, IPs, minorities and youth verifiably involved, and have they understood the issues?
- *Credible community assimilation and processing of REDD+ offering and trade-offs* – i.e. are communities able to present costs, benefits, and risks of REDD proposals to project validators and auditors?
- *Credible community-level institutional capacity to make decisions and sustain engagements* – i.e. can these certifying agents determine that key information has been fully vetted and community decisions are "representative"?
- *Credible community-level (and institutional partner) capacities to monitor, evaluate, and adaptively manage REDD engagements to comply and adapt as needed* – i.e. have skill levels been verified as adequate, and are technical support measures as needed in place?

- *Credible ability to assess and advocate the need for policies, enabling conditions, and technical support to achieve REDD+ permanence objectives to be sustained in the face of evolving threats* – i.e. can communities verifiably demonstrate analytical capacity for sufficiently astute policy analysis to merit award of FPIC?
- *Credible community institutions and mechanisms to enable equitable and timely benefit sharing to be achieved* – i.e. have communities identified and negotiated favorable outcomes, or have they simply accepted what was offered to them?
- *Determination if land tenure is enabling or disabling[2]* – i.e. have external stakeholders together with community representatives assessed land tenure as a possible constraint, and identified what bundle of formal or informal rights are needed to justify moving forward with REDD?

If REDD projects now being implemented in the Readiness Phase were to be evaluated using these criteria, few pilot projects would likely respond successfully. Moreover, it is not clear if pilot projects are even working to be in the position to answer these questions at the end of the Readiness Phase, when the roll-out strategy for full REDD+ implementation is presumed for 2015.

Overview of basic distinctions between feasibility and safeguard approaches

The social dimensions in REDD are largely addressed through SIAs as well as social safeguards. SIA is primarily a tool that is useful to assess if projects have lived up to what was anticipated for social impacts in design. It is not used to determine specifically what needs to be designed from a social standpoint so that projects work. That said, use of SIA is one way in which social risks associated with REDD projects can be mitigated.

The distinction between social feasibility and safeguards is important. Integrating social objectives at an early stage in REDD leads beyond safeguards (Vira 2012). Understanding that REDD involves much more than safeguards for protecting the rights of front line forest communities, and programming in social feasibility analysis to enable getting past it, will be a major pivot point for REDD in the near term.

Lessons learned from the CMP

The CMP experience is *highly* relevant to REDD. Paradoxically, its most profound lessons do not seem to have raised the bar for establishing best practice in REDD for the better. In particular, the CMP has yet to decide how to "deliberately incorporate social strategies and human welfare targets into the Open Standards for the Practice of Conservation". The National Audobon Society et al.'s (2011) publication *Tools of Engagement: A Toolkit for Engaging People in Conservation*, provides some sense, though, of where the process is headed.

Most notably, social feasibility is not a consideration in the five-step CMP framework. Social feasibility *could* be addressed somewhere under situation analysis

Table 7.1 Simple comparison of social feasibility and social safeguards

Social feasibility assessment	Social safeguards
• Determines what is required for projects to work and become sustainable	• Determines what is required to protect peoples' economic, social, and cultural interests
• Presupposes that safeguards must be met to be feasible	• Emphasizes protecting status quo versus discerning either minimum or optimal development requirements to achieve sustainability
• Considers minimum capacity and action requirements for successfully addressing resource tenure, existing stakeholder conflicts, and identifies prospective benefit sharing	• Establishes an implicit "defensive" posture where local peoples are perceived as a constraint to REDD, somewhat similar to the notion of communities as a target of threat analysis in conservation programming
• Presupposes that REDD can only work sustainably if people play a lead and central role in design and implementation, and receive equitable benefit sharing from REDD	

in phase 1, once critical threats pertaining to people's resource-use patterns had been identified.

The framework from the *Tools of Engagement* publication illustrates how central threat analysis, and the behavior of people, shapes all aspects of conservationist solutions to problems pertaining to biodiversity status and, in the case of REDD+, deforestation responsible for habitat destruction, biodiversity loss, and the impact of climate change on biodiversity. The framework themes were further underscored by the CMP (Conservation Measures Partnership 2012).

If threat assessment begins with the understanding that threats can be human activities that directly degrade the target, for example, habitat loss because of development (National Audobon Society et al. 2011), this effectively precludes a social feasibility orientation.

The National Audobon Society et al. (2011) suggest the following of practitioners:

> As you're digging into the root causes of the problems, you might find that you still need to understand more about the community or issues that you're focusing on… The root cause analysis helps you make sure you're accounting for the biological, social, economic, political, and cultural factors of a situation… In some cases, you might want to do a community assessment to find out more about the situation.
>
> For example, you *might* [emphasis mine] want to find out more about:
> • Community boundaries.
> • Community capacity and activism.
> • Community interaction and information flow.

- Demographic information.
- Economic conditions and employment.
- Education.
- Environmental awareness and values.
- Governance.
- Infrastructure and public services.
- Local identity (and culture).
- Local leisure and recreation.
- Natural resources and landscapes.
- Property ownership, management, and planning.
- Public safety and health.
- Religious and spiritual practice.

This non-prescriptive flexibility has limitations. True, flexibility can be useful, but at the same time it opens the door to extreme relativism – no one's methods are best, as comparative analysis and judgment is never undertaken. Information and methodology consumers are allowed to choose their preferences, as if walking down a supermarket aisle and making selections based on branding and packaging. As this information on methods has oftentimes not been peer reviewed, or compared, and with little to any data on how results correlate to impacts, objective evaluation becomes difficult to impossible (Brown 2010a).

Others have expressed similar misgivings. In a review of US$7 billion in World Bank participatory development programming, Mansuri and Rao (2004) noted that "handbooks, guidelines, and terms of reference all use the concepts [participation, community, social capital] uncritically, assuming that they are widely and uniformly understood. What each of these concepts implies, however, is quite controversial". The fact that there has been no subsequent standardization or consensus on minimum threshold requirements for terms, and when concepts such as participation or, in the case of REDD, FPIC are obtained, this leaves the door open to relativization and abusive use.

It is unlikely that there will be many instances "when potential beneficiaries make key project decisions" enabling participation to become "self-initiated action" in what has come to be known as "the exercise of voice of choice" (Mansuri and Rao 2013) or empowerment. Little from the empirical record suggests "voice" or "empowerment" associated with REDD program and project success is being engendered by best practice guidance and standards deployed.

An example of how best practice has led to REDD controversy: Cambodia's Cardamoms

The following is one well publicized example of why social feasibility issues are so important in REDD, and why current best practice is insufficient. Moreover, it is an example of how PES' models can empower buyers of services "to define the nature that they want to save and how, while leaving little scope for participatory or bottom-up natural resource management" (Milne and Adams 2012). This is one way in which social feasibility can be precluded by planners.

In early 2012, a jaw-dropping, highly controversial, protracted story began unfolding in Cambodia's Cardamom Mountains. It has been well publicized in the public domain (Lang 2012a). It involves one of the thought-leaders in REDD, CI. By leveraging its world renowned leadership in biodiversity conservation, CI has been assuming a lead in REDD+ programming globally. The Cardamoms case involves numerous stakeholders: the Royal Government of Cambodia (RGC), communities in three provinces, numerous large donors and foundations, several respected academic anthropologists, the *Phnom Penh Post* reports, and one of the leading critical sources of information on all things REDD, REDD-Monitor.

The case study epitomizes the gamut of perceptions that different stakeholders may bring to bear from the same set of events and data. When there is no unifying framework for defining and appreciating what is happening on the ground pertaining to the social dimensions of REDD, controversy can be amplified

At its most basic: CI markets a successful program and receives donor funding for Cardamom forest conservation activities. A national newspaper breaks a story about corruption and illegal logging. CI is implicated in the *Phnom Penh Post* story because of alleged support to government employees it is paying. Social scientists, one of whom had reportedly worked for CI in Cambodia, look closer at the facts and interpret a wholly different, far more complicated and less sanguine story for forests and social welfare trends in the Cardamoms than that portrayed by CI. CI disavows any negative implications. Critics of CI and REDD attack and counter-attack. The rosewood of the Cardamoms continues to disappear. REDD monies continue to flow into the project zone all the same.

Finally, a former CI employee and director of the Natural Resource Protection Group in Cambodia, Chut Wutty, is murdered by Cambodian military police on April 25, 2012. Leading journalists from the *Cambodia Daily* then investigate details of further illegal logging purportedly undertaken by a private company in the Cardamoms that perhaps are pertinent to the killing (Lang 2012b).

Regarding the murder, a former CI staff member in Cambodia notes:

> It is this failure of mainstream and "official" conservation efforts that pushed the battle for Cambodia's forests to the fringe. This is what drove Chut Wutty and his colleagues at NRPG to risk their lives gathering data on illegal logging operations in the Cardamom Mountains and elsewhere. The work of NRPG revealed not only the culpability of government officials who abuse their powers to profit from logging, but also the hypocrisy of NGOs like Conservation International that have denied the existence of logging altogether, in order to maintain the façade of effectiveness, along with their government and donor relationships.
>
> (Milne 2012)

Government rangers paid by CI have been directly profiting for years from illegal logging of the Central Cardamom Protected Forest (CCPF) (Boyle and

Titthara 2012). The reason for failure, despite CI's claims for project success, are that the definition of the problems in the Cardamoms and the PES' solution "were drawn up by economists in Washington, expatriate staff in Cambodia, foreign biologists and government staff" (Milne and Adams 2012; Lang 2012b). Community perspectives were thus excluded, while the role of elites reinforced (Milne and Adams 2012).

CI's "communities" were created around the requirements of the PES' model. "The outcomes", Milne and Adams write, "were inevitably shaped by pre-existing power structures", while it was the "failure of mainstream and 'official' conservation efforts that pushed the battle for Cambodia's forests to the fringe" (Milne 2012).

What makes this story important is that the CCPF has received direct support from some of the world's most ardent supporters of forest conservation and now REDD. Agence Francaise de Developpement, USAID, and the Gordon and Betty Moore Foundation all supported the CCPF.

The Cardamoms' example illustrates how thought-leaders in PES and REDD are using TMAs, along with public relations and privileged government relationships, in manners that may simplify and camouflage social complexity so as to preclude social feasibility.

To avoid similar outcomes, different strategies based on social feasibility methods are needed. Unfortunately, REDD planning in Prey Long, Cambodia, by CI Japan is predicated on the "success" of the Cardamoms.

Extending mistakes from the Cardomoms into Prey Long?

Prey Long is a forest estimated to be around 400,000 hectares. The reference region is "the area where the project analyzes its deforestation rate, driver, and pattern as a geographical reference" (CI Japan 2012).

The relationship between CI's Cardamom activity and the evolving Prey Long REDD+ program is summarized by CI Japan (2012) as follows:

> In Cardamom, Conservation Agreement with local communities by CI and law enforcement for forest management by FA had been implemented together [sic]. Since Cardamom project had shown success in the region, FA tasked with CI [sic] to explore the opportunity to implement the same scheme for the Prey Long Area by utilizing the REDD+ mechanism.

The objective of the feasibility study is as follows: "In summary, to assess the feasibility of combining Conservation Agreement at the community level and the forest management by FA at each province in the Prey Long Area in the longer term by involving community to reducing emissions from deforestation in the area [sic]."

The approach proposed by CI for the Cardamoms and CI Japan for Prey Long is implausible. There is no logical relationship between what the Feasibility Study presents, and its recognition that: "It is especially important that the project to follow the Free, Prior, Informed Consent (FPIC) process to establish the governance encompassing the large area of four provinces [sic] (CI Japan 2012).

Box 7.1 Subsidiarity principles and participation are key to success in REDD

Ostrom (1990) showed the importance and potential of community participation in rule-making in management of common pool resources such as forests.

Although achieving desirable outcomes across potentially competing social and ecological objectives is complex, Persha et al. (2011) suggest that when forest users participate in rule-making aspects of forest governance, the probability for positive outcomes are enhanced. "Participation is associated with a lower probability of less desirable outcomes (unsustainable forest systems and those characterized by trade-offs) and a higher probability of sustainable forest system outcomes, across smaller and larger forests." They conclude that "working toward formal participation of local forest users in rule-making processes for use and management of forests from which they draw their livelihoods is an important way to increase the probability of obtaining more positive outcomes across social and ecological dimensions". Clearly, these findings are of relevance in REDD. While the premise of participation is strong, the levels of participation and thresholds for what will be needed in terms of rule-making and benefit sharing mechanisms remain to be clearly defined and negotiated in each case.

There is reason to believe that this approach is representative of how feasibility issues are approached by conservation NGOs in REDD generally, not just CI. This is because there is neither a standard nor a framework for addressing social feasibility. As CI is part of the founding members of CCBA, clearly it is within their purview to review their assumptions of the adequacy of CCBA social standards in the future.

The Cardamoms is marketed by CI Japan as a "successful" model – "The scheme to protect the forest after turning it into the Protected Forest along with local community and FA is already proved to be successful in CI's activity in the Cardamom Protected Forest [sic]" (CI Japan 2012). Given contrary reports from the Cardamoms (Milne 2012), there is legitimate cause for actually considering *failure* as part of lessons learned being assimilated and capitalized upon in the adaptive management of REDD policies, standards, and TMAs employed not just in Cambodia, but more broadly.

Social feasibility in central African REDD

In 2009 and 2010, Satya Development International (SDI) LLC,[3] a consultancy managed by the author, undertook a series of REDD strategic planning activities in Central Africa under contract to Forest Trends, a subcontractor on USAID's TRANSLINKS project implemented by WCS. Forest Trends is well known for jumpstarting the Katoomba Group, as well as EcosystemMarketplace.com. The work involved an assessment of opportunities to impact positively on REDD activities in the Congo Basin.

Several clear conclusions emerged from the desk and field studies:

- No one in the field of REDD+ in central Africa had been addressing "social feasibility" *directly* in design and implementation of REDD+ projects.
- Project developers implicitly appeared to believe that what they are doing was getting at social feasibility in some manner.
- Project developers and standard setters appeared primarily interested in SIA to safeguard community-level stakeholders from negative impacts attributable from REDD+ activities.[4]
- Risk mitigation through SIA appeared to be the priority for REDD project developers.

Project developers were interested to *hear about* social feasibility. Yet there was no perceived urgency or value addition in focusing on it, again, as it is not an explicit design and evaluation criteria for getting REDD projects approved through the CCBA. At present, CCBA is clearly the leader in standard setting for social and biodiversity impacts.

Despite presenting it at numerous fora, SDI's conclusion was that social feasibility was likely seen as being redundant and already accounted for under SIA and REDD safeguard approaches. The issue appeared to SDI to be a red herring fallacy; by suggesting that since the CCBA or any other standard does not require social feasibility to be explicitly verified does not mean that REDD proponents and project developers should not undertake social feasibility. Across central Africa, social feasibility analysis is not being done. Nor was it seen to be a priority for projects in development for either voluntary or ultimate compliance market purposes.

The impact of this gap at program and project levels is already being felt.

Can REDD+ work in the absence of demonstrated social feasibility?

The probability for REDD projects working in the absence of demonstrated social feasibility is minimal. Projects could, of course, be lucky and work out. They could theoretically also succeed by applying PRA methods in design. Based on experience, and the evolving empirical record where voluntary projects and national-level programming are evolving, the probability for this occurring is nonetheless low.

Review of the REDD literature suggests that it is challenging to impossible to find organizations concerned with social feasibility as an organizing concept in project design. IUFRO (2012) comes close to recognizing that a whole new capitalization of forest sector experience is needed that attends to social issues in REDD, and highlights the importance of tenure and social and economic factors in REDD success.

Even if one considers carbon rights and forest tenure issues alone, addressing social feasibility requirements for each is arduous. This stands in contrast to how tenure and carbon rights issues are currently handled in the PDDs reviewed.

Addressing the broad array of social and economic development issues that factor into feasibility is, furthermore, consistent with Vira's (2012) observation that, some years ago, talk in REDD was "about low-hanging fruit, [but] it's not low-hanging, it's tough. You've got to climb many ladders to get to REDD to get to that fruit, you've got to work really hard and fall down a few times along the way". In full agreement with Vira, there in fact is no such thing as low-hanging REDD fruit.

Various authors consider how carbon rights (Knox et al. 2010; Vhugen et al. 2012) are addressed in REDD project design. While they are comprehensive in assessing the range of conditions and assessment of probability of success in achieving different options for assigning carbon rights (Knox et al. 2010), they are less clear in identifying the tools and sequencing needed to satisfy the assessments that would actually be needed to comprehensively clarify respective rights. While Karsenty et al. (2012a) refute the logic of even attempting to pursue carbon rights, it is clear that the policy urgency of getting a compliance, carbon market mechanism in place may trump rationale for a cautious approach.

Taking the DRC as an example, deciding how carbon and tenure rights should be tackled that would consider social feasibility is challenging. While the new Forest Code in the DRC was introduced in 2002, the Congolese Parliament has yet to approve the bylaws specifically regulating community forestry that have been drafted but not yet implemented ten years later (Benneker 2012). Optimism on broader, timely tenure policy reform in the DRC (or most other central African countries) in this context is therefore unrealistic.

Current forest use in the DRC is "the result of constant hassling and negotiations between local actors, including government officials and politicians, loggers, local associations, entrepreneurs, local communities, traditional chiefs and occasionally, NGOs, negotiation processes are therefore endless and complicated" (Bennneker 2012). To successfully operate, this is the context in which the Mai Ndombe Project or Disney's support for the Union of Associations for Gorilla Conservation and Community Development in eastern DRC (UGADEC) through CI and Diane Fossey Gorilla Fund International (DFGFI) in North Kivu needs to be understood. In the Kivus, layers of additional complexity must factored in for to the enduring and virtually impossible-for-outsiders-to-comprehend conflict in north Kivu. This involves militias, lingering Interahamwe rebels who fled Rwanda after perpetrating the genocide there, the Congolese military, the Rwandan military, and the unique position of the Banyamulenge, a Tutsi group living in eastern Congo for over 100 years. This is a tapestry that makes any type of technocratic approach to a REDD project in the area far-fetched

To figure out how a REDD program or individual project can work in a country like the DRC is extraordinarily challenging. Without beginning with a social feasibility framework, it is implausible to perceive how the social complexity on the ground can be approached at all coherently, to assess if and how to move forward.

Why social feasibility is the lever REDD planners and developers avoid at their peril

CIFOR has provided its 3E+ criteria for REDD+ project development (Angelsen 2009). CIFOR focuses on climate *effectiveness*, cost-*efficiency* and *equity* outcomes, in addition to their generation of *co-benefits*: biodiversity and other environmental services, poverty reduction and sustainable livelihoods, governance and rights, and climate change adaptation. Social feasibility criteria appear to be implicit and cross-cutting. But are they?

While the 3Es presupposes that certain conditions be present for each "E" to be met, effectiveness, efficiency, and equity serve more as normative principles to aspire to. No information is provided on what needs to be done to actually reach these end states, or to determine if attained. As is the case with REDD methods generally, aspirational principles are left to individuals and agencies to define and justify. This leads to a universe of relative interpretations.

Employing a social feasibility framework *will enhance the probability* for achieving the CIFOR 3E+ criteria. By not advocating that social feasibility be attended to, policy centers like CIFOR will likely continue to extol the virtue of the 3Es while bemoaning continued disappointing REDD results on the ground.

Land tenure, benefit sharing, and negotiation are central to social feasibility

A shift from formal legislation to emphasis on collective seeking of workable solutions is needed to pilot REDD initiatives. By enhancing carbon values and its trade, the poor may end up inappropriately subsidizing the very process they are meant to benefit from, and ultimately help sustain.

Increasing the value of nature based goods and services may result in their capture by politically powerful local actors, thereby excluding the very poor from access to potential benefits (Vira and Kontoleon 2010). To avoid this, political decision making and social inclusion issues need to be tackled at the same time, otherwise REDD can do little to help the resource dependent rural populations who are among the intended targets.

This shift has not occurred to date. Projects continue to be parachuted top-down in REDD as they have been in both the conservation and development sectors where the guidance on best practice informing them originates. Absent any feasibility framework, this approach perpetuates the likely unfeasibility of these REDD projects.

In regard to land tenure, the absence of adequate legal frameworks for REDD will not likely be resolved in a timely manner by legislation on carbon rights. Good laws are typically the work of years, not months (Bruce 2012). Moreover, hasty legislation on carbon rights, where even possible, is unlikely to frame and resolve tenure issues in a sustainable manner. Legislation on carbon rights is a legitimate longer term objective, but may be counterproductive – if the root causes of tenure insecurity and conflict are not well factored into legislation. In the short term, the main tool in creating and realizing expectations with regard to benefits is *a carefully negotiated and thoroughly understood agreement* between the REDD sponsor, the governments, and local communities or individual beneficiaries (Bruce 2012). The principal objective of the agreement is to identify the resource, record the basic intention, and reach fundamental understandings, rather than trying "to decide land tenure issues beyond its ability to decide" (Alcorn 2011), as local perception of the agreement as legitimate is more important than its legal elegance or even its enforceability.

Yet despite the fact that domestic legal frameworks form the "backbone" of REDD+ implementation, "it is widely acknowledged that, the relatively recent development of REDD+ at the international level combined with the small number of pilot projects around the world has left a vacuum in terms of how existing law is applied to implement successful REDD+ projects" (4CMR 2012). This constrains feasibility across REDD implementing countries.

Negotiation as key

In REDD, similarly to conservation and development, projects as a rule are identified and designed externally to the specific contexts in which they are implemented. Chambers (1983) identified the principles and basic methods for PRA to avoid this. At the time, these were highly progressive.

Thirty years later, while the rhetoric of conservation, development, and now REDD has become mainstreamed to the point where it would seem like participation has become *everyone's* priority, participation has too often become a euphemistic ploy. In practice, little has changed fundamentally since 1983. Top-down planning continues on unabated.

Projects are still, by and large, designed by external agents. Moreover, PRA is used as a checklist tool to validate that consultation has occurred, though the consultative process may resemble manipulation on Pretty et al.'s (1995) scale more than it does empowerment.

Over the past twenty years, I have visited more projects that were ill conceived, poorly designed, and poorly implemented despite the fact that PRA was employed. While not the fault of PRA, the tool has simply become a misunderstood panacea in planning situations whose complexity far exceeds what PRA can reasonably be expected to do. In none of these projects, was negotiation premised as part of a social contract linking project developer and beneficiary groups. In most cases, small scale development projects addressing basic needs was presumed sufficient in return for improved local practices.

In addition to PRA, which is a fine step 1 tool to initiate a much longer and extensive capacity building and participatory, potentially collaborative planning process, negotiation is a pivotal component that can make or break social feasibility. Without negotiation (or "bargaining") it is impossible to speak of level playing fields, stakeholders "seated around the table", or even FPIC, save where the meaning of consent is suspect.

While negotiation is key to establish effective benefit-sharing mechanisms (Bruce 2012), openness to bargaining is also pivotal to achieving feasibility and sustainability in REDD overall. Many stakeholders such as IPs and local communities are destitute and lacking in sophisticated human and social capital to negotiate with Canadian, American, or European agencies. For example, monthly household revenues in sample sites across Bandundu province in 2004 were assessed at the time at US$3/month (IRM 2006).

Human and social capital remains highly constrained in places like Bandundu province. Negotiation capacity in most of its communities is limited. The northernmost reaches of Bandundu, including the Mai Ndombe, were colorfully described in *King Leopold's Ghost* (Hochschild 1998) as a major rubber producing area. These communities have never *negotiated* anything in good faith with government, let alone with private sector shareholder companies whose shares are traded on Canadian or American stock exchanges. More recently, communities like those of the Mai Ndombe have simply had to accept what logging companies deigned to pass on through SRCs that often failed to comply with requirements.[5]

Meanwhile, opportunities for negotiation with conservation NGOs has never been common practice, as BINGOs provide information, environmental education, and communication *to* communities, conduct PRAs *in* communities, but do not negotiate *with* communities as partners. Capacity building is therefore needed to help communities assess realistic negotiation targets and methods, prior to engaging in exercises where results may practically involve several cases of whiskey and a bit of money in exchange for a forest if and where the "carbon cowboys" operate. Who though in the development community or conservation is funding this capacity building?

In contexts where negotiation is not best practice, as is now the case in REDD, what is *the guidance* for approaching negotiation? Where is it done, who is involved, and what are the threshold criteria for "genuine" or "effective" negotiation?

Or because this is not a part of recognized best practice under the voluntary CCBA standards, for example, how can negotiation processes move forward in the absence of the recognition of the need for a new approach to stakeholder engagement?

These are some of the questions that resist being tackled by policy makers and project developers as the trinity of REDD finance, MRV, and safeguards continue to attract most attention. Based on the PDDs that have been verified and validated, the low bar established by the CCBA standards that enables virtually any meeting to meet consultation standards, and while a chief's signature may satisfy FPIC requirements, be it representative or not, how things will change without the CCBA itself recognizing deficiencies in its own standards is almost unimaginable.

Not surprisingly, *none* of the big conservation NGOs or consulting firms working on REDD provides capacity building for decentralized governance entities at community or local jurisdictional levels to negotiate REDD outcomes. And why should they, when standards and donors do not demand differently?

Yet, how exactly are communities with prevailing illiteracy rates, and prevailing *functional illiteracy* of the theory underpinning REDD, to negotiate with thought-leaders driving REDD at national R-PP policy levels? Or with international corporations who have negotiated contracts with governments for conservation concessions, the details of which are not made public? Or with non-profit organizations with annual revenues in the US$150–200 million range up to near US$1 billion a year in the case of the largest environmental non-profits, who are to varying degrees involved in REDD?

Communities need more sophisticated tools to defend their rights. Simply advocating based on UNDRIP and other charters that are either virtually disregarded, or interpreted as they may by governments and the private sector, is no longer enough safeguard for IPs and remote forest communities.

COAIT and social feasibility in AD

One effort specifically designed and field tested in Cameroon and the DRC from 1998–2005 to address social feasibility in biodiversity conservation already exists.

Whether it is COAIT (Innovative Resources Management 2005), or a sister methodology yet to be developed with similar objectives and proven effectiveness, and is adapted to AD and REDD, this type of methodology is worthy of piloting.

From 1998–2007 in Cameroon and the DRC, COAIT was used as a tool for land use planning, conflict mitigation, anti-corruption, and sustainable agriculture programming. COAIT was designed to mobilize communities around the twin economic development goals of improved local resource management and enhanced local participation. In the DRC, it was tested across Bandundu, Equateur, and Orientale provinces in several project contexts.

Originally developed in response to the lack of on-the-shelf methods that enable communities to be proactive in their own development, COAIT facilitates intensive community participation in local institutional analysis. It leads to the assessment and strengthening of human capital resources, and more mobilizing effective community initiative in development. Feasibility analysis of both social and technical issues is stressed. Lessons learned have shown that COAIT helps communities contribute to planning and oversight more than external, expatriate-driven planning processes (Innovative Resources Management 2005; Brown et al. 2008).

COAIT appears to be one of the few toolkits that could be available for REDD design and implementation that specifically seeks to strengthen community level stakeholder capacities enabling communities to be proactive or collaborate effectively on an informed basis. This is key to the FPIC criterion in REDD, and is also fundamental to the carbon permanence objective that REDD planners and market participants hope to achieve.

COAIT addresses the challenges articulated in Merlet and Bastiaensen's (2012) framework. It stresses capacity building of the poorest, local marginalized stakeholders to enable more level bargaining spaces such that equitable outcomes emerge. While clearly building on information-gathering strategies developed in PRA, it extends significantly beyond the scope of PRA by focusing on establishing local stakeholder definition for social feasibility in projects.

Using institutional assessment, detailed villager driven participatory mapping, and stakeholder analysis, COAIT extends beyond PRA by incorporating innovative tools that promote community involvement in information gathering and CBA. Its end product is identification of sustainable enterprise development and resource management options that can be spearheaded by local communities, and packaged into a sustainable development prospectus. This prospectus can be marketed by an individual community, or groups of communities aggregated for economies of scale at broader "landscapes".

The process takes anywhere between six and twelve months to implement, depending on the baseline and the ability for communities to identify partners to support the training. Where PRA processes can transpire on the basis of a few perfunctory meetings perhaps extended over several months, the capacity-building component of COAIT together with the analysis and decision-making components for land use planning exercises indisputably will take longer. This is because multiple phases – awareness raising, training, group planning sessions, community validation, refinement final validation – are involved. Much depends on the initial level of social capital,[6] human capital,[7] and prevailing social contract[8] between communities and the entities to which they recognize and/or accept overarching prerogatives.

Companies with CSR programs, foundations supporting community empowerment programming, or donors seeking to create an enabling environment for feasible and sustainable development requiring meeting FPIC standards may logically support COAIT or another similar process. The determinant in any support is the desire for the donor to wish to empower communities with analytical and decision-making capacity for both the present and the future.

Based on development rhetoric, this type of approach would likely enjoy wide support. Empirical evidence showed that, in fact, few agencies have been interested in replicating COAIT, despite the verified success it was able to achieve in practice. This may be because community-driven problem identification and solutions key to social feasibility remain marginal priorities to most agencies with explicit performance-based deliverables to account for.

Possible steps for incorporating social feasibility into REDD+ programs and projects

For social feasibility to become a driver in REDD program and project design, versus an incidental by-product as is the case now, a number of policy and practice level factors would need to come into play:

- REDD program and project developers would need to perceive the need for a social feasibility agenda.
- REDD program and project developers would need to distinguish why social feasibility is distinct and a value addition from SIA.

At the level of standard setters such as the VCS and CCBA, were social feasibility to be identified as an important stand-alone indicator in project audits and overall validation, this would obviously create a demand for it. At present this is not the case. Here, conceptual support from a high-level panel such as IUFRO, the CDM's High Level Panel, or some other body within UN-REDD or FCPF would help.

At the practice level, were one or more thought-leaders in the sector to identify social feasibility as key to their strategic approach to REDD project design, this could perhaps create precedent and demand for standard setters (VCS, CCBA, Plan Vivo, etc.) to incorporate it. These could also be BINGOs or private sector shareholder firms working on REDD as a profit-making practice area. So too, institutional investors like investment banks such as Goldman Sachs or BNP Paribas may see it in their own ultimate interests to do so.

Should data demonstrate that the empirical record is showing that key aspects of REDD+ projects are proving stubbornly problematic (such as FPIC, participation, land tenure security, institutional factors enabling equitable revenue sharing), and that these are somehow linked to biophysical indicators such as permanence of a project's carbon assets, this could lead industry thought-leaders to rethinking the need for explicit attention paid to social feasibility in REDD design.

Finally, and perhaps most poignant of all, should voluntary buyers of carbon demand that *project developers prove* that social feasibility issues are being addressed, this could establish demand for social feasibility as a standard set of issues requiring attention as well.

Currently this is not the case.

The urgency of the feasibility agenda

If *either* market/incentive-based REDD programming, fund-based REDD programming, or a hybrid is to root and flourish, projects will obviously need to work. Therefore, they must be designed to do so. Currently, they are being designed principally to meet standards set to enable the measurement of carbon. This can be mutually exclusive with feasibility on the ground at local levels from a community perspective.

Transparency and accountability regarding results and impacts of REDD projects will likely become increasingly expected in coming years as scientists and donors (USAID 2012) are recognizing historical deficiencies from biodiversity and other sectors. REDD projects will need to demonstrate they are avoiding deforestation or sequestering carbon on a sustainable basis. They will need to do so without negatively impacting on communities. And they will need to contribute to poverty alleviation and development.

Conversely, there will be a need to show that REDD is actually benefiting communities, *not just safeguarding them from harm*. Any strategy that enables benefits to primarily be captured by external REDD agents and local elites at the expense of front line, impoverished forest communities who have deferred customary rights and subsistence practices, cannot be judged as viable *regardless* of the value of carbon offset credits generated. In short, there will be scrutiny of whether REDD works as originally advertised. Absent this, why would both subsidized and private REDD funding continue to flow?

Many analysts have recognized that local communities have few incentives to enforce forest resource use rules when their own rights are unprotected. Thus, clarification and increased security of rights has been suggested as a key first step toward REDD readiness. To date, while most national Readiness Preparation Proposals (R-PP) acknowledge this need (Sommerville 2011), few layout strategies achieve these goals. Absent guidance at international and national levels, enabling conditions continue to slip.

8 Capacity building

Often discussed, rarely implemented

This chapter presents a basic overview of capacity building issues and some lessons that have been learned that are pertinent to REDD. The topic is important because in most tropical forest contexts, the refrain from international organizations and governments remains consistent for the need for NGO and community capacity building. Yet, there is little evidence that investments in capacity building to empower communities to participate both effectively and proactively has ever been taken seriously.

Background

Adherence to NGO and community capacity building as an objective is as predictable among development and conservation planners and practitioners as is respect for participation and empowerment. Its need is exemplified by a simple googling of "the need for capacity building"; for example, the National Council of Nonprofits (2013) states: "Capacity building is vitally important, now more than ever."

While the quote is in reference to capacity building for NGOs, the same dictum has been applied to capacity building for national NGOs as well as communities in development settings where a variety of functions are anticipated to be fulfilled by the NGOs and community groups.

That said, comprehensive compliance or follow-through on the capacity building remains largely unfulfilled across the developing world. Invariably, development assistance is rarely directed to capacity building at local NGO and community levels. This leaves glaring gaps when it comes time, for example, for communities to creditably fulfill FPIC requirements in a credible manner.

Forebodingly, for communities to negotiate social contracts and REDD agreements with sophistication, dedicated capacity building will be needed. Are there development agencies, REDD proponents, or voluntary project developers prepared to step up to the task?

NGO and community capacity building in Africa: lessons from the 1990s

Virtually the same capacity building challenges exist today in Africa regarding REDD as existed in the 1990s for natural resource management (NRM). The only qualification to this is that situations are more complex today due to exacerbated resource degradation, population growth, demand for land and resources in Africa from Western countries as well as China, persistent poverty, enduring administrative corruption, and a growing middle class creating resource demands of its own. To add to the complexity, an ever expanding array of programs and projects continue to descend on rural citizens, of which REDD is markedly one.

The lessons the author learned from seven years of NGO capacity building in NRM in Africa are of direct relevance to REDD today. These are summed up in an analytical assessment (AA) of the impact of training, technical assistance, and information support (Brown et al. 1996). The AA looked at cognitive, analytical, behavioral, and biophysical level indicators. The learning that was generated has implications for a range of NRM activities, of which REDD is no exception.

Three points from the AA are relevant to REDD:

1. When it comes to NRM, of which AD and REDD are subsets, lax acceptance of overinflated assumptions about stakeholder capacities – government, BINGO, community – is a sure-fire recipe for failure.
2. Even when local leadership appears strong, and when it appears that technical assistance (TA) provided is strong, inattention to the most basic design assumptions will lead to failure.
3. It was systematically shown to be the case that NGOs were adept at proposal formulation to meet checklist criteria and extremely weak in understanding the distinction for design needs to address technical and social feasibility.

In that particular capacity building project, it was clear that capacity building for larger NGOs was almost as important an issue as for small NGOs. This was because large NGOs, international NGOs included, were superiorly adept at proposal preparation and submission, but very weak in attending to feasibility analysis and sound project design. This meant that hard to refuse proposals were submitted, which very often were poor or unsuccessful projects when implemented.

In REDD, instead of small African NGOs, the need to strengthen design capacities also involves BINGOs and consulting firms, who have the wherewithal and self-interest to advance REDD projects to capitalize on funding opportunities. If however they are not properly designed either becuase the activities are inappropriate or unfeasible, failure will ensue.

The most interesting, albeit difficult lessons during the 1989–1996 period working in four Sub-Saharan focal countries, involved assumptions critical to the design of appropriate and feasible activities that would enable sustainable resource management practices to be realized. What we learned was startling: even the most apparently simple intervention was in fact far more complex to design and successfully implement than we had first anticipated.

In practice, we learned that activities that were passing successfully through the filter of our national proposal evaluation processes were in fact neither necessarily *appropriate* projects, nor *feasible* projects. Our project design and proposal submission standards were simply too lax. In fact, the standards for project design that we were demanding of small African NGOs, while perhaps more rigorous on social issues than those demanded by the CCBA of REDD project developers, are similar enough to suggest that a sharing of lessons learned is warranted.

In Private Voluntary Organizations-Nongovernmental Organizations' Natural Resources Management Support Project (PVO-NGO/NRMS), "appropriate" was defined as a condition that meets certain objective requirements or needs. For example, a hillside on the border with Rwanda inside Uganda may be bare, stripped of vegetation. Planting of a particular tree species on the hillside may be deemed appropriate by project developers from a pedological perspective, and a proposal is drafted and submitted to the project for validation and funding based on a perceived opportunity by an NGO.

"Feasibility" was referred to as the conditions required for a potential activity deemed to be appropriate to work in practice – were there people willing to plant trees on the hillside? Would someone from outside the presumed beneficiary group attack people planting trees on the hillside due to conflicting perceptions of tenure rights on the deforested hillside? And so forth.

To our surprise, we found after seven years of implementing project activities that we had been underestimating constraints in proposals. Due to the urgency of getting funded pilot projects initiated, we were letting too many half baked, on-the-surface plausible ideas out the door with funding. We had not done a good job of training NGOs to assess appropriateness and feasibility in project design. What actually was passing our litmus test for a successful proposal too often bore little relation to what was needed to demonstrate project feasibility. And here, pressure from constituents to receive funding, along with pressure from the donor to disburse funding in a pilot, capacity building initiative, led to numerous inappropriate pilot activities being funded. Yet because we systematically were keen to learn from failure, as well as success, the lessons have proven to be useful to some at least over the intervening years.

REDD is currently going down exactly the same track. Hopefully, the presumptive learning that has been factored into the Readiness Phase is actually occurring, both from mistakes, as well as from any reported successes.

Implications for CCBA standards and appropriate imprecision

Continued adherence to the principle of "appropriate imprecision" (Richards and Panfil 2011) opens the door to fuzzy thinking on social issues impacting feasibility in REDD. To move to a social feasibility framework, planners will need to consider if the degree of appropriate imprecision that is enabled by adhering to a theory of change project design framework undermines feasibility by enabling planners to rationalize virtually any imprecision as good enough for their planning purposes.

If third party evaluation will be allowed, this weakness will likely be exposed in most CCBA validated REDD+ projects. If not, ambiguity and discontent may well degrade both the CCBA and REDD brands on the voluntary side, and all aspects of the R-PP process that serves as an umbrella at country levels for individual REDD project activities.

At present, there is no systematically imposed penalty for poorly designed projects in REDD. This is coupled to lack of transparency or systematic accounting of the relationship between taxpayer or individual donor investments in REDD with results and impacts. All the same, anecdotal evidence, brand recognition, and public relations combine to push projects through the pipeline. Outside the REDD context, Trefon (2011) refers to this as "the political culture of aid inefficiency" with Moyo (2009) simply calling it *"Dead Aid"*. The design of socially feasible projects continues to evade international aid providers. What is odd in the case of REDD is that insufficient or ineffective practice is shielded by the euphemism of "best practice".

The following examples drawn directly from Brown et al. (1996) show how complicated effective project design can actually be at local levels. These anecdotes show why apparently *"simple"* projects are more complex than meets the eye, are difficult to design, and why superficially elegant proposals often belie an absence of feasibility that can prove fatal during project implementation. If PDDs are an indication, the same vulnerability will definitely be experienced in REDD projects now being tested in the Readiness Phase.

Proposal elegance is, ironically, a noticeable facet that is apparent in REDD+, CCBA validated projects. CCBA auditor validation serves, meanwhile, as a shield for developers to avoid having to ask the difficult question: *What design is needed for the REDD project to achieve permanence, non-leakage, and meet biodiversity and poverty alleviation objectives without damaging present and future development prospects for local people?*

The following three examples suggest why social feasibility cannot be taken for granted in REDD project design, and what types of capacity building may be considered at project levels to address these issues.

Example 1: live fencing for a women's group in Mali

What could be easier than planting and growing a live fence, inside a protective metal fence, in a 120 linear meter perimeter of a vegetable garden?

North of Segou, Mali, at a women's agricultural cooperative, an apparently sound live fencing initiative ran into problems because project designers had not adequately considered the social feasibility of planting and sustaining the fence. Prior to onset of planting, half of the women believed the live fence would grow, while the other half did not. The result was that about half the fence was maintained, and therefore grew, while the other half did not.

This project was very appealing because it was a pilot effort testing an innovative NRM technology with high replication potential in south central Mali. Villagers were poor; social capital appeared to be strong. The live fence was planted aside a metal fence that was to be rotated from group to group under a credit program format, village to village. The activity as proposed had more to do with that experimentation, and with supporting the collaboration between a Northern and a Southern NGO, than it had with probing for social feasibility. This was because for the project proposers, the latter appeared to be straightforward so that it was taken as a given.

Whether the women *wanted* the project, or believed that it was *technically feasible*, was never at issue during project selection. Whether the two NGOs possessed the design skills required to maximize the probability of successful implementation also was not questioned, as they both were well respected.

However, the inability of the project developers to identify constraints along with mitigation measures – that is, TA or extension services to address the fact that, as determined in retrospect, half the women did not believe the live fence would work in their three linear meters of live fencing because they believed the soil was not as rich as their immediate neighbors (which, in fact, was false) – directly led to failure.

Had the PVO-NGO/NRMS project mandated that all NGOs in Mali participate in project design training workshops that emphasized feasibility analysis prior to implementing project-funded field activities, would this have made a difference in the results of this pilot project?

Example 2: a public garden in Cameroon

In Cameroon's Northern Province, the establishment of a public garden in the city of Garoua proved far more complex than expected for similar reasons. The idea of a public garden in a dry Sahelian town, which had broad support from the mayor and representatives of a well-placed local NGO, seemed like a sure winner for the people of Garoua. All that was needed was the land, to be given by the city council; funds, to be provided from the project; and the know-how, labor, and some other contributions from the NGO. The land to host the garden was obtained from the town council, albeit with no formal transaction to legalize the transfer. A barbed-wire fence was erected around the site. Trees were planted. Drainage trenches and a well were dug. The project disbursed half the funds allocated.

Although the town had long enjoyed government favor as the home of Cameroon's late president Ahidjo, activities were stopped abruptly upon the demand of the state government. Numerous parties were claiming tenure rights or ownership of the same garden land, including the state (through the divisional officer), the city council (through the mayor and NGO), and a private individual (the son of the late president).

Thus, while this project appeared ecologically and conceptually sound and feasible, it could not be fully executed because of the ambiguous land tenure situation.

Had land tenure been examined during project design, it is unlikely that the proposal would have moved forward. But an analysis was not done expressly because the project seemed so eminently feasible: it had the full support of local authorities, in particular the mayor, so how could it possibly not have been viable?

The project learned that, when land issues are introduced in project proposals, the social feasibility of even the most seemingly innocuous initiative must not be taken for granted. The ability to assess feasibility is, however, a skill that generally can only be obtained through training.

Had PVO-NGO/NRMS mandated that all NGOs in Cameroon participate in project design training workshops that emphasized feasibility analysis prior to implementing project-funded field activities, with emphasis on tenure would this have made a difference in the results of this project?

Example 3: the craze of passion fruit in Uganda

Passion fruit was all the craze in Uganda in the early 1990s. One final example involves a Ugandan CBO that *had* gone through project trainings in project design, financial management and accounting, and agroforestry sought to capitalize on opportunities. It designed an initiative that appeared, on the surface, to represent the priorities of a local women's association based 100km from Kampala: establishing an agroforestry activity on a parcel of land allocated to the women's group by a local group of men.

During later site visits, it was clear that the integration of trees into a field of food crops had succeeded; the trees were growing nicely, as were the food crops. The women claimed that productivity on agroforestry fields was higher than on fields that did not integrate tree crops, and all seemed very happy. Women were beaming as they clearly were making money from passion fruit sales.

On closer questioning six years after initiating the "successful" project, however, it emerged that the agroforested field area had in fact previously been a forest patch. The women had cleared the natural forest, then planted trees to secure the passion fruit vines, along with food crops. In the process of clearing, moreover, two women had been bitten by snakes and died. Were this anywhere in francophone Africa where tenure is expressly conferred through cutting down forest and subsequent planting – known as *"mise en valeur"* (creating value) – the logic would have been understood. Here, land for planting and income generation had been granted, with details of the grant and any encumbrances on the women by the men's group unknown.

Several lessons sprang from this experience. While the women's group had secured funding for an agroforestry activity at a time when agroforestry was being hailed by many in Uganda as a development panacea (1990–1991), the need to consider feasibility for even this popular kind of undertaking had not been fully impressed upon the group. The women assumed that as long as they proposed agroforestry, their proposal would be approved, and the project had, in fact, "worked". In that, they were right. Had they fully described their plan in their proposal, the activity would not have been funded, because cutting down a natural forest to replant tree species integrated with food crops could not be considered appropriate from an NRM perspective.

The point is that the women's group had learned the proposal-writing skills necessary to present a project successfully. They had not, however, assimilated how to design a project that was appropriate and feasible.

Had PVO-NGO/NRMS mandated that all NGOs in Uganda participate in project design training workshops that emphasized feasibility analysis prior to implementing project-funded field activities, would this have made a difference to the results of this project?

Pertinent lessons learned for REDD from NRM capacity building

In each of the three cases, a basic element in social feasibility had been sacrificed in the design phase: determination in case 1 if all women had bought into the live fencing objective; assessment of any overlapping claims to land and potential tenure conflict in example 2; and incomplete due diligence to reconise a fraudulent proposal based on unsustainable design principles in case 3.

The need to attend to feasibility is true of any project in the world, regardless of its size. Whether it is the manner in which the major institutions moving REDD forward at national levels are addressing REDD policy, or individual project developers who are employing pre-compliance or voluntary funding to initiate projects, the expressed need to "keep transactions low" is directly impacting feasibility. By saving money in the short term, the approaches undermine feasibility over the longer term, leading to increased risks on investment. Proposals are successful in receiving funding, but this likely belies even medium-term feasibility. REDD will never work unless it radically changes its policy regarding how feasibility is approached across the board in project design.

9 Financing issues

This chapter provides a basic overview of financing issues that have been heatedly debated over the past seven years concerning REDD. It presents basic information about market and fund options, and why a hybrid approach to financing AD activities may make the most sense at this point.

The REDD market premise

REDD is premised on the need for major amounts of finance coming from traditional ODA sources. The anticipation has been that markets – both compliance and VCMs – will provide the principal source for funding.

Delivering REDD finance has taken more preparatory work, capacity, and tailoring than initially envisaged, as the twenty-four Northern, Annex II countries have been trying to balance climate and development objectives, and most REDD finance continues to be directed through development assistance budgets (Creed and Nakhooda 2011). As of mid 2013, prospects for sustainable REDD finance remain unclear.

Some argue that the global community has wasted ten years trying to get carbon markets right, and that we still have "poorly performing carbon markets (from the point of view of seriously reducing emissions) and other alternatives that could have been more effective have been successfully sidelined" (Boyd et al. 2011). Meanwhile, pricing and carbon trading mechanisms will in the end prove to be just one small component of a much broader transformation that is required in capitalism if the worst effects of climate change are to be averted (Newell and Paterson 2010).

Others suggest that to break the current gridlock and scale up REDD+,

> what is required is a strategic, performance-based financial intervention that will make market outcomes (such as price and demand) more certain; provide adequate incentives to forest countries and communities to enter long-term, results-based agreements; and enable them to attract finance for economic development that will benefit forest dependent communities.
>
> (Kanak and Henderson 2012)

For this to happen, effective social contracts will be needed.

An estimated annual investment ranging from US\$17–33 billion (Stern 2006) or US\$22–38 billion (UN-REDD 2011c) is required to halve the current rate of tropical deforestation and support global efforts aimed at holding global temperature rise below 2°C by 2050. It is improbable that these estimates reflect the gamut of institutional and transaction costs that will be needed if REDD is to work and become sustainable, if social feasibility requirements are considered. Nor will the funding level be correct if the 2°C temperature is already on track to being surpassed, as REDD's contribution to ER may be viewed with greater urgency.

The belief is that "without immediate action taken, it is estimated that the global economic cost of climate change caused by deforestation alone could reach USD 1 trillion/year" (UN-REDD 2011b). The benefits over time of actions to shift the world onto a low-carbon path were first projected in the order of US\$2.5 trillion each year, with the shift to a low-carbon economy bringing huge economic opportunities, as markets for low-carbon technologies were first seen to be worth at least US\$500 billion, and perhaps much more, by 2050 (Stern 2006).

The Eliasch (2008) review suggested that the inclusion of the forest sector in global carbon markets would lower the costs of reducing emissions, and could provide the financing and incentive structure for the reduction of deforestation rates by up to 75 percent in 2030, even higher than the 50 percent that Stern (2006) predicted.

The Environmental Defense Fund subsequently developed an analytical model considering forest carbon finance. It stated that international carbon markets are the first, and possibly last, chance to create economic value for forests at a level commensurate with large-scale deforestation (Schwartzman et al. 2008). As Isenberg and Potvin (2010) suggest, this statement was clearly an exaggeration, as either funds or a hybrid approach involving market incentives, along with fund mechanisms, could be structured to address REDD. Others suggest that "sustained investment", as opposed to performance based payments, is the most logical approach to the challenge (Karsenty et al. 2012a). This approach appears consistent with those who focus on sustainable land management (SLM) and sustainable forest management (SFM) by integrating SLM/SFM priorities into government decision making and the political culture (Asen et al. 2012).

As of mid-2012, the global community is *sixteen times* behind where it should be to reach an abatement target of 5.5 billion tons of CO_2 by 2015 (WWF 2012). WWF suggests that "what is currently being asked of forest nations is to ignore lucrative alternatives to conserving the forests, and to engage in what might be a 30 year activity where the rewards beyond year 2 or 3 are completely unknown". The question then is, why exactly would forest nations, or financiers, participate in this type of program if the economic opportunity costs are potentially so high?

Yet while there were modest gains in VCM activity to the US\$150–200 million range in 2011, carbon markets hit the wall in 2012, with CDM carbon *prices* tumbling 70 percent year-to-year from 2010–2011 to around US\$3/ton (High Level Panel on the CDM Dialogue 2012). Projections are for further price plummeting through 2020. The relation between broader compliance market price plummets and REDD voluntary prices is unclear. As the climate division of France's Caisse des Dépôts estimates that voluntary market absorptive capacity for all types of ER activities is at 30 Mt CO_2e/year, and the potential quantity of carbon credits in REDD+ project pipelines can reach up to 100 Mt CO2e/year (Karsenty 2013), this will likely impact both REDD credit pricing and thus project financing.

These falls are attributed to overly modest mitigation targets that "no longer create strong incentives for private international investment and local action in developing nations, with governments, private investors, and financial institutions losing confidence in the CDM market" (ibid.). Without a compliance mechanism in place, voluntary markets are left on their own to pick up the slack.

Despite this, multi-stakeholder support for REDD from governments, BINGOs, international financiers such as Goldman Sachs, BNP Paribas, EKO Asset Management Partners (EKO), along with UN-REDD, the FCPF, and bilateral country programs such as DfID, USAID, and the largest, Norwegian Agency for Development Cooperation (NORAD), appears stable if not growing. Creating value through PES for carbon offset credits remains a substantial area of interest for a variety of institutions, no less so than big conservation NGOs in partnership with private sector entities, and supported by ODA funding. Pronouncing thumbs up or down on REDD, therefore, remains difficult as of 2013.

The one thing that can be said is that, for the foreseeable future, REDD's growth appears relatively modest in light of the carbon finance challenge for the overall climate change mitigation problem posed by deforestation, as framed initially by Stern and Eliasch. As REDD is yet to become an official compliance mechanism, and despite prospects for continued growth in voluntary markets projected to surpass the US$150 million mark in 2012 (see Diaz et al. 2011), the gap between actual carbon finance and actual needs to coherently address deforestation and forest degradation remains tremendous.

REDD and the green economy

Currently, REDD and its latest REDD+ incarnation are at the heart of the new green economy that was marketed at Rio+20 by UNDP, UN-REDD, the World Bank, and other thought-leaders as the approach to sustainable development and ACC mitigation. While it is possible to speak of AD without referring to market mechanisms, one cannot refer to REDD without the immediate implication of carbon markets and, therefore, the green economy.

Yet, the technical mechanisms for the commoditization and transaction of a commodity and securities exchange at its infancy faces great challenges if it is to be taken to scale. Debate over "carbon rights" is just beginning to heat up in 2013 (Knox et al. 2012; Karsenty et al. 2012a), with no basis for reconciliation for what such rights may be in countries like Indonesia or the DRC. In Indonesia, IPs' rights are not yet even recognized (Survival International 2012). In countries where rights are either not recognized or ambiguous, how will REDD function as part of the new green economy?

Carbon markets

Newell and Patterson (2010) provide the most comprehensive work on carbon markets as an approach to climate change mitigation. Pricing and carbon trading mechanisms are, in the end, just a small component of a much broader transformation that is required in capitalism if the worst effects of climate change are to be averted (Newell and Paterson 2010).

Market-based solutions aligned closely with prevailing ideologies regarding regulation and the "primacy of efficiency" have spurred carbon markets as the most politically acceptable solution (Boyd et al. 2011). The preferences of private sector operators in leading economies for trading schemes, over taxes or regulation, have been particularly true in the US; while a national cap-and-trade program failed to materialize in 2008, California's AB 32 attempts to legislate just that for REDD in 2013.

Some see the attempt to regulate a carbon market as an endeavour in essence cannot be regulated. It is suggested that this is an indicator of the utmost corruption at highest levels, with attempts to regulate simply entrenching its status as a locus of international corruption and exploitation (Lohmann 2009). The illusion of offset regulatability is sustained, it is argued (Lohmann 2009) partly because climate policy has been captured on both national and international levels by an elite alliance comprising big business, commodities traders, financial firms, neoclassical economic theorists, and an influential group of professionalized, middle-class environmentalists. All are bent on seeing offset trading expand rather than be abolished.

With the overarching multi-stakeholder push for a green economy, which reached its apogee in Rio+20 in 2012, conventional wisdom has it that markets are the most efficient means for generating the levels of finance needed to broadly scale up REDD, among other GHG mitigation activities. When compared to the difficulty of securing even the few billions of dollars committed by Northern industrialized countries for REDD Readiness Phase activities through 2012, the argument that markets offer the most efficient mechanism for leveraging the funding needed is compelling. For most global development agencies and private sector investors, along with BINGOs active in REDD, private sector resource mobilization outshines potential for securing similar funding through government in the current, still difficult, economic climate.

Climate change falls under the category of "super wicked" policy challenge "because of the enormous interdependencies, uncertainties, circularities, and conflicting stakeholders implicated by any effort to develop a solution" (Lazarus 2009). Even with the lure of markets, however, in three 2012 US presidential and vice presidential debates totaling six hours, climate change was not even mentioned once. The projected costs to jobs, thus presidential politicians, were perceived as toxic in their own right. And then hurricane Sandy blitzed the northeastern US. An estimated US$50 billion in damages was incurred. Former Republican and current billionaire mayor Bloomberg of New York endorsed Barack Obama's presidential candidacy on the basis of Obama being a leader on climate change. While four years earlier Obama had advanced cap-and-trade legislation that appeared, incorrectly at the time, to be perched to pass successfully through the US Congress, little tangible progress on emissions reductions has been achieved.

The preferences of powerful stakeholders in leading economies for trading schemes versus taxes or regulation (Boyd et al. 2011), with private sector support joining with BINGO advocacy for a carbon economy that can produce supposed win–win outcomes may well resurface in the US. It is within this framework that REDD markets as a subset to carbon compliance markets, is best understood.

Yet there is scant evidence from the experience in the principal climate change markets that effective mitigation is actually promoted through market mechanisms. For example, the EU ETS fails to make countries take responsibility for their own emissions, instead allowing them to offset emissions by buying permits from countries from the South.[1] And the failure of the CDM, the longest standing and largest market for mitigating Western industrial source carbon emissions, led one critic to note: "Because of the CDM's structural flaws and cheating by project developers, billions of dollars worth of credits are being sold by projects that never needed assistance from the CDM to be built" (International Rivers 2008). The CDM experience, though distinct from REDD, is thus relevant to assess REDD's own prospects as a market mechanism.

Pursuing market approaches to mitigating climate change has implications for achieving social feasibility also. There are particular challenges to social feasibility when investors in REDD projects must report on profit and ROI to shareholders, including potential for cutting corners to reduce transaction costs.

Böhm et al. (2012) suggest that disappointing carbon market statistics are clear: "the only drops in carbon emissions that have occurred over the past two decades were during recessions or some other type of serious economic collapse". The 1990s contractions in the ex-Eastern bloc countries, and the 2008 global financial crisis and resulting recession correlated with considerable GHG emission reductions. Meanwhile, "carbon markets have had near to no impact whatsoever, other than creating new business and profit opportunities".

Voluntary and compliance markets

Recent trends suggest that voluntary mechanisms may have a bright future, particularly if as Zubair Zakir, Head of Carbon Sourcing for the CarbonNeutral Company suggests voluntary markets are "taking faith in the fact that forests are so important that any compliance schemes that exist ultimately would take these credits – and if not a compliance scheme, there will be other investments available" (quoted in Peters-Stanley et al. 2011). Said otherwise, many investors may be banking on credits generated through ongoing voluntary pilot projects being ultimately creditworthy under a REDD compliance mechanism.

Disaggregating speculative REDD investments, from CSR investments, has yet to be attempted.

Over the years, standards and registries that guide and track voluntary GHG reductions have multiplied. The "unofficial" message from VCM participants is that voluntary mechanisms are as rigorous as their compliance market counterparts, and arguably more innovative. Many voluntary actors now aim to dispense with distinctions between voluntary and "compliance-grade" offsets as "they scramble to write the rules that will fill various regulatory vacuums" – like that which some nations will face until 2020, when they may join in a new international legal framework for GHG mitigation (Peters-Stanley 2012).

The International Institute for Sustainable Development has come up with a useful checklist of issues of concern to private sector and governments respectively as part of the enabling environment for REDD+ (International Institute for Sustainable Development 2012). These relate to clear price signals in the marketplace, regulations, payments for performance criteria, clear national level regulations on carbon, forest and land rights, an agreement and benefit-sharing mechanism developed for use with communities, etc.

WWF (2012) suggests, however, that prospects for private sector finance at the scale required is unlikely under 2012 conditions. With few incentives for developing countries to commit to the necessary investments required to create adequate enabling conditions for investors when the opportunity costs from other land uses are considered, inevitably makes them speculative.

Ecosystems Marketplace suggests in a report commissioned by several industry trading associations (Peters-Stanley 2012) that voluntary markets are pushing the envelope in terms of methods and accountability. An assumption is that this will help to enable compliance markets to advance and function. While voluntary REDD market activity reached US$170 million in 2010 (Peters-Stanley et al. 2011) this declined by over 60 percent in 2011 (Peters-Stanley et al. 2012), and is a far cry from figures that *could be* extrapolated for REDD market valuation from Sukhdev's multi-trillion dollar characterization of the total value of ecosystem services (see Sukhdev 2011).

Hajek et al.'s (2011) suggestion that the "bottom-up" construction of REDD+ as a strategy be adopted to encourage innovation and flexibility – key for success in nascent carbon markets – appears to be happening. As over twenty national or sub-national governmental agencies have incorporated voluntary mitigation activities in their climate change strategies, VCM activities are in a position to influence REDD's future. This will do little to satisfy critics who argue that carbon offset markets cannot be constructed on the basis of step-by-step technocratic improvements, while only addressing governance improvements at the margins (see Lohmann 2009).

Market rhetoric meets administrative muddle

At present then, the relationship between finance for avoiding deforestation and associated activities under REDD+ is muddled.

Box 9.1 Abyd Karmali, Managing Director, Global Head
of Carbon Markets, Merrill Lynch on carbon markets

> Those who assume that the carbon market is purely a private market
> miss the point that the entire market is a creation of government policy.
>   Moreover, it is important to realize that, to flourish, carbon markets
> need a strong regulator and approach to governance. This means, for
> example, that the emission reduction targets must be ratcheted down
> over time, rules about eligibility of carbon credits must be clear etc.
>   Also, carbon markets need to work in concert with other policies and
> measures since not even the most ardent market proponents are under
> any illusion that markets will solve the problem.
>
> (cited from ClimateChangeCorp 2009)

"Fast track finance" currently under discussion (Vieweg et al. 2012) lacks
an underlying conceptual framework which would define clear objectives and
structures to guide implementation. This means that while developed countries
have committed to long-term mobilization of US$100 billion in climate change
finance, of which REDD activities are a subset, criteria and definitions for the
sources of funding remain vague, predictability for funding in distinct commitment
periods remains ill defined, and common definitions for key issues like "new and
additional" and "balance" between mitigation and adaptation activities also
remain to be defined (Vieweg et al. 2012). Taken together, the framework for
financing REDD activities remains highly uncertain, and represents a significant
constraint to participants at all levels of the process. Nonetheless, through 2011
the European Commission (2011) maintained optimism that US$100 billion in
finance could be raised.

Meanwhile, the argument for or against carbon markets as a financing
mechanism to avoid deforestation remains squarely at the heart of the debate
over policies and incentives in REDD.

The two basic approaches to REDD financing are government funding and
market-based instruments. There is considerable debate over which is best. Some
have suggested some form of hybrid is needed.

Climate investment and green funds

Climate Investment Funds totaling US$7 billion were designed to be a transitory
carbon finance mechanism. It may be replaced with a new multilateral body,
presumably Governors' Climate and Forests Task Force[2] (GCF), which is poised to
become the main channel for climate finance (Polycarp and Patel 2012). The GCF
is expected to deliver large-scale finance to developing countries to address climate
change, though its funding sources and mechanism for moving monies, and its
operating and accountability rules remain to be defined (Polycarp and Brown 2012).

While much with the GCF remains unclear, it is reasonable to assume that a REDD compliance mechanism could well be associated with the GCF for all aspects pertaining to creating enabling environments for compliance market operations at international and national levels.

As the experience of World Bank funds for creating win–win outcomes for the environment and people have proven problematic (Dooley et al. 2008), fund mechanisms in and of themselves provide no guarantee for efficiency or equity above and beyond markets either. The GCG will be no different.

Technical arguments for and against markets and funds in REDD

Funds are seen by technical critics of REDD and PES markets as offering a more realistic mechanism to reduce the drivers of deforestation and forest degradation than markets (Karsenty 2009, 2010). Markets suffer from critical challenges involving additionality, leakage, and permanence of carbon sequestered, along with social equity issues pertaining to fair market value for carbon. Karsenty (2008) argues that an alternative REDD architecture which relies on a "special fund" would protect against massive flooding of carbon markets by non-additional credits, and could also help finance potentially efficient policies and measures.

If REDD+ involves a large part of a country's forested area, it is necessary to establish a good link to the general forest policy, with the suggestion being that a national REDD fund organized under a country's national administration, or conditional budget support could work in conjunction with an international fund (Karsenty et al. 2012a), linking it directly to a nation's general forest policy could be authorized to issue CERs sold to firms with reduction responsibilities.

If REDD activities were to be financed by a fund, then demand should be determined by the cost of reducing emissions from deforestation rather than by estimating developing countries' supply in terms of carbon credits (Isenberg and Potvin 2012). The cost of reducing emissions from deforestation and forest degradation based on the average of a range of estimates is proejected to be US$14.2 billion per year. Regardless of the variation in the estimates, REDD will necessitate an amount of money that is significantly higher than the current level of official development assistance (ODA) (Johns et al. 2008).

In 2013, increased ODA for REDD remains politically untenable. This provides REDD proponents with further justication for relying on market based approaches to carbon finance.

Unlike Schwartzman et al. (2008) who claim that only through a carbon market could sufficient financial resources be mobilized to tackle REDD, Isenberg and Potvin (2010) argue that global health financing of global health can serve as a model for REDD. The fact that recipient countries are accountable for the results of the action undertaken and that donor countries are likewise accountable for the support pledge (Schieber et al. 2007) would creates a basis of mutual accountability that could leverage public funding, private donors, and philanthropic support similarly for REDD. This optimism does not jive well, however, with the persistent unwillingness, as WWF pointed out, of Northern countries to abide by their commitments under REDD.

Climate change remains off the negotiating table in the US. This is despite mounting evidence of the galloping rate of Arctic ice melt that may in a short span of time generate impacts globally (including in the coastal US due to rising sea levels), and the wallop that 2012's superstorm Sandy paid to New York and the east coast. Expectations for more than the US$481.5 million allocated by the US State Department for climate change[3] in 2012 is probably overly wishful thinking for many years to come in the context of US fiscal cliffs and debt.

With the initial premise that wasto take AD to scale, considerable private sector originated carbon finance would be needed – with projections up to US$33 billion annually (Eliasch 2008), inclusive of transaction costs – to mobilize such sums of carbon finance, clear profit incentive would be needed. This underpinned the logic of the marketplace that has come to predominate. From there, next steps to derivatives trading of carbon securities (Sandor 2012) were supposed to work through the CDM, the EU ETS, and the now defunct Chicago Climate Exchange (CCX) and other exchanges. These would provide the foundation for this cornerstone of the evolving green economy.

With the linear descent of carbon prices in the CDM and EU ETS to the US$1.50–4/ton range in 2012–2013, pleas for strong government intervention to prop up prices energed, lest the basement for carbon prices continue to lower, even while the logic for support of a REDD mechanism was being promoted (High Level Panel on the CDM Dialogue 2012). Despite this, the market approach to both avoiding deforestation and sequestering carbon prevails as conventional wisdom despite growing dissonance. In October 2012, the projected glut of CERs in the marketplace was such that price plunges for CDM carbon were anticipated to fall to €0.50/ton (approximately US$0.67/ton) by 2020. One industry leader, Point Carbon, had suggested that "the low price for CERs may mean that investment in emission-reduction projects under the CDM will dry up over the coming years".

This leads to the obvious question: So where will all the finance for AD and demonstratively needed mitigation of deforestation actually come from?

Some analysts (Karsenty 2010) are optimistic that the types of "fast start" fund commitments in the US$4 billion range for the 2010–2012 period from Northern governments may be most apt to deal with the types of reforms needed to address the massive deforestation prevailing in fragile states. But is this realistic? As the need for *the same* reforms clearly exists in 2013, it may be that another round of funds will be needed to finance these enabling reforms that must target the forest sector, agricultural sector, and governance and tenure reform in countries where REDD is being implemented. Without these reforms, REDD cannot work. For one thing given the state of carbon markets is clear: how different mechanisms will help unleash the necessary sectoral reforms to avoid deforestation, through a fund mechanism, a market mechanism or a hybrid, is to be clarified.

Accepting that the wide-ranging estimate of US$15–33 billion a year in funding is accurate for addressing the deforestation component of climate change, in the political and economic environment of 2013 a new urgency and strategy for resource mobilization will be needed. Yet, where is this to come from? And until policy and technical issues are worked out, along I believe with social feasibility issues that will undermine investment over the longer term, it is difficult to see where the levels of private sector finance will realistically emanate from either.

As with markets, funds will also need to be evaluated on the results and impacts they generate in REDD. If the record of results and impacts from fund based mechanisms in development were available for the past thirty years, there would be an objective basis to argue for funds being a viable mechanism to address development needs and, by inference, REDD too.

Where funds puport to have been most successful, it is reasonable to assume in the absence of causal data that the criteria for funding projects had some relationship to the appropriateness and feasibility of activities proposed. The same will be true for REDD under a fund mechanism – activities will need to address technical and social feasibility, in the same manner as they will for REDD activities financed through market mechanisms.

While many critics argue that REDD has completely lost its way as one of the more implausible instruments in the evolving green economy due to its adherence to a market mechanism (Lohmann 2006; Sullivan 2011; McAfee 2012b). This, however, does not provide a solution as to what requires doing to mobilize finance for addressing deforestation, beyond Lohmann's imploring of radical, structural transformation of unsustainable Northern economies. While conceptually appealing, politically for the time being this is clearly a non-starter as well, though perhaps a repeat of hurricane Sandy type events in the US and Europe may begin to tip the balance towards some deeper structural changes.

At this point, the burden of proof for avoiding deforestation through carbon markets, without inducing massive restructuring of rural economies to the detriment of rural peoples, needs to be made by proponents. If it cannot be, the market premise should be rethought and potentially abandoned in favor of the best rational bet for mobilizing resources through an appropriate mechanism to promote better forest governance, with lower externalities to rural communities. This will likely lead to funds, which while they cannot raise close to the initial levels of finance envisioned as needed, may well mobilize sufficient sums to begin to address needed reforms at national levels. This is a first step to measurable reductions in deforestation. Here too, however, the role of citizens in driving home the rationale for the reform process, as well as overseeing its compliance, will be pivotal.

Fund mechanisms of disbursement for sectoral reforms to impact AD do not in themselves reduce risks pertaining to feasibility criteria any more than market approaches would. The demand for feasibility as part of the design and provisioning of funds should therefore be conditional on feasibility criteria being met as well.

Thus Karsenty et al.'s (2012a) proposal for a "sustained investment" strategy in REDD also faces an uncomfortable question: Why will results and impacts from sustained investment in REDD prove any different than development initiatives over the past thirty years that have largely relied on government planning, multilateral organizations, and large international research institutions for program design, implementation, and funding? What are the indicators if so? Some analysts argue that "[t]he best financial approach for REDD would be a flexible REDD mechanism with two tracks, a market track serving as mitigation option for developed countries and a fund track serving as mitigation option for developing countries" (Isenberg and Potvin (2010). Vatn and Vedeld (2011) challenge the assumption, as many have argued, that it is only through a market mechanism that requisite funding levels to tackle REDD objectives at scale can be addressed.

Payment for environmental services and REDD

The PES model proposed for REDD should result from a voluntary, contractual agreement between parties which sets out the service expected, corresponding payments, and length of service (Karsenty 2011). The amount of a PES for carbon sequestration would not therefore depend on the monetary evaluation of natural assets, but would be the subject of negotiations which in principle would cover the opportunity cost linked to the usage restrictions or changes (Karsenty 2011). This might undercut the logic for market participation for investors.

Yet, how prepared are most communities to assess opportunity costs and negotiate fair agreements? The argument can be made that if opportunity costs are limited only to the present, while discounting costs in the future, negotiations will be unfavorably skewed against communities. If transaction costs do not consider what communities will need to get to the point where they *can* negotiate on an "informed" basis, and thus *can* provide "informed" consent, this type of contractual model based on present opportunity costs would not necessarily be equitable for many parties receiving payments. In short, contractual agreements are worth shooting for, but only if interlocutors in negotiations begin from some semblance of an equal footing.

Why a hybrid financing approach is inevitable

Several reasons enhance the possibility for a hybrid approach to AD finance. Current carbon price signals coupled to the weak performance of the CDM raise the question of where carbon investment will come from. ODA funding prospects will at best be stable, while bilateral or government funds are paltry in comparison to projected finance needs. At the same time, an emerging movement with significant capital assets for investing in ventures where *some level of profit* mixes with social and environmental objectives could be possible, if the detailed conditions for how this could work are established and found credible by investors.

Nor do enough people trust that a market mechanism for carbon can work or, indeed, should even be attempted. While in countries like PNG, media coverage of REDD has been politicized and influenced by government and an optimistic industry (Babon et al. 2012), community level people are suspicious about the carbon trade objectives embedded in REDD. This suspicion underscores the need for addressing the underlying issues – the causes of unsustainable deforestation and forest degradation – be it through market mechanisms, funds, or some hybrid.

Finally, the conundrum over carbon rights is not one that is likely to be resolved to the satisfaction of stakeholders in countries like the DRC or Indonesia, such that investors can feel confident that investments in REDD carbon will not go up in smoke once their backs are turned – i.e. that permanence can be achieved. This will inevitably limit total investment.

Therefore, a possible solution for carbon finance at the levels and scale required is a mechanism incorporating both concessionary, ODA-type fund mechanisms, with a special form of progressive, market based mechanism. The latter would be developed through the emerging industry practice known as "impact investing".

While critics will argue that early carbon market investors could claim to be impact investors, as well as massive pension funds like Teachers Insurance and Annuity Association – College Retirement Equities Fund (TIAA-CREF) who recently set up a US$2 billion fund for farmland investing with projected returns of 8–10 percent (Gillam 2012; Or 2012) which has been loudly criticized for farmland grabbing, if impact investing frameworks are appropriately designed and negotiated impact investing, could, in theory, make tremendous amounts of financing accessible. This is, of course, a very big "if". For this to become an option, the willingness of key stakeholders to open up to negotiation processes regarding principles and terms will prove key. This idea, and conditions for success, have yet to be tabled.

If a platform can be created under a new framework approach to negotiating social contracts for REDD that brings three key groups of stakeholders together to negotiate the terms of such a hybrid – front line communities and their associations, national and sub-national governments, and the potential impact investment community open to discussing AD under a rejuvenated REDD label or some new term – a novel mechanism that would respond to the criteria established by Lou Munden (2011) for rationalizing private investment could *perhaps* be created. This is where the inter-governmental community and development banks such as the World Bank's FCPF could be most facilitating and innovative.

Characteristics of a hybrid platform

The key to a hybrid platform that is *acceptable* to a wide array of stakeholders will be that the framework for the nature of engagement between investors and stakeholders in countries is actually subject to negotiation under conditions of transparency and accountability. Of course this sounds unlikely, but it will be pivotal for establishing the basis for a new social contract for avoiding deforestation, through REDD or another mechanism to be named.

Any negotiation will inevitably limit the ceiling on potential profits that investors will make, and thereby limit the number of interested investors. This will impact on the potential scale that avoiding deforestation programming can operate at the outset.

On the other hand it will establish and pilot a mechanism that is pivotal for developing new social contracts linking natural resource stewards with governments often enjoying *de jure* ownership over resources to be stewarded in AD programming. It also could be credible enough to lure responsible "impact investors" who wish to operate, and impact at grander scales. Impact investors could be able to provide US$500 billion in finance before 2020 (Freireich and Fulton 2009). Whether REDD would be an investment area of preference would need to be determined.

If indeed "scalable" activities can be "aggregated into a portfolio from the various 'local solutions' to deforestation" (Munden 2011), this type of bottom-up approach could create the basis for a negotiated hybrid plan for financing REDD in which all key stakeholders would be involved in bargaining over the architecture and terms. It is possible that the investors Munden envisions would likely represent that emergent category of investors keen on supporting social and environmental causes, but still interested in earning some level of profit. To work at scale, the profit generated from any impact investing and the rationale of grander carbon finance than ODA funds will ever offer, will need to be negotiated. A framework for doing so is needed.

Should weak market signals be foreboding for REDD, or does it represent an opportunity?

As a function of the broader push for compliance markets, REDD prospects will be tied longer term to frameworks and incentives that the UNFCCC process is trying to put into place. Unless, of course, stakeholders become so fed up and abandon it for less ambitious, bilateral or sub-regional initiatives. Nonetheless, two short-term issues that are difficult to predict need considering: (1) the role that private investment will play in the evolution of VCMs (see Bayon et al. 2007); should the CDM market and UNFCCC COP 21 in 2015 fail to increase clarity, market prices remain weak, and the move to a compliance market remains ambiguous; and (2) the opportunity that this failure creates for REDD, under the CDM itself, to enable pilot activities that inject momentum for alternative "REDD+" mechanisms, and possibly accelerating their development in turn.

Therefore, the jury is still out on how significant, or insignificant in market terms, REDD can become. That is why it is all the more important to enhance the chances of undertaking the *right kinds of pilot activities*, so that learning can feed into a future UNFCCC supported roll-out of REDD. The present round of learning will fall far short in terms of environmental governance policies, incentives for stakeholder engagement, benefit sharing mechanisms, and feasible REDD project design in general. Myriad issues with national MRV capacities remain too.

Private sector preconditions for investment

Luttrell et al. (2012) argue that getting groups on board to bring change at different political scales in REDD+ requires equitable sharing of benefits. Yet the answer to the question of who should benefit from REDD+ remains highly debated.

For example, should benefits flow to people who have legal rights to the land and carbon-storing resources, or to good forest managers, who might not have a clear legal claim to the forest? These include REDD+ implementers such as private companies or NGOs who bring the investment capital to the table.

Until there is a logical framework for assessing and negotiating fair outcomes regarding benefit sharing, REDD will be constrained. If benefits are arbitrarily assigned, versus satisfactorily negotiated as part of new social contracts, permanence of REDD carbon assets and the sustainability of PES initiatives will be compromised. Both are critical to rationalize any impact, or blended investment strategies.

Therefore, while the appropriate social and environmental safeguards must be developed to surround private sector engagement, they may increase transaction costs for REDD+ projects to the point where investors are dissuaded. Unless the process employed is socially feasible, the risk to investments may become undue, with investment curtailed as a result.

Box 9.2 A rationale for carbon concessions – the weak performance of the logging sector

The dismal state of affairs in Congo Basin forestry from a development perspective has generated much attention over the years. It also helps rationalize the push to conservation concessions that the Disney Foundation supports with UGADEC and CI/DFGFI in North Kivu, that attributed to ERA in the Mai Ndombe, or Bonobo Conservation Initiative's activity in Sankuru under its CBFF supported project.

The jury is still out as to whether conservation concessions premised on REDD carbon finance will prove any more beneficial to local communities than the older model based on "retroceded" taxes from the forestry sector. Too often, these failed to arrive as promised in the provinces, and where negotiated social responsibility contracts (*cahiers des charge*) did not deliver tangible benefits to communities, and where poor forest people stay poor, or get poorer.

> I recently returned to a village where I worked as a young forester. When I left 25 years ago, it was with the promise that logging would bring a future of social and economic development. Timber was the only resource these villages had to ignite development. And today, commercial timber is gone. The same families are there. They were poor 25 years ago and they are poor today. But today, they have less forest and less hope. They feel cheated by the government, the private sector, the local chiefs and by me. They feel let down. And I believe that in many ways they are right. We are all responsible for letting them down.
>
> (Giuseppe Topa, Africa forest specialist, the World Bank (2002), quoted in Greenpeace (2007a))

The Munden challenge to markets

The Munden Project takes a very different view of carbon markets, and REDD markets in particular.

The Munden Project has called the current market-based model for REDD "hopeless" (Munden 2011), not because it was "irrationally exuberant", but because the model appeared "more manifestly dysfunctional" in October 2011 than when the original Munden Project report surfaced in March of the same year. In the interim, the Climate Markets and Investors Association (CMIA 2011) challenged the Munden Report's analysis and conclusions to add fuel to the fire.

While vague, Munden's (2011) proposal is for a broad performance-oriented framework. It is premised on a rights-based approach to *avoiding deforestation*. Munden argues that offset markets are not the best way to avoid deforestation. Rather, performance is to be measured to determine whether there are scalable activities that can be "aggregated into a portfolio from the various 'local solutions' to deforestation". This bottom-up approach, without stating it as such, appears to be premised on social feasibility principles as a means to convincingly attract private capital. It also may be akin to what is otherwise referred to as impact investing.

10 Risks related to REDD

This chapter portrays the range of risks confronting stakeholders to REDD. It considers risks from the perspective of action or inaction.

The nature of the risks

REDD faces risks both upstream and downstream. Upstream risks pertain to how avoided deforestation through REDD approaches is addressed. Downstream risks pertain to the verifiability of what has been accomplished, including costs incurred, to enable carbon asset permanence and leakage to be verified.

Risks in REDD can be analyzed in terms of both commission and omission. These pertain to addressing deforestation as a subset of climate change issues. They also pertain to the manner in which deforestation is attempted to be mitigated.

If REDD is poorly designed, dislocations and lost livelihoods for local peoples can ensue. So too, potential opportunities for investment may be forsaken as investor confidence degrades. For donors and investors, actual return on investment may not pan out. For governments, development could be set back leading to greater financial burdens on already often impoverished nation states.

If REDD is taken to scale after the current pilot Readiness Phase, there will be technical issues that remain extremely challenging to address. These present situations for inducing error. Inaccurate measurement of emissions reductions through REDD, improper allocation of rewards for what may in reality be "hot air" as opposed to any "additional" reductions from what would have occurred anyway are among the possiblities. The following presents a conceptual framework for assessing possible risks from failure to address social feasibility.

Climate risks amplifying

To appreciate the relevance of REDD as a GCC and GHG mitigation response, it is important to understand the extent and pace of climate change. This enables a degree of "calibration" of REDD as a policy response, i.e., given the scope of the challenge, what will REDD accomplish if it works, and what will be the downside risk *if it doesn't work?*

Keeping with Manzi's (2008, 2010a, 2010b) counsel to be objective about actual benefits and costs of GHG mitigation, comparing potential carbon offset benefits through REDD, with potential costs in the event of failure, is also logical. This is particularly true if REDD is increasingly relied upon as an interim policy response while a more robust climate change mitigation measure is enacted.

Startling data from the National Snow and Ice Data Center (2013) add force to Hansen's statement that the data substantiating ACC is, for all intents and purposes, statistically incontrovertible (Hansen 2012). Childs' (2012, adapted from Smith et al. 2009) updated "burning embers" assessment of different types of escalating risk posed by climate change corroborates Hansen's position, and illustrates that reasons for concern in 2012 compared to 2001 are significantly higher. This applies to unique and threatened systems, risk of extreme weather events, along with the distribution of impacts, aggregate impacts, and large-scale discontinuities.

Over a ten-year period, the risks from climate change appear to have worsened significantly. This may be due to "positive feedback" in which the very effects of climate change in turn exacerbate negative trends. This may explain the increased incidence in extreme weather events globally in 2010–2012, culminating in the northeastern US in October 2012 with hurricane Sandy. The fact that New York's Mayor Bloomberg endorsed Barack Obama on the sole basis of his climate change policy, which was not even enunciated during the 2012 electoral process, perhaps foreshadowed the inevitable return of GCC as a political issue that even the US would be forced to more coherently deal with.

Looming behind the extreme climate events scene stands methane. The evidence of massive Siberian methane gas releases reported (Connor 2011) in the Arctic Ocean in 2010–2011 could involve hundreds of millions of tons of methane gas locked away beneath the Arctic permafrost, and may partially explain some of the National Snow and Ice Data Center's latest findings on Arctic ice melt. While positive feedback due to reduced reflection of sunlight is known to contribute to warming, one of the greatest fears is that the trapped methane could be suddenly released into the atmosphere, leading to rapid and severe climate change. With methane twenty times more powerful as a GHG agent than carbon, this risk is as or more consequential than what is posed by terrestrial carbon.

As theoretical risks materialize into measurable and verifiable global warming, sea rises, changing weather patterns, transformation of once-in-a-generation climate events into annual or every other year occurrences, the pressure for acting will increase. This could pose a risk if in the case of REDD, it is prematurely rolled out as a mitigation strategy before it is ready for prime time.

There is no evidence that the risks depicted in Figure 10.1 are perceived or being addressed by anyone through current programming. As IUFRO's Global Forest Expert Panel (IUFRO 2012) noted, there is *still* time to rethink assumptions about how social and economic issues are to be integrated early on in REDD planning before it is too late. The question is though: do thought-leaders and those driving programming see the need for this? And if so, will they, and can they, take action?

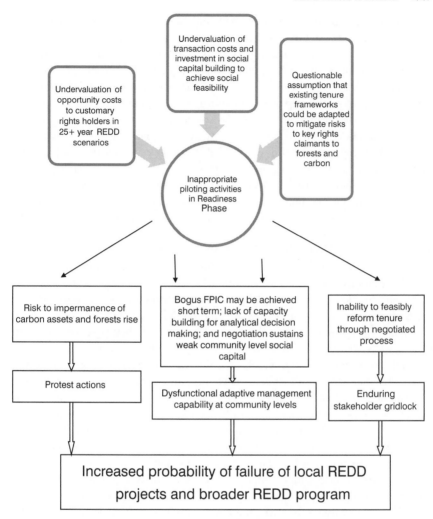

Figure 10.1 How inattention to social feasibility increases risks of REDD failure

The risk of not moving beyond safeguards

Forest governance is a political, social, and cultural problem, not a technical one (Trefon 2011). In this regard, the way the global community has approached REDD, aside from the political dimension at government levels, has been backwards.

REDD continues to be treated first and foremost as a finance and technical problem of measurement to enable carbon trading. Unintended risks are to be "safeguarded against". Were AD the focus of attention, a suite of activities to reform policies and provide local communities with the tools to mitigate deforestation could offer a more direct and plausible solution. This remains contrary to conventional wisdom on strategies and approaches.

The emphasis on safeguards in REDD to date has created a defensive posture, as people are seen as problems somehow to be mitigated. If, alternatively, people in communities were posited as the source of solutions to deforestation, and provided with a context of enabling policy reforms, capacity building, and tools to analyze and negotiate feasible outcomes, a new social contract for addressing deforestation could be established. This social contract would offer greater feasibility than what is currently being pursued through difficult-to-achieve market mechanisms and PES approaches that do not feature sufficient attention to social feasibility.

At present under REDD, social issues are treated as ancillary issues to be mitigated with non-binding principles. FPIC can be acheived through simple community analysis of the REDD offering; a country's signing on to UNDRIP for example is seen as providing a framework for IPs' rights. Weakly applied FPIC and a non-binding UNDRIP that ends up being more aspirational than effective, are safeguards that may well not advance community welfare in practice in many REDD situations. For REDD to be sustainable, more seriousness is needed.

As participation is not clearly defined, nor clear thresholds for determining if it has been attained or not established, at national R-PP or voluntary project levels, ambiguity is perpetuated. This weakens the credibility of "best practice", which in turn injects systemic risk into program and project design processes.

While "tweaking" safeguards at the margins will be possible and likely supported by REDD planners, transformational change to defining and operationalizing participation is what is needed. That said, it may not be welcomed, as this could lead to the type of paradigm shift that could threaten the principal driver underpinning REDD – the need for massive carbon finance to avoid deforestation at scale. For this, control over design and management is a key criterion, and more flexible participation may be seen as jeopardizing that.

There are several simple indicators for the relatively weak evolution of applicable policy and practice key to credibly addressing the social dimensions in REDD. These are (a) the surprisingly weak quality of analysis available in the public domain for already validated CCBA projects as previously discussed, (b) the proliferation of one-off major UN-REDD Readiness Phase workshops on the social dimensions of REDD+ in an environment where activism opposed to REDD is growing, and (c) bilaterally supported technical meetings such as the one sponsored by USAID in October 2011, in which social scientists from around the world involved in REDD explored how IPs' issues, gender issues, human rights, democracy, governance, land tenure, benefit-sharing, and distribution could be improved upon in terms of policy, guidelines, and practice in REDD+.[1]

The preoccupation of REDD planners with safeguards, paradoxically, obfuscates progress in developing TMAs that can be used in design phases to enhance feasibility. Unless the mix of issues presented in Chapter 6 is addressed upstream in planning, social impact assessment will be relegated to its traditional use – as an evaluation tool to assess implentation and adaptively manage progress.

By placing emphasis on non-binding safeguards, current REDD policy makers inadvertently sent two wrong signals from the start. First, instead of creating a framework for strategically approaching the development aspects of REDD to induce improved environmental service provision from local stakeholders that sufficiently factors in community visions, the focus on safeguards has put local peoples in the position of being objectified, presumably, as negative impacts are impending. As Vira (2012) observed, this creates a defensive posture that subtly becomes the accepted modus operandi of planners and project developers – attempt to reduce impending harm.

Yet why should the starting point and operational priority be to attempt to control against harm, versus maximizing (or at least optimizing) gain for local peoples through policies and practices that promote sustainability based on negotiated outcomes? And does this MO inadvertently contribute to harm being inevitable, with its mitigation then the desirable outcome?

Because there is no verification process for safeguards actually being implemented, nor to any requisite degree or threshold, safeguards become a rhetorical device to project a *notional* standard for REDD project developers. When no *material* standard of safeguard practice has been structurally established in project contexts, reaching objective consensus among stakeholders about situations and events is compromised. The need for greater clarity and establishing higher safeguard standards has preoccupied all levels of stakeholders to REDD at UNFCCC levels down to projects.

As the emphasis on safeguards undermines the focus on development priorities through its defensive posture (Vira 2012), the trend in this regard is not surprising. To a point, safeguards can be understood as the flip side of the coin for threats. Just as *threats* are the focus in biodiversity conservation to design mitigation measures around, threats-based analysis and programming have kept external project developers squarely in the driver's seat in project design (see Conservation Measures Partnership 2012 for how threat analysis is employed). While clearly the driver of the current biodiversity conservation paradigm, threats-based analysis and programming has put local stakeholders into the backseat as consulted peoples as opposed to planners (see Salafsky et al. 2003; CMP 2007; Conservation Measures Partnership 2012), behind governments and BINGOs in all aspects of PA planning and implementation. External agents decide on the threat, and develop mitigation responses in turn.

With safeguards, external agents identify the threat to maintaining *current* social welfare, as opposed to a negotiated future vision incorporating development objectives. The objective is to maintain status quo levels, as opposed to improve conditions based on measurable indicators.

The way in which conservationists employ threats analysis to limit responsibilities in addressing social welfare issues provides a window into how conservationists may assess responsibilities to respect safeguards in their REDD projects. In the manner that conservationists prioritize human well-being targets dependent upon biodiversity conservation, there is a real risk they will only consider safeguards that are a function of what they determine their responsibilities to be in REDD projects. As REDD projects may often attempt to induce changes to production systems and resource use patterns, how safeguards are balanced with activities that reduce perceived unsustainable land use practices seen as detrimental to project objectives, becomes pivotal. If imposed, as opposed to negotiated, both the efficacy of safeguards, as well as the legitimate development aspirations of local communities, may be placed at risk.

The Conservation Measures Partnership explains (2012) that though a conservation team may care about all aspects of human well-being, if its ultimate aim is conservation, it should focus on human well-being as it is derived from, or dependent upon, conservation. So, for example, a team might choose human well-being targets of fisheries livelihoods or forestry livelihoods, as these are clearly connected to the health of fish species or forest conservation targets. In contrast, the team would probably not focus on human well-being targets related to literacy or religious freedom. While acknowledged as important elements of human well-being, the CMP (2012) suggests they are not directly connected to biodiversity conservation.

Thus, if in planning and monitoring activities teams use diagrams like conceptual models and results chains, they should ideally only include human well-being targets clearly dependent upon biodiversity conservation. The CMP (2012) suggests further that the categories of human well-being are not important to display in a diagram and could even lead to confusion.

> We recommend that teams simply use these categories to make sure that what they are identifying as a human well-being target is indeed an aspect of human well-being – and not, for example, an ecosystem service or a socially beneficial strategy or result.
>
> (Conservation Measures Partnership 2012)

Technical solutions to avoid harm can thus be applied, it is suggested, to the root causes of threats to biodiversity conservation. Using the same logic, technical solutions to avoid unintended harm to people caused indirectly or directly from REDD projects can also be established. In both cases, the sense is conveyed that externally proposed technocratic solutions to these problems can be obtained. Scant public domain evidence suggests, however, that this is in fact plausible.

From a development perspective, local peoples may well oppose the emphasis placed on safeguards if they understood that stagnation could result for them, that the potential for increased risks to subsistence may occur without clear solutions guaranteed, or that foregone future development was a possible outcome. If given the choice, they likely would prefer not to stagnate as citizens of countries where hundreds of millions of dollars destined for development often arrive, to remain in the same place, or worse, to slip back "only somewhat" (through "reasonable" application of safeguards to prevent abuse), this would probably be an option many communities would reject if given the chance.

It is unrealistic to expect that all risks can or will ever be eliminated from REDD. It is thus unreasonable to expect any agency to eliminate all risks, or be able to control for them.

Nor can feasibility be assured given the myriad factors at play in REDD. That said, many of the projects that are being designed and approved by the principal validation and accreditation standard for REDD, the CCBA, appear from the Project Development Documents (PDDs) available online on the CCBA website – www.climate-standards.org – to be based on superficial understandings of sociocultural complexity and "basic needs only" plans for community development. This creates risks.

Because the bar has been set so low, erring on the side of facilitating project validation as a way to increase the value of proposals to generate financial backing for projects, risks increase. The alternative is to design and implement activities that have a higher probability of being appropriate and feasible. Where local peoples either fully and verifiably collaborate in design, or else own the design process, the risks for impermanence, leakage, and weak sustainability indicators will be lessened. That the bar has been set low to date is not a surprise; NGOs most vested in the CCBA are some of the same who have established best practice under the "Open Standards for the Practice of Conservation".

To reverse course, policies, practices, and incentives need to align more around the components of what it will take for REDD to work. This involves issues such as standards and principles, as well as how the bundle of issues constituting the social dimensions of REDD can be more credibly dealt with both substantively and procedurally.

Is overemphasis on social science expertise a risk?

The heterogeneity and contested nature of climate stresses for different groups and ecosystems highlights the unusually important role of social science in the analyses of climate change and potential responses to climate threats (Adger and Barnett, 2009). Yet, a disconnect between need and engagement prevails.

While there is no doubt that social sciences have much to offer planners, because conservation and REDD are in essence political processes, the major onus of responsibility for portraying or representing local visions and risks must in fact fall to ever-more-empowered and capable communities as social scientists are not community representatives. Communities must be able to drive participation, and demand that programs and projects become socially feasible. Not surprisingly, social scientists have yet to prove capable of doing this.[2]

Social scientists have traditionally been employed by development and conservation agencies to undertake social and stakeholder analysis upon which programs and projects are designed. Social scientists' position in conservation projects is described as follows (Welch-Devine 2009), and has clear implications for REDD.

Even among natural scientists who are quite favorable to including social science in their [conservation] projects, there is often the conception that the natural scientists will design and execute the project, while the social scientists will 'talk to the people' and get their 'buy-in' – in essence, *the social scientist is*

> *there not to provide a more thorough understanding of complex political and social contexts, but to make the people 'behave'* [emphasis mine]. When social scientists are asked to provide information, the data asked for are often limited to such things as 'how people are a threat and what incentives will make them change their behavior'.

Accepting Welch-Devine's description, it is clear that social scientists may find themselves in the position of being co-opted and instrumentalized by conservation organizations. This can lead to the disempowerment of the very people that, philosophically, social scientists as a rule of thumb are intent on supporting.

Co-optation of social scientists clearly represents a risk to both local peoples as well as the AD objective underpinning REDD. This is all the more reason why empowering local people to analyze, negotiate, advocate, and lobby for positions that are appropriate and representative of "community" needs, versus relying on external social scientists to explain and defend community positions, is vital in REDD. At the same time, so long as conservationists and REDD proponents rely on social scientists to "represent" local interests, this perpetuates and enhances the structure of risk faced in REDD projects.

MRV level risks

The risks of generating incommensurate data sets in REDD are high. To develop project-based MRV systems at the level of detail needed during the data collection stage, especially within the VCS standard, is time-consuming, costly, and "sometimes nearly impossible to obtain" (Cracknell 2012). Due to the differences in national data sets, accounting methodologies, and associated definitions of forests and deforestation, costly data collection activities will not necessarily produce comparable and reliable information, which may compromise the credibility and effectiveness of a nested or national REDD+ mechanism. This poses a major risk for generating "hot air", a preoccupation for many observers (Karsenty 2009; Poynton 2012).

While generating hot air is problematic for national and global REDD accounting purposes, it also creates risks to people, as REDD activities are occurring on the premise that mitigation of GHGs from deforestation is crucial. If in fact air is what is being generated, the risks to people as well as to financiers of the activity are high.

The fit between the technical accuracy required by standard setters compared with the reality of data availability in REDD+ eligible developing countries is uncomfortable. This raises the question of how this issue can be managed as REDD+ moves toward nested systems of accounting. Whether national and sub-national monitoring can be delivered "at a level of detail that certifiers are happy to certify will be a key issue" (Cracknell 2012). If overly rigorous, the risk of eliminating weak MRV national systems is high. If lenient, lots of "hot air" will be created.

The means to establish baselines against which ERs are measured and verified is fraught with technical challenges (Karsenty and Ongolo 2012). The counterfactual extrapolation of what *would have happened* in the absence of REDD is impossible to know. This can open the door to generation of hot air, as historical reference emissions levels (RELs) against which future emissions can be measured, and carbon credits issued, is subject to large margins of error. These margins of error are due to incomplete historical data sets, questionable technical means of accurately detecting deforestation, subjective bias in interpreting data, and the potential for political lobbying for inflated RELs along the way.

The cumulative sources of error in data generation and its management can cause very large uncertainties. In some cases deforestation has to be reduced by over 50 percent to allow for a clear detection of ERs (Pelletier et al. 2011). This illustrates that the technical risks in REDD accounting remain daunting. These risks will remain daunting so long as national technical capacities remain weak, and the leverage of governments to set RELs against which MRV will work remains high. Unless improved, suggestions of rigging will persist, while data quality of ERs will be of questionable veracity.

REDD managers are *currently* facing some of the core technical challenges in MRV, as a recent WWF publication (Cracknell 2012) indicates. The WWF capacity building guide makes recommendations on what will be needed to move from REDD+ projects to projects embedded in nested systems, emphasizing that the literature notes that developing MRV systems capable of accurately detecting ERs or gains remains a big issue (Cracknell 2012).

Some suggest that uncertainty levels are so high that detecting changes in emission rates could remain hidden in "error margins" for years to come (Bucki et al. 2012). This is a major issue for the UNFCCC, which needs consistent benchmarks to allocate REDD+ incentives to nations on a fair and equitable basis. It is important too for other organizations that are developing standards for emissions certification. Moreover, due to the differences in national data sets, accounting methodologies, and associated definitions of forests and deforestation, costly data collection activities will not necessarily produce comparable and reliable information, which may compromise the credibility and effectiveness of a nested or national REDD+ mechanism.

It is reasonable to conclude from the WWF's guide that far more is *unknown* than is known, and that the argument for "lowering the MRV bar" is nonetheless, somewhat paradoxically, gaining steam (see Bucki et al. 2012). This reinforces that the technical risks posed by REDD+ projects are aslso very significant.

While the WWF guide (Cracknell 2012) provides insight on why it is important to undertake feasibility analysis for identifying which MRV system should be used in nested REDD+ programs – "Feasibility and cost benefit analyses should be undertaken before any REDD+ project is initiated" – it is silent on how feasibility analysis of social and institutional factors needs to occur such that REDD+ projects in evolving nested contexts *actually* work, for risks on that side of the ledger to be effectively mitigated.

Emphasis currently placed by REDD+ project designers and managers on mitigating risks posed by outstanding MRV technical issues is understandable. To enable project validation and sustainable financing under the current framework, MRV systems must work.

Nonetheless, there is a risk of overemphasizing MRV issues to the detriment of risk mitigation for incompletely, and unconvincingly attended to, social dimensions in REDD. For REDD to work – both require attending to.

Risks from overconfidence in industry risk management tools

The VCS has compiled a risk analysis tool (2012) for project developers conducting non-permanence risk analysis and buffer determination required for Agriculture, Forestry and Other Land Use (AFOLU) projects' procedures. There is, however, no analogous tool VCS employs to objectively assess risk of project failure, or potential costs of such failure from the perspective of the social dimensions of REDD projects.[3]

While the VCS tool sets out the requirements for project proponents, implementing partners, and validation/verification bodies to assess risk and determine the appropriate risk rating for a project, the risk rating depends on the availability of difficult-to-obtain data. Moreover, the data that is obtained can, arguably, readily be fudged to fit the needs of project proponents and developers. This leads to questions of confidence and reliability.

The challenge that REDD certification bodies will potentially face retrospectively, if and when projects fail, will be in determining how adequate risk assessments actually were in objectively gauging the gamut of risks faced in projects. While a different activity area, as the scathing critique by the World Rainforest Movement of FSC certified monocultures in Uruguay attests (Carrere 2006), the most hallowed of certification bodies do not necessarily employ the most objective tools in assessing the certification merits of certificate holders.

When it comes to certifying more qualitative and complex issues such as FPIC, which is central to both FSC and REDD, all bets are off as to whether this can or will be done credibly. The risk of getting this wrong, based on the reading of information on the social dimensions of REDD that auditors are using to validate projects under VCS and CCBA standards, appears significant and worrisome.

Corruption risks

Risks pertaining to corruption and fraud present huge areas of doubt among REDD critics. Countries with weak transparency and accountability legacies pose great risks in REDD that may prove challenging to mitigate in practice (Global Witness 2010, 2011, 2012; CDM Watch 2012). TI, meanwhile, has created a manual to deal specifically with corruption related to REDD+ (Transparency International 2011), looking specifically at the relationship between weak forest carbon accounting, and potential for corruption and fraud in REDD (Barr 2011).

Some suggest that corruption risks extend beyond governance to encapsulate every aspect of human endeavor due to commodification, "where social relations are made over in the image of the market, and where there is a price to pay in terms of our moral, spiritual, and civic values" (Sandel 2012).

An example of corruption and mismanagement from Uganda (Cavanagh 2012) suggests that many countries may be "unready" for REDD+, as opposed to ready. In this case, forest and conservation sector deliverables have systematically been mismanaged and misappropriated. The implication is that the MRV systems used in REDD+ may be susceptible to BAU practices that will enable corrupt practices to be replicated.

Meanwhile, it is unclear whether the REDD community is systematically and comprehensively investing in creating firewalls so as to minimize risk from this type of corruption. While the TI guide, for example, is exemplary for providing guidance on issues that civil society groups in-country can apply to assess entry points for corruption and proactively plan for it, many practical and non-operational questions remain as to who will take up the guide, and with what political and financial support.

There are examples where civil society has taken the lead in combating corruption in remote forest areas of developing countries that has led to measurable reductions in corruption impacting trade and commerce (Brown et al. 2004). Learning lessons from similar experiences should determine the role that civil society groups can logically play in strategizing for anti-corruption activities in REDD.

International level corruption risks through organized crime

In a joint 2012 report, UNEP and INTERPOL (Nellemann 2012) cited ten different methods that organized criminals employ to access timber illegally. Each is facilitated by grand and "lesser" administrative corruption. These methods combine to create and fuel an illegal timber trade valued at least at US$11–13 billion/year. This represents as much as 20–50 percent of the value of logging worldwide when laundering of illegal wood is considered (Nellemann 2012).

While noting that both REDD+ and the European Forest Law Enforcement, Governance and Trade Action Plan (FLEGT) program are crucial mechanisms to support sustainable forestry and emission reductions, the work and methods required to investigate and combat organized crime may be underestimated.

> With the billion dollar investments in REDD+ and a developing carbon trade market designed to facilitate further investments in reducing emissions, illegal international cartels and networks pose a major risk to emission reductions and climate change mitigation through corruption and fraud, while also jeopardizing development goals and poverty alleviation in many countries.
>
> (Nellemann 2012)

Companies working illicitly now in the forest sector launder illegal logging under fraudulent permits for ranches or plantation establishment schemes. Under REDD+, this could open an additional conduit for exacerbating risks and impacts to IPs *and* local communities already burdened by high levels of corruption and weak governance. Illegal issuance of permits and inappropriate zoning could occur under REDD, much as it has historically in the tropical forest sector.

Proposals to mitigate these risks remain aspirational versus operational. Moreover, the potential for empowering local communities as an integral component in combating corruption in the forest sector appears to be being missed by major donors. A USAID-funded piloted model for fighting administrative corruption along the major trade arteries of the forest sector in seven DRC provinces may offer a basis for sub-national and local level oversight of REDD in central African countries at a minimum (see Brown 2007; Brown et al. 2004).

Reputational risks

While worries about inflated RELs endure (Cracknel 2012; Karsenty et al. 2012a), the potential for reputational risks associated with REDD implementation have yet to become a centerpiece issue in REDD debates.

The principal factor impacting reputational risk could be due to human rights abuses, unacceptable livelihood and development costs incurred, or failed REDD programming. Four categories of actors who have driven REDD are most at risk:

- *The intergovernmental system* led by UN agencies including UNEP, UNDP, and the FAO who created and manage UN-REDD, the leading driver of REDD programming, along with the World Bank and regional development banks such as the African Development Bank, InterAmerican Development Bank, etc.
- *National governments* (through their development agencies) who have taken the lead in backing international efforts to make REDD the leading option for mitigating climate change through a net-zero offsetting strategy – e.g. the UK government through DFID; the Norwegian government through NORAD; the US government through USAID and Overseas Private Investment Corporation (OPIC).
- *BINGOs* who have tied their programming horses to REDD as a major funding source to replace/augment biodiversity conservation funding that has reached a plateau, with prospects unclear in the enduring financial crisis facing the world in 2012 and beyond – e.g. TNC, WWF, Flora and Fauna International, CI, and WCS, along with smaller NGOs such as the Union of Concerned Scientists, Forest Trends, and others.
- *Major research centers* whose agendas are tied to climate change and REDD, such as CIFOR and ICRAF.

Other stakeholders facing scrutiny, and potential reputational risk, include technical experts, and financial institutions that have backed, or will be backing, REDD. These include private sector groups listed on stock exchanges, and limited liability companies (LLCs) involved in different aspects of voluntary and compliance carbon marketing. The second may be financial institutions such as Goldman Sachs or BNP Paribas that have funded REDD pilot activities for NGOs or invested in developing the sector. So too, OPIC, in providing insurance to Terra Global Capital, LLC on its fund investments in perilous investment environments like Oddar Meanchey, Cambodia, for example, is another agency potentially facing reputational risk.

Donor reputational risk

REDD projects in Cambodia, Madagascar, the DRC, Mexico, and Indonesia have been mentioned in previous chapters where disputes or some degree of conflict between project developers and civil society groups has been placed in the public domain. While problematic publicity appears to be being generated at a rapid rate, comprehensive studies on conflict in REDD projects are lacking. Several issues on reputational risk in REDD emerge.

A reasonable sense of how one of the principal REDD donors perceives risks to its reputation comes from The World Bank through review of its FCPF by its own "Independent Evaluation Group" (IEG). In noting that "expectations have been raised at both the country and local levels by the FCPF about future rewards of REDD+", the IEG goes on to say that "The Bank faces a risk to its reputation in case financing does not materialize on the scale envisaged", and that "[t]he Bank, therefore, needs to make sure that it stands fully behind the REDD+ agenda" (Independent Evaluation Group 2012). Quite reasonably, the IEG goes on to suggest that a strategic reflection on the Bank's overall approach to REDD+ is necessary, particularly regarding how country-generated REDD+ strategies will be aligned with Country Assistance Strategies (CASs), Poverty Reduction Strategy Papers (PRSPs), and the corresponding operational portfolios. This would appear to offer an excellent opportunity for the Bank to assume leadership in promotion and negotiation of new social contracts that will enhance REDD's overall feasibility.

Increasing negative publicity that "the story of REDD is just a lie"[4]

If the narrative on REDD increasingly becomes one of people exaggerating success for pecuniary reasons – either for purposes of land grabbing or carbon marketing – this may well have a spread effect across development sectors and regions that will damage donors and their implementing partners.

In response to a series of highly critical publicity in Australian papers on the Kalimantan Forest Carbon Partnership (KFCP), the Commonwealth of Australia's Hansard Commission reviewed its REDD programming in Indonesia.[5] The Green Party's representative on the Commission, Senator Milne, asked a series of questions about the project stating that the "KFCP is quite a serious issue here because this project has been a total failure compared with what was claimed for it and what has actually happened".

The Senator alluded to initial claims that work was to avoid deforestation of 50,000 hectares and rehabilitate an additional 50,000 hectares of degraded peatland, yet only one-third of the AUS$100 million allocated had been spent, and just under 1,000 hectares replanted. She suggested that the project "is a total failure", requesting explanation of "what went wrong, why, when it started to go wrong, and what this means realistically for REDD in Indonesia".

The response to the question from Mr. Comley of AusAID was reasonable, however unfortunate.

> I think the issue that has been found in Kalimantan has essentially revolved around the fact that land tenure issues have been more complex than first thought and resolving the land tenure issues has taken longer than first thought… What does it mean for REDD? I think it means that until such time as further work on those land tenure issues is resolved, REDD will take longer than first thought. That is the principal reason.

The response reflects that social feasibility issues were not seriously considered up front in the project design phase, leading to unrealistic expectations raised. Yet, it is implausible that the social constraints relate to *land tenure alone*. This experience could well be repeated in other country contexts.

Potential damage to thought-leaders of repute

Jane Goodall is the face of gorilla conservation, and of biodiversity conservation as a whole. In an interview on public radio's "Democracy Now!" in the US, Goodall's support for REDD was critically assessed by Anne Petermann.[6] Goodall was quoted as saying that it is important to protect forests through programs like REDD because of the protection they afforded biodiversity while enhancing livelihoods.

When Petermann was asked by the interviewer, Amy Goodman, about the issue with Goodall's support for REDD, Petermann's response was that she thought that "she and many other people are grasping at REDD because they're so desperate. I mean, the world's forests are falling at an alarming rate, and something has to be done. Unfortunately, REDD is not going to be the solution, and that's what I think people don't understand yet".

To Petermann, "REDD is going to benefit the largest forest destroyers in the world, while displacing the people that have traditionally protected the forests for all their lives. And that is a serious problem."

While Goodall is clearly justified in seeking a solution to deforestation on multiple grounds, should REDD fail, this may come with possible reputational risks to renowned thought-leaders, which also include philanthropic icons like George Soros who sees REDD as "transformational".[7]

Potential new sources for REDD critique

There is an outside chance that a convergence of factors could turn REDD into an intriguing learning laboratory for public protest in the first quarter of the twenty-first century. The convergence of factors involves (a) the perception of injustices, however right or wrong, by local peoples vis-à-vis government, BINGOs, and the private sector over loss of forest lands key to livelihoods, (b) new alliances between civil society groups in REDD countries and Northern NGO activists, and (c) the proliferation of cell phone technology, most strikingly in rural areas of Africa.

The critique of REDD may increasingly become an activity that involves citizens of involved Southern countries, not simply international NGOs who traditionally have stepped up for the voiceless. With 450 million mobile phones already in Africa, one for every two people, and with absolute figures growing (Mayers 2012), rapid dissemination of information no longer requires a CNN correspondent for alerts of conflict, as was the case in the late 1980s. Citizens have the ability to disseminate information as they choose in real time.

Critique by rural Africans now possessing cell phones will of course not be limited to REDD. Unrelated administrative corruption, or determination of market prices to identify optimum timing for sale of produce, will occur. This will simply create greater fluency with technology which, in the case of any REDD failures, could lead to negative publicity. This could create models that will be replicated in places where, historically, dissent was maintained at very local levels.

So too, educated African diasporas may become a source of public REDD critique as well. While limited in number, these may be inspired by how the Malagasy diaspora played a crucial role in successfully combating large-scale transnational land acquisitions involving Dawoo and Varun (Merlet and Bastiaensen 2012). It is simply suggestive of what *might be expected* from anti-REDD activists in years to come.

Finally, migrant association-based activism using internet tools (blogs, discussion forums, and webpages) has been reported in Madagascar (Teyssier et al. 2010; Burnod et al. 2011). These associations, a subset of the diaspora, may become sources of anti-REDD activism.

Given its relative abundance compared with Africa, web-based anti-REDD activism from Asia or Latin America may become an increasing phenomenon. As groups no longer depend on media outlets to get their messaging out, with the proliferation of mobile phone technology particularly to developing countries at ever more competitive prices, this may impact REDD in unforeseen ways in the coming years.

Information dissemination could therefore become a focus for ever broadening alliances of anti-REDD and anti-PES activists who do not accept the premises of the proposed push for a global green economy. They may well model communications on examples like climate-connections.org, or redd-monitor.org, providing information directly to the global community without passing through Northern portals.[8]

Risks of trivializing gender

Nominally, REDD is inclusive of IPs, communities, and minorities. It is also intended to be gender balanced. In practice, these objectives are hard to achieve, no less so for gender balance than effective inclusion of IPs in REDD processes. Institutionalized exclusion of women in the forest sector is a global phenomenon, largely because forests are managed and policies are designed for and by men, women's literacy rates are low, and while men's work is often linked to timber and markets, and is thus more visible, women generally use forests to support their families (Onta 2012).

While these factors *clearly* impact on the feasibility of achieving FPIC in REDD, it is less clear how individual projects and standard setters systematically address this as part of the validation and verification processes.

Checklisting women's participation in stakeholder meetings, directing questions to women; and enabling women a couple of days' opportunity to reflect on REDD project proponent propositions as part of FPIC, while necessary, are not sufficient to constitute adequate attention to gender. Like participation itself, gender inclusion is a topic that historically has received much publicity and attention, yet has remained resistant in many instances to significant progress.

The risk of exacerbating existing gender imbalances and further skew benefit distribution for forest sector revenues has been described in Oddar Meanchey, Cambodia (Boudewijn 2012). Here, the introduction of REDD into a context where community forestry is already operative, risks increasing intra-community stratification between men and women, as monetary benefits from REDD only go to men. In this regard REDD projects could find themselves unwittingly supporting regressive gender imbalances through its revenue sharing strategies unless vigilant.

The risks of not conducting risk analysis

The fact that risk analysis is not demanded of PDD developers and proponents may contribute to perpetuating an "announcement culture" (Olbrei and Howes 2012). This is where "aid announceables rather than results are given prominence", as has been the suggestion for REDD policy implementation in Australia and the aforementioned KFCP. While being "quietly downsized", the failure to conduct upstream risk analysis is seen as having likely precluded sustainability, prevented the sharing of lessons through lack of transparency in reviews and evaluations, and represented a setback for REDD in Australia and the subregion Australia most works in (Olbri and Howes, 2012).

Were risks of both commission and omission to be factored into the analysis of projects like KFCP, or other projects being piloted in the Readiness Phase, it would be likely that many projects would not move forward. Yet, current donor standards of intergovernmental agencies or multilateral development banks like UN-REDD or FCPF, for example, do not demand this. Nor do voluntary market standard setters demand rigorous risk assessment upstream either. This leads to simplistic risk identification, along with simplistic risk mitigation plans. The thorniest challenges to REDD projects involve risks that relate to social capital and institutional issues. Neither common definitions nor thresholds for the quality of community participation in decision-making meetings, the credibility of decisions made leading to FPIC or concurrence of objectives presented top down by project developers to communities for concurrence, exist at the level of standard setters. This means that, at present, these challenges are not approached realistically, systematically, or convincingly. No attention is paid to the minimum social feasibility requirements for REDD to be sustainable. This amplifies risks.

Social and institutional issues customarily are treated as required add-on elements in program and project design. Too often design in development and conservation has been based on untested assumption versus evidence-based data and facts. This approach in development is not dissimilar to the use of leeches by doctors in the medieval period (see Duflo 2010); by acting as if the risks do not exist, as the KFCP illustrates, impacts from poor risk assessment can come back to bite.

The risk inherent to carbon commodities and securities

Instead of being a tangible product that can be measured, valued, and traded, carbon "is an assemblage of agreements, conventional practices, durable artefacts and rules held among people who operate in very different contexts around the world" (Whitington, 2012). It is this uncertainty about the very nature of *"the thing"* that is traded in carbon which feeds speculation and climate opportunism, hindering the establishment of international agreements (Böhm and Siddartha 2009; Böhm et al. 2012). This uncertainty triggers a series of risks, ranging from failed environmental policy, to project failure. Carbon commodification thus undermines carbon sequestration and AD, the argument goes, which of course is the very objective of the carbon commodification to begin with.

These failures create risk for global society. At local levels failure is experienced most acutely. Impacts are from opportunity costs to disrupted livelihoods (Lohmann 2006; McAfee 2012b; Sullivan 2011), evictions modeled on those experienced in the biodiversity conservation context (Dowie 2009) or externalities on other aspects of governance and culture may each occur.

That said, these risks continue to prevail. Either they are under-appreciated or simply ignored by those most in control of the REDD agenda. For example, the High Level Panel on the CDM recommends that the CDM, despite absorbing tremendous criticism over the years, suggests that it "[d]evelop and test project-based and/or national/subnational REDD+ programmes, while implementing appropriate controls to mitigate risks" (High Level Panel on the CDM Dialogue 2012). It rationalizes this because the CDM is "the world's only truly global carbon market today" and "is collapsing for reasons outside its scope, and nations must intervene to avert this downfall".

The Panel sees REDD as a possible path to help avert the CDM's downfall. As REDD itself is highly risky, and as the CDM has a weak success legacy of its own to bolster REDD prospects, this marriage would seem to suggest a doubling of risk, for both REDD and the CDM (Lederer 2011).

The CDM may well succeed in assuming a role in eventual scaling up of REDD programs. It may be asked to house an eventual international compliance mechanism negotiated through the UNFCCC process. If so, this will be all the more reason to maximize efforts and urgency to reconstruct REDD's strategic and methodological paradigm. In particular, getting TMAs in place that enhance environmental governance prospects and overall social feasibility will help reduce risks associated with the predominant focus that MRV and carbon finance currently maintain in REDD. The CDM would need this.

Risks of misreading equitable benefit sharing for communities

REDD feasibility may well hinge on how successfully benefits are shared, if and once they are generated. Benefit sharing is not, however, the only basis for defining equity. The criteria used to define equitable benefits, who decides on benefits, and how this is done, is pivotal. The risks of misreading what equitable benefit sharing comprises for communities impacted by REDD is significant and can upset projects and sustainability.

Current participatory tools employed as best practice do not mitigate these risks. Tools such as PRA or appreciative inquiry (AI), for example, are designed for the ranking of basic needs and are useful for community visioning exercises. They are not designed for more exacting CBA of trade-offs from different benefit sharing arrangements. Nor are they designed for analysis of how or what to negotiate for, based on assessment of the value of REDD carbon that external partners may be transacting.

Carbon commodities are securities that communities have as-yet-defined rights to and potential financial stakes in. Basic participatory toolkits normally employed in conservation and development are ill designed to make nuanced assessment and sophisticated judgment for dealing with analytical issues pertaining to land use or engagement decisions involving community carbon. In situations where multiple stakeholders may lay claim, tools that enable assessment of costs, benefits, and risks are needed.

"Benefit-sharing" is the general response to questions of equity in REDD. Yet clarity on the mechanisms and accounting specifics remain ambiguous in the REDD Readiness Phase. When REDD project developers suggest that 50 percent benefit sharing with local populations will take place with local populations, on the surface this appears to be a great opportunity for local communities (see Poffenberger 2009).

One factor constraining implementation of benefit sharing programs that increases risks of REDD failure involves uncertainty over what form of benefit sharing mechanisms will be ultimately classed as legal (Luttrell et al. 2012). Many REDD projects are currently operating in a vacuum in this regard, which of course creates major risks as communities really do *not know* what they will have legal recourse to in the event of dispute. This ultimately will threaten the legitimacy of, and support for, REDD.

Until the accounting base for the calculation is clear, however, the 50 percent share attributed to local communities in the case of projects like WCS's Makira in Madagascar (Burren 2012), BCI's Sankuru (Hurley 2012), Terra Global's Oddar Meanchey (Poffenberger 2009), Conservation International's REDD Pilot Project in the Tayna and Kisimba-Ikobo nature reserves of eastern DRC that is financed by the Disney media and entertainment company (World Rainforest Movement 2011), or ERA and Wildlife Works' Mai Ndombe REDD project in the DRC (2011, 2012) retains an uncomfortable degree of ambiguity over what the 50 percent ultimately will translate to in practice. Nor is it clear as to how well communities have internalized what such a 50 percent offering will confer, as well as appreciation of the broader context of opportunity costs through any deferral on resource use as part of a PES agreement.

The risk of community perceptions changing over time, if not initially grounded in analysis and full understanding, may be high. These could create problems for REDD projects down the line if questions of equity and full disclosure of implications of commitments are not clarified up front to communities.

To mitigate these risks, REDD project developers should ensure that communities understand details of how market prices, actual project costs, net revenues, and net revenue sharing with communities will work. It is only on this basis that communities can factor in opportunity costs in the present and at different stages in the future, and reach informed consent and agreement.

The risk that benefit sharing schemes prove to be little more substantial than the spinoffs promised to the peoples of Josephstaal in PNG in 1992, will undermine the status of REDD. But it will also undermine the rolling out of major natural resource sector initiatives in the future. Here, the risks posed by uncertain price signals for carbon, poor or questionable accounting practices, the integrity of the community partner, the oversight capacity of communities, and the ability for communities to seek judicial system recourse in the event of non-compliance with formalized or informal agreements, will all come into play. This will be true for individual projects, and for REDD in the aggregate.

Clearly then, local stakeholders who engage in REDD+ in late 2012 are taking on great risk in situations where states and jurisdictions have yet to pronounce on the legality of benefit sharing arrangements for carbon market transactions. The reason for delays may have to do with administrative dysfunction, or in state indecision over how much rent they wish to skim off carbon transacted through the levying of fees or taxes. These rents will reduce the "net 50 percent" profit communities would ultimately receive. In addition, then, to risks posed to any ultimate revenue shared at community levels, there is also a risk that legal clarity on legal *recourse*, in the event of dispute, may remain unclear.

So too, transparency in reporting requirements may continue to remain unclear. Equally, participation and FPIC *thresholds* that identify when legally binding community consent is reached is notional at best. For example, in situations where customary chiefs or administrative chiefs hold sway in relations with government, as is the case in the DRC, how disputes will be settled when communities maintain they did not provide consent, while customary and administrative chiefs conversely may provide consent for REDD projects, creates great risk of project failure if not conflict. This is the type of risk worthy of considerable mitigation investment upstream.

Additionally, there is a risk that international crediting mechanisms may not work in practice to unleash a benefit sharing stream. While some argue that for a given quantity of carbon credits granted to a country, REDD *projects* should first be credited (Karsenty et al. 2012a), this depends on management and the efficiency of revenue sharing mechanisms. Where will the locus of revenue sharing administration for this reside? Who is providing leadership at international and national levels to credibly clarify this in a manner that communities, in addition to project developers, can be comfortably de-risked?

The risk o benefit sharing from non-representative community decisions over benefit sharing

Local elites are often well positioned to capture benefits in decentralized governance settings (Dutta 2009). These are situations that may be increasingly encountered in REDD. How communities decide what to do with any community level development benefits generated from REDD projects – schools, clinics, etc. – which in the logging sector would be termed SRCs, has no clear model to build upon in the REDD context. Though not exclusive to Africa, frequently in the African context, SRCs linking logging companies to communities have been a resounding failure. Depending on the specific situations there may be detrimental residual impacts for any other activity involving resources and expatriate institutions.

In central Africa for example, conservation organizations have often failed the expectations of communities for development activities that communities had understood would be a part of the agreement they had with the NGO. This often places elites in complicated situations where they may negotiate agreements that are personally advantageous with external agencies, but which are not perceived as useful by communities. In situations where acrimonious carryover from failed delivery of development promises exists, this poses heightened risks in REDD.

For example, in 2003 CIFOR presented an analysis of options to WWF-Cameroon for how resettlement of illegal villages inside Korup National Park, at the time over twenty years after its gazetting, should best proceed (Diaw et al. 2003). The CIFOR team noted that "communities' perceptions of the KNP are dominated by the sentiment that they have been deceived and abandoned" and recommended that "the only solutions that can work are the ones that will come out of sincere negotiation and collaboration with the communities of the Korup National Park. The whole range of legitimate stakeholders will have to be part of this process."

Resettlement is not always the issue. In northern Republic of Congo (Nouabale-Ndoki), far western Central African Republic (Dzanga Sangha) and eastern DRC (Kahuzi Bihega), PA managers face issues of community disaffection. Cross-cutting each case are years of resentment that development benefits had not been delivered by conservation organizations as communities had understood or anticipated they would be. While generalizations always have their exceptions, the backdrop for REDD in central Africa is complicated by the legacy of both logging and conservation at the community interface.[9]

Yet despite this legacy, how many conservation NGOs in central Africa are learning and capitalizing from it to reach out to negotiate new conservation and

REDD agreements with *communities*? This would mean negotiating bilaterally with individual chiefs who may or may not represent their constituencies, a proposition easier said than done.

From a methodological standpoint, relying on PRA or AI to decide on community development priorities may work, but it also can be simplistic, particularly if revenue sharing plans are put into question. The manner in which decisions are made *internally* amongst communities could skew benefit sharing and undermine feasibility.

This means that NGOs or private sector conservation concessionaires would need to develop skill sets to understand internal community dynamics for assessing if decision making is representative of ethnicity and gender for example, or whether conflict is being generated and, if so, how to diffuse it. There is every reason to believe from the information provided in CCBA-validated PDDs that these skills currently are not well developed among voluntary project developers.

Livelihood risks

As forest peoples living in potential REDD-eligible landscapes have so little margin for error in their livelihood security, any program that could worsen their plight, despite the advertisement of strong "safeguards" being built into the program, brings dubious ethical implications *if* the risks are well envisioned in advance.

Calls for different policies to address a range of shortcomings in REDD, while improving prospects, have been solidly made now by numerous NGOs, activist groups, and researchers over the past five years (see Global Witness 2012; RRI 2008, 2011; AIDESEP 2010; Lang 2011). Impacts of this advocacy are, however, unclear.

The three UN agencies responsible for framing REDD and REDD+ (FAO, UNDP, and UNEP) acknowledge that REDD+ risks "decoupling conservation from development", enabling "powerful REDD consortia to deprive communities of their legitimate land-development aspirations", undermining "hard-fought gains in forest management practices", and eroding "culturally rooted not-for-profit conservation values" (Ribot 2010; FAO et al. 2008). Clearly, the array of risks has been foreseen.

Operationalizing CIFOR's "3Es", along with adding a fourth and fifth "E" extrapolating from Ribot (2010) – enfranchisement and emancipation – would go a long way to securing social feasibility in analysis that would help in de-risking livelihood security losses. Yet, operationalizing this is not a small challenge. This is an area where pressures from the private sector may come into play.

As more realistic transaction cost assumptions are factored in, and carbon prices in the broader compliance markets where REDD carbon is for the time being not traded, demand for more robust appraisal of the transaction costs associated with achieving the "five Es" could, and should, be demanded by investors sympathetic to CSR and keen on ROI too. Reducing social risks will need to become higher prioritized. Incentives for stakeholders willing to pay for a REDD with higher probability for success, and reduced risk for failure, will need to be conceived.

Yet as Robinson (2012) points out, assuming that working with corporations for the sake of financing advantages, even those with avowed CSR agendas, will not guarantee generating biodiversity conservation results. The same will logically hold for REDD, where legal requirements and enforcement of the CSR aspects of REDD – equitable revenue sharing, delivery of social benefits, etc. – are unlikely to change corporate behavior if it decreases profitability (Knorringa and Helmsing 2008). Therefore, corporations must see that comprehensively attending to social risks in REDD is pivotal, along with credible MRV, to justify investment from a business standpoint.

Unrealistic deferral of agricultural land-use underpinning livelihood security will be problematic for most local stakeholders to accept. So too, over-inflating the value of shorter term revenue generation against longer term investment returns that could possibly be obtained outside REDD could backfire; if stakeholders at a later date feel they have been duped, this could create permanence risks that planners may not have accounted for. Therefore, any additional burden created through physical displacement impacting reduced livelihood security, or in terms of opportunity costs to development aspirations that go beyond simply safeguarding the little that many people now possess, is risky. This has led analysts to consider other approaches to achieve forest conservation, reduce GHGs, and address local environmental security at lower levels of risk (Karsenty et al. 2012a). Consensus on alternative approaches has not been achieved.

The following set of questions (Ribot 2010) can help guide thinking about what could likely go wrong when it comes to trade-offs between REDD and local peoples, and that can stimulate reflection on how to strengthen TMAs:

> What does it mean that 'REDD benefits in some circumstances may have to be traded off'? Who is doing the trading? And what about the more likely converse alternative: that local needs will, in fact, be traded off for REDD benefits? These trade-offs involve people's lives and histories at the edge of the 'legal' world. How will they be protected? How will REDD proponents, in the place-based realities that these real people inhabit, ensure that trade-offs are just—especially as REDD+ is being dictated by national agendas? How will REDD strategies take their needs and aspirations into account?

There is no indication that the most innovative projects developed for VCS and CCBA validation are able to answer these questions. Therefore, the risks currently posed to REDD livelihoods due to insufficient preparation are high.

Risks from politicizing indigenous perspectives

The position of IPs on REDD is easy to politicize and polemicize. This creates risks, as Chief Tashka Yawanawá (2012) has suggested is occurring in Acre, Brazil. In requesting the time and space to decide if REDD is beneficial or not to the destiny of his people and the planet, Yawanawá is arguing against a priori, politicized positions that are simplistic and highly risky, either pro or con. This is logical.

The extremes of public opinion on REDD are found in Latin America and Amazonia. The role of California's AB 32 legislation that could make carbon offset credits from Brazil and elsewhere eligible in its compliance market, are of major interest.

The October 25, 2012, debate in REDD-Monitor[10] over whether Brazilian IPs endorse or reject REDD+, and how this factors into inclusion of REDD+ as eligible carbon offsets for California's AB 32 legislation, offers a fascinating case of how diversity of opinion may be respected or simplistically politicized.

Environmental Defense Fund President Steve Schwarzman in an October 17, 2012, article on the Environmental Defense Fund website argues that there is great diversity of opinion in indigenous communities on REDD+ in Brazil. He offers that "there are some 385 indigenous peoples, speaking about 300 languages, living in 2,344 territories that cover about 2.1 million km^2 – an area four times the size of California – in the Amazon, as well as some 71 isolated groups". Schwarzman expresses the opinion that the NGOs and IPs visiting California to advocate against REDD+ offset credits being included in the AB 32 program, "offer one set of perspectives on REDD+, but their views should be considered in the context of the spectrum of indigenous organizations currently engaged on these issues, many of whom view REDD+ quite differently".

Schwarzman is correct. The opinion is also misleading. Data shows that 6 percent of Americans deny global warming is happening while 70 percent believe it is, while 97 percent of scientists agree that climate change is real, and is mainly due to the actions of people (Leiserowitz et al. 2012). Ninety-seven percent of scientists agree that climate change is real, and is mainly due to the actions of people, while 3 percent do not.

Were the issue about climate change deniers, whether the public or scientists, someone in Schwarzman's position may arguably belittle the credibility of the climate denial proponent.

Relativizing the diversity of opinion on climate change says little more than acknowledgement that different viewpoints exist, but that those denying the science of climate change may be wrong and are clearly in the minority. Should climate change deniers therefore be given equal weighting to express their opinion simply because they can?

In the absence of weighting, and subject to various assumptions, appreciating what Schwarzman is arguing for or against is difficult since no one knows the opinion of *most* Brazilian IPs on REDD+. If one of the most articulate Brazilian IPs himself doesn't have a fully formed opinion on REDD (Chief Yawanawá), what do we make of other IPs who purportedly do? Have they gone through full cost–benefit and risk analysis of REDD versus other options? Have they been brainwashed or bribed?

Reaching Amazonian IPs is difficult. Most Brazilian IPs in all likelihood do not know what REDD+ is. For those who do, outside perhaps the Suruí who have had capacity built through the Katoomba Group and which is favorable to REDD+, and "owns the carbon trading rights to their land" according to Baker and McKenzie law firm (Butler 2009b), it is impossible to know whether most IP opinions are "informed" and represent solid intra-community analysis of costs, benefits, and risks from REDD+, or not.

In this context, where a framework for objective situational analysis is wanting, it is likely that the opinion of umbrella organizations representing numerous IPs may hold more weight than individual IPs with special relationships with pro-REDD international NGOs. It is quite possible that the Brazilian IPs that travelled to California represented the opinion of most IPs who have thought REDD through and concluded they wished to reject it. Yet even here, it is impossible to know whether these rejectionists have gone through a credible CBA either, or whether their position has been simplistically politicized.

This is why social feasibility in REDD depends on enabling IPs, and other remote forest communities, to have capacities built as a first step to be able to objectively assess REDD on an informed basis. This would provide means for appreciating the diversity of opinion. It is also fundamental to building a new social contract in REDD. Credible social contracts, beyond perhaps the case of the Siruí, do not exist

Is OPIC's REDD insurance a risky precedent for countries?

In 2012, Terra Global Capital LLC won the *Environmental Finance's* (www.environment-finance.com) 2012 award for its "sustainable Forestry Deal of the Year" with the USG's OPIC in Cambodia.[11] The rationale for the project, along with comprehensive analysis of the range of risks faced in the project,[12] is described in what may be a standard setting report for the private sector involved in REDD (see Poffenberger et al. 2009).[13] As Oddar Meanchey may be a possible model pilot project for those hopeful for the future of REDD, the role of OPIC insurance in evaluating Oddar Meanchey as a representative REDD pilot project is worth considering.

Environmental Finance noted that the project is considered unique in the manner in which it is addressing "mosaic deforestation", and the many drivers of deforestation. The article quotes Terra Global's Leslie Durschinger who notes that "it is also unique in that it involves working with the Cambodian government, a much larger and more bureaucratic entity than most private sector companies, and one for which something like REDD is brand new".

The value of having political risk insurance as a mechanism to reduce the exposures for investors cannot be overstated, Durschinger says. "It is a very, very powerful risk-reducing financial product for investors in the sector" (www.environment-finance.com).

Pacific Environment et al. (2012) and Lang (2012f) provide a critique of the OPIC political risk insurance policy issued to Terra Global, focusing its attention on the implications for the RGC in particular in the event of default. Two main issues are raised:

1. Pacific Environment et al. (2012) point out that under the UNFCCC, countries may be required to regulate REDD projects in order to participate in compliance carbon markets. Depending on the regulation implemented,

this could affect Terra Global Capital's ability to profit from the sale of carbon credits on the voluntary market.

2. The report questions the due diligence that OPIC carries out.

OPIC and the project developers claim that the Oddar Meanchey REDD project will have significant climate, and local environmental and development benefits. But the viability of REDD projects rests largely on carbon markets, which are not reliable, thereby creating the risk that OPIC-supported REDD projects will fail to provide the revenue stream needed to deliver the promised benefits to implementing partners and local communities, and the high returns to investors (Pacific Environment et al. 2012).

From a policy and liability standpoint, "OPIC's use of political risk insurance to protect against the rightful application of nesting regulations turns the concept of political risk insurance on its head, and suggests, inexplicably, that the US Government is providing insurance against other countries fulfilling their future international obligations" (Pacific Environment et al. 2012).

In scenarios where there are insufficient financial incentives to keep trees standing, say if the price of carbon dropped too low, and promised benefits would not materialize, this could trigger deforestation. If so, the insurance would be needed as sections of the forest could be cleared for mining, agro-industrial plantations, and other environmentally destructive projects.

It is also conceivable that in the absence of adequate revenues and benefits, provincial or national government authorities could shift to other land use priorities. This could lead to "a perverse result in which one of the project 'beneficiaries' ends up being the entity that triggers the political risk insurance and becomes a target of US government pressure in an attempt to avoid an insurance payout" (Lang 2012f).

Given corruption in Cambodia's Cardamom forests (Milne and Adams 2012), the investment costs actually needed to achieve an acceptable threshold for social feasibility in environments like Oddar Meanchey, similar to the Cardamoms, may be much higher than what investors have allocated for their investment or transaction cost calculations. How well OPIC understands this, and whether in fact this creates any higher level of risk to OPIC for payout under the terms of the agreement, will be interesting to monitor. At the least, greater attention to social feasibility requirements in Oddar Meanchey would inevitably impact on the return on investment.

The risks of employing win–win scenarios

The inevitability of achieving "win–win" outcomes in REDD and PES programs was contested by McShane et al. (2011) in the Advancing Conservation in a Social Context project. The analysis is germane to REDD.

Basing their conclusions on empirical evidence, the authors argue that instead of expecting and marketing win–win outcomes in complex missions like biodiversity conservation, and now REDD, it is more appropriate to view them as one involving hard choices and trade-offs from the outset. This represents an important shift in how issues are framed.

This conclusion is also consistent with analysis of win–win scenarios projected in CBNRM projects in Madagascar and biodiversity conservation (Hockley and Andriamarovololona 2007). Clearly, the marketing of overly facile win–win scenarios, while potentially effective for fundraising and public relations, bears risks in the event of failure as the profile and stakes of the REDD activity in question may have unrealistically been raised. This could undermine future investments in the sector. It also impacts mitigation costs.

The risk that investors may demand more due diligence on social issues

It is possible that investors may demand higher due diligence standards of project proponents and developers in voluntary and pre-compliance project activities. This is a risk that the UNFCCC, standard setters, and project developers alike may face. Inarguably, it would have highly salutary effects while short-term costs would rise, higher probability for ROI would be secured.

That said, it is presently a remote risk for stakeholders driving REDD processes. This is because the global community has bought into the belief that best practice, and ever improving MRV, coupled to safeguards, will lead to private-sector-driven carbon finance that will couple with more modest publicly financed ODA programs for REDD. This finance in turn validates the methods and approaches employed. Something would be needed to dislodge this belief.

The only possible reason this may change is if those funding REDD through taxpayer funded ODA programs, along with investors in it for CSR, climate, and ROI reasons, challenge thought-leaders on the viability of the prevailing TMAs.

Given the strength of public relations programs of those stakeholders most vested in REDD, however, it is unlikely that awareness raising for the need for revised TMAs will originate from within the REDD industry. It would seem that discretionary investors will require evidence of whether the strategies employed in REDD are viable. If and when they understand that what is presently being advanced may not prove good enough from an investment standpoint to address permanence and leakage over the long term, higher standards may be demanded. Whether investors up the ante on this remains to be determined. IUFRO's recent work (2012) may help here.

Still, as transparency and accountability are increasingly demanded, evidence that investments are meeting their goals is likely to become a priority. Voices for accountability will be coming from multiple sources. The current inclusion of indigenous communities, activist academicians, social justice and environmental lawyers, and even bloggers from the lay public, suggests this is occurring.

Activists, though, may well have the last word on where REDD should be going. The combination of livelihood security, coupled to available technology, is offering marginalized groups more leverage than ever before, along with a new voice (see Lang 2012g).

In today's interconnected world, top-down programming based on dubious assumptions will be increasingly challenged by those segments of "the bottom billion" who are becoming more aware, and able, to respond. This subtle shift in power, or "pushback" (Rights and Resources Initiative 2011), should not be underestimated for its impact on REDD or conservation and development more broadly. That said, the potential for REDD investors to wake up to what practices will really be needed to derisk the social and institutional elements of projects, should not be underestimated either.

Risks to implementation after the current preparatory Readiness Phase

If the Australia experience is representative, the global community of proponents and practitioners is far from "ready" to scale up REDD programming. It further confirms the suggestion that the use of economic instruments such as PES and REDD can only succeed under certain governance contexts (Karsenty et al. 2012a). National institutions must work. Public goods need to be tied to a justice system that is *accessible to all* – and which provides for tenure security at individual and collective levels. Finally, these rights must be able to be defended by authorities and citizens. Few developing countries appear to meet these criteria.

The Proof Committee Hansard testimonies from Australia illustrate that it is impossible to conclude that either Indonesia, one of the premier forest countries involved in REDD, or the Australian government, both institutional leaders and proponent of REDD, are actually "REDD ready" as of 2013. A second important conclusion, which is a subset of the first, is that lack of preparedness to address social feasibility issues is constraining readiness at all levels of planning and programming. While unpreparedness to meet land tenure challenges was highlighted in the testimony, it is fair to assume that a number of associated issues pertaining to weak community capacities prevails.These result from lack of adequate policy, and how corruption is to (1) be avoided, and (2) mitigated if it occurs.

The risk, therefore, of rolling out full-fledged implementation programs on the premise that stakeholders are "ready", simply because the preparatory Readiness Phase is winding down, would appear to be *very significant*. What is really needed to be tested in the pilot phase – policies along with TMAs for negotiating REDD viable social contracts at national levels – has not been thought of, let alone addressed.

REDD could have a significantly higher probability of meeting the many challenges it faces *if* a social feasibility framework were applied from hereon out. At present, it is not on the radar. This reduces, or virtually eliminates, any chances for REDD to work. On the other hand, the range of risks, and their consequences, is ramped up.

The risks of conflating biodiversity conservation needs as a driver for REDD programming

Carbon and biodiversity are often different ecosystem attributes that represent separate policy concerns (Phelps et al. 2012). While in practice siting REDD in some ecoregions, as opposed to others, may make sense – i.e. the Seram rain forests of Indonesia, Borneo, and Sumatra lowland rain forests and Sumatran peat swamp forests, Niger Delta swamp forests, Madeira-Tapajos moist forest, and the Isthmian-Atlantic moist forest have been suggested as REDD-sensible (Buchanan et al. 2011). Still, REDD programming alone will not be what will resolve loss of biodiversity in most of the world's tropical forests. Thus, believing that REDD and biodiversity conservation should be implemented in the same spaces may increase challenges and risks counterproductively for both programming objectives.

Risks to farmers of eliminating slash and burn

Finding the right incentives that will convince farmers and resource users in tropical forests to abandon practices that contribute to deforestation will not be easy. Impoverished peoples do not often have the luxury to risk adopting new practices that may fail. While many would prefer to avoid the grueling labor of clearing primary or secondary tropical forest for subsistence purposes, until affordable technologies exist that rationalize abandoning tried and true practice, the latter will continue.

New approaches to develop credible agricultural technologies that offer incentives enabling farmers to avoid the difficult task of primary forest clearing using shifting agricultural techniques to produce food for subsistence, remains a priority in many countries where productivity and innovation are low, and food insecurity is high. Taking a country like Madagascar where for decades policy makers have been trying to reduce *tavy* and *tanely* – two forms of slash and burn agriculture on secondary and primary forest respectively – have largely failed as viable alternatives are lacking.

Working on the assumption that slash and burn is bad for the environment, and is an inefficient technology for farmers, international research centers in Asia, Africa, and Latin America have tried to replicate the green revolution in the past thirty years. Progress has been incremental as opposed to transformational, and, even there, has not been enough to lift the poorest farmers out of poverty.

Recently, research from Biodiversity International in Quintana Roo, Mexico, over a fifteen-year period has shown that slash and burn in fact has salutary effects on forest regeneration, particularly of high value timber species like mahogany. By creating major forest gaps that mimic hurricane effects which facilitate natural regeneration (Cherfas 2012), high value species like mahogony may benefit.

Misunderstandings about how slash and burn actually works to promote natural forest regeneration often results in inappropriate conservation and forestry policy measures, with small foresters and farmers losing out from these policies (Paddoch 2012 in Cherfas 2012). Yet it is fair to say that while a minority of scientists like Paddoch or Fairhead et al. (2012) may argue the salutary benefits of swidden agriculture, with Alcorn (1990) going so far as to call swidden "managed deforestation" as agriculture is built around removal of trees and not the forest, conventional wisdom has it that swidden, much like nomadic pastoralism, is at the root of natural resource degradation and is to be transformed.

Yet, are REDD planners about to offer viable incentives to substitute for slash and burn practices, or will alternatives to slash and burn be presumed to be developed under complementary programs? For all cases where deforestation drivers are attributed to unsustainable local agricultural practices, a first priority will be to identify viable technological and economic alternatives for farmers with few subsistence alternatives. Simply "sensitizing" farmers to the purported nefariousness of practices, in the absence of alternatives, will be ineffectual.

For the time being, as opportunities for transformational change of agriculture are not realistic in much of the developing world, small incremental changes will need to suffice. This will place a greater demand for REDD to optimize all aspects of its collaborative programming with agencies promoting proven agricultural innovation to help reduce land use pressures on forests. It also will push REDD projects to see how more distant, arguably important deforestation drivers are addressed through policies at national and even international levels. If REDD expects that behavior change is to be primarily borne by poor farmers, its prospects will indeed be dim unless strong, viable livelihood alternatives are offered them.

Opportunities

Most of the analysis to date has been critical of the direction REDD has taken. The "low bar" of standards established, and associated best practice employed to achieve mitigation and poverty alleviation REDD objectives has been underscored in *Redeeming REDD.*

That said, there is no reason that the bar cannot be raised, the rigor of the standards and auditing processes improved, and the quality of the TMAs employed to be more attentive to the social dimensions of REDD. If so, this may well be a moment of opportunity in avoiding deforestation and sequestering carbon. It all depends on how decision makers and investors particularly, assess the present and future, and adaptively manage the course from hereon out.

If approached realistically, with a strategy that empowers people by providing realistic incentives to escape poverty through AD, REDD *could* offer an opportunity to generate "triple win" outcomes. This, however, is conditional on negotiating a framework for approaching such a triple win, and underpinning it on a formalized social contract that outlines stakeholders' obligations and risks assumed. Absent this, zero sum outcomes, where for each winner there will be one or many losers, will be more likely.

If feasibility analysis drives the planning process, this type of multiple win can be achieved. It is almost impossible for it to result from the current approach to REDD. Paradoxically, while "triple win" outcomes are implausible to impossible as REDD is now structured, *if* the right types of policy setting, capacity building, and overall attention to social feasibility were to be prioritized, achieving "triple wins" would not be out of the question. "*If*", however is a big word.

11 A new social contract for moving forward

> What we have ignored is what citizens can do and the importance of real involvement of the people versus just having somebody in Washington make a rule.
>
> Elinor Ostrom, Nobel Economics Laureate

Throughout this book I have argued that REDD programming has been technocratically obsessed and overly abstract. People are placed second in practice, despite the prevailing rhetoric of participation and, philosophically, the primacy of people in REDD.

I have asserted throughout the book that the privileging of MRV over social dimensions may provide comforting conviction that the REDD world makes sense, and rests on a secure foundation. Yet, I believe that this reflects on what Kahneman (2011) notes when he suggests that we often exhibit a collective ability to ignore our ignorance. Or as Trivers (2012) may argue, we oftentimes lie to ourselves the better to lie to others, which I believe continues to be the case in REDD. Specifically, we simply are acting in the REDD world as if things are under control, and that social and institutional issues are subsidiary to MRV. While consistent with conventional wisdom of industry thought-leaders, the empirical basis for the strategy remains, arguably, weak.

As empirical experience of REDD lessons learned in Indonesia, illustrate (Indradi 2012; Milne 2012), whatever learning has taken place, fundamental issues have yet to be addressed, the approach remains unconvincing, and is, ultimately, disappointing. This goes far beyond Indonesia, as hopes for significant learning that were to have been achieved during the soon-to-be-ending Readiness Phase have yet to address key policy issues fundamental to operational success and sustainability.

When Rousseau (2004) warned the intellectuals of the day to beware of listening to imposters, and that things become undone once one forgets that "the fruits of the earth belong to us all, and the earth itself to nobody", for all intents and purposes he could have been presaging the chorus of critique leveled by those opposed to REDD today. For rather than being an exercise of the people as Ostrom (1990) clarified is necessary for resource management to work in places where property is collectively owned and where competing claims may exist, which is at the heart of those opposed to REDD's evolution to date, REDD continues to perpetuate a top-down planning process. While it is expected that people on the ground avoiding deforestation will participate wholeheartedly due to the strength of incentives on offer, I have maintained that the inattention to social feasibility and the range of issues comprised within it undermines REDD from the outset. By focusing on what we say *will* happen, versus creating the conditions for what we hope *to* happen, we delude ourselves to believe that an urgent cause, coupled to good tools and money will save the day. Perhaps, but more likely, not.

This chapter presents a vision for what a new REDD social contract might comprise, and who may be involved in its promotion. Framing REDD through social contracts is a necessary first step to introduce feasibility into the planning process. This will need to be done at international, national, and sub-national levels. While of course idealistic, unlikely given the lack of precedents for negotiated social contracts framing major initiatives like REDD, there is no reason it could not be accomplished if a critical mass of citizens from around the world got meaningfully involved, and demanded leadership and accountability. The current approach is simply not working. Stakeholders are aware of this, and understand the risks of inaction.

Rationale for a new approach

Millions of citizens across developing countries could potentially be mobilized to support AD and proactively sequester carbon to mitigate GCC if they were given a good reason for doing so. While REDD has failed in this regard to date, the issues have not gone away, nor has the interest in avoiding deforestation among millions globally.

"Average" people in developing countries at this point have more reason to resist REDD or be indifferent to it, than to be supportive; albeit some remote, misinformed people still believe REDD represents the gateway to wealth and prosperity. Justifiable or not, this remains the trend. While international negotiations through the UNFCCC process strive to be participatory, most people dependent on tropical forests for livelihood security have limited sense of REDD, its risks, and potential rewards. In this regard, formal negotiations do not reflect on the type of social contract needing to be established at national and sub-national levels for REDD to gain traction and longer term sustainability.

There is no indication whatsoever that this situation will change. While REDD remained formally in limbo through the Doha UNFCCC COP 18 negotiations, it nonetheless retains interest for investment for planners, investors, and those opposed to it. While marketed early on as a livelihood game changer for the poor, REDD represents a sine qua non case study for what Mosse and Lewis (2004) describe for international development policy, as "the convergence of ideas of neoliberal reform, democratisation and poverty reduction within a framework of 'global governance'".

Meanwhile, the opportunities for learning during the current preparatory, REDD Readiness Phase have been largely squandered. In paying inordinate attention to the subset of MRV issues at the expense of social issues, and instead of the overarching challenge – creating an appropriate framework at different scales for successful environmental governance – REDD's wheels continue to spin. Ironically, REDD planners have not been able to see the forest for the trees.

This leads to several questions to raise and answer, prior to counterintuitively suggesting that REDD can in fact be redeemed.

Why after twenty solid years of sustainable development rhetoric is a major initiative like REDD continuing to repeat mistakes that have been muddled through over the past decades concerning elementary issues like local participation and capacity building? Equally, how can the global development community continue to allow itself to dance on the head of a technocratic REDD pin when long-standing critique of the challenges and implausibility faced in establishing workable MRV systems, coupled to the implausibility of carbon trading as a solution to climate change, have yet to be credibly answered (see Lohman 2006; Lang 2010; Llanos and Feather 2011)? Is there the slightest indication that the opportunity is being seized upon for establishing REDD as "an emancipatory program" given its global scale and potential power (see Ribot and Larson 2012), or one in which both redirected benefit streams and redistribution of forest tenure are recognized as the legitimate rights of forest peoples (see Sikor and Stahl 2011)? For "emancipation", recognition of rights, and their operationalization, the tabling and negotiation of new social contracts at multiple levels cannot, obviously, be avoided. For the time being, this is not yet REDD..

The moment is therefore ripe for rethinking the logic and strategy for avoiding deforestation before the window of opportunity fully closes. The goodwill of the global community to address this issue is clearly fatiguing with the inability of the UNFCCC to close a deal on a compliance mechanism after seven years of trying. Yet, oddly, while critics of the current REDD trajectory argue on moral and philosophical grounds for respecting rights and empowering indigenous communities in REDD, few have focused on operational options that could serve as the focus for triggering more constructive coalition approaches to the impasse that has largely prevailed for seven years.

While it is wishful thinking that planners and REDD thought-leaders will step up and admit to failure in the current approach, if deforestation is to be avoided, and its contribution to climate change mitigated, there is enough evidence to suggest that a new strategy is needed. So long as the process remains dominated by external agencies, and local ownership of those living closest to forest resources stays limited, there is no reason to believe that recycled stratagems from conservation and development, along with inadequate forest sector policies, will stand up to the scale of challenges presented by deforestation.

Somehow, the global community must move past the current impasse if viable solutions are to be found. On its present pathway, deforestation will be exacerbated, climate change indicators from deforestation contributions will worsen, standards of living for tens of millions of forest peoples subject to REDD in developing countries will plunge further, and the contribution from deforestation to African droughts and the next hurricane Sandy will amplify.

Furthermore, corruption, displacement, and overall opportunity costs resulting from REDD will inordinately impact politically marginalized, poor local peoples in tropical forested countries. Meanwhile, externalities will also be absorbed by poor communities in polluting countries too, as offsetting carbon emissions through compliance ERs' legislation will not eliminate the emissions but, simply, will enable them to continue. The cost will be what polluting companies, or other investors, pay for the offsets in tropical forests. It is safe to assume that most shareholder-owned companies will try to pay as little as they can.

However predictable, illogical, or immoral continuing down the current REDD scenario may be, delusional climate change policy continues to be exercised to the detriment of the global community (Helm 2012). REDD is simply a microcosm of failure of the global community in its approach to climate change overall, as planners refuse to own up to the real investments required to implement feasible strategies to address a problem that is only worsening as indicated by recurrent, now "normal", freakish climatic events. Furthermore, these investments will need to go beyond establishing new policies and laws that implementing agencies will, as has often occurred in the past, simply ignore, however progressive they may appear to be.

A social contract generated through multi-stakeholder negotiation is the first step

Avoiding deforestation successfully will require that an effective and equitable social contract be negotiated between the core group of stakeholders who must establish the framework for a viable global initiative. Negotiation among key stakeholders to define the terms for avoiding deforestation has never been a premise under REDD. Key groups of stakeholders have been left out of the framing of REDD and its priorities. This has undermined its viability. While there is wide recognition that global policy regimes are deficient (see Cadman 2013), few focus on how to practically change this.

The notion that local stakeholders are pivotal to the future of avoiding deforestation is supported by several decades of common pool resource management research (Ostrom 1990), community forestry research (Agrawal and Angelsen 2009), a decade of management transfer experience in biodiversity conservation in Madagascar (Hockley and Andriamarovololona 2007), and an array of other evidence from the literature. Communities with strong management authority and sense of security tend to conserve forest resources, carbon, and biodiversity, as well as enhance livelihoods (Blomley et al. 2008; Ellis and Porter-Bolland 2008; Chhartre and Agarwal 2009; Ojha et al. 2009; Nelson and Chomitz 2011; Porter-Bolland et al. 2011). And they apparently can do this far more cost-effectively than BINGOs, if the research in Madagascar is to be believed (Hockley and Andriamarovololona 2007).

Yet, why do programs like REDD not establish frameworks whereby communities are given the chance to design conservation programs and manage resources they live closest to, of which REDD is clearly one? This question was most clearly asked nearly a decade ago (Chapin 2004), and has yet to be satisfactorily answered.

Is it because of an incommensurability of values – that Northern funded programs must be run by Northerners, even if they have never demonstrated the competence to do so – as measured by results and impacts? With compacts between Northerners and national elites at all levels enabled, to the detriment of democratic representation of local peoples in the process (Ribot and Larson 2012), REDD perpetuates a familiar BAU approach. Is it because planners believe, absent empirical data, that results and impacts are distinctly better under externally managed programs, even though the data is clearly equivocal in this regard? Or does it simply come back to the usual suspects – power and money?

Clearly, if the Isuseño people were able to take the lead in managing the Kaa Iya National Park in Bolivia, with support from the WCS and USAID throughout the 2000s in the second largest national park in south America, this suggests that IPs and local communities elsewhere can take the lead in managing AD and REDD projects. Moreover, BINGOs and research centers can play important *support roles* in advancing MRV systems as needed. This itself would be transformational and long overdue for all those NGOs that for decades have philosophically espoused working themselves out of a job.

If for no other reasons than the scale of its efforts and the configuration of its management and funding, Kaa-Iyaa (Guaraní for "the spiritual owners of the forest") should be looked at as a possible model to learn from for REDD projects, not just in Latin America, but globally. That the model may work *because of* support from the WCS and USAID in this case, is precisely the point – if given appropriate support, local communities *can* take the lead.

Avoiding deforestation is not something that IPs or local communities in the Ituri forest of the Congo, the Chiapas in Mexico, the highlands of PNG, the forest–agricultural frontier of Brazilian Amazonia, or the peatland swamp areas of Kalimantan Indonesia necessarily wake up to in the morning as their focus for their day. While the Isuseño do so in Kaa-Iyaa purportedly because they are paid for the service, and not for the nature underlying the service (Lowrey 2008), an important distinction if true, in their case they are happy to do so. Just as BINGOs cannot do their work without payment, so too local communities deserve payment that reflects a negotiation process, if resources are to be sustainably managed (Brown et al. 2008).

Nor is it something that will readily become their priority for the years or decades to come, as the presumption in REDD would have it, unless a solid rationale is provided to them. The current logic in REDD is several papers thin. That will remain so, unless the conditions are created for it to become robust.

While local stakeholders at the front lines of where tropical deforestation occurs are not the only players who need to be successfully mobilized, they do hold leverage over the short- and long-term success of any initiative seeking to avoid deforestation. Figuring out how to involve them coherently at international, national, and local levels in the framing of a new social contract for avoiding deforestation must become a first priority. While the specifics of the contract will vary from one site to the next, a core of generic parameters – rights, roles, responsibilities, etc. – will guide the process everywhere.

Clearly, and with no illusions, the quest for a new social contract for avoiding deforestation can *only* happen, however, if current decision makers and thought-leaders own up to the inadequacy of the current approach to REDD to date, and the unlikelihood that lessons learned will substantively change these prospects. While this type of leadership may be difficult to obtain, it can be sought.

However, adherence to the sunken cost fallacy in respect to REDD – "we've invested so much in REDD to date and hence must simply redouble our efforts as the mission is of such importance" – will not help make the currently unfeasible approach work either. More dubious investments, or losses, will follow.

Inordinate externalities to communities must be averted

Given the range of risks for local peoples associated with failure of the current pilot test of REDD, where impacts may reach tens of millions of forest peoples, it is not surprising there is resistance.

Is there enormous resistance? No.

Is it grounds for worry? Yes.

The prospect of loss of livelihood, exacerbated poverty, loss of future development opportunities, relocation, intra-community conflict, and other risks is inordinately high for front line communities to REDD. REDD pilot projects, and the guidance under R-PP processes at national levels, have done little to assuage concerns. Standards established for voluntary REDD projects do not consider feasibility, save indirectly or by chance.

Why forest communities would buy-in to projects where benefits are ambiguous at best, the probability for detrimental impact on livelihoods and lifestyles is high, and feasibility is not demonstrated, is unclear. Until the costs for mitigating deforestation are perceived as no longer inordinately high for those front line communities, the program stands no chance of sustainable scale-up.

The *Stern Review* (Stern 2006), which has framed the evolution of REDD from the outset, argued that mitigating costs of climate change would be low overall. However, the inability of the global community to treat project level mitigation costs incorporating social feasibility, which would increase transaction costs and impact short-term investor profitability, has created an unfeasible framework for avoiding deforestation. So far, the burden for offsetting carbon emissions in Northern countries and emerging economies through avoiding deforestation is to be borne, primarily, by poor, politically marginalized communities. This needs rethinking, not simply because it is morally questionable. It simply will not work in practice.

How UN agencies can redeem themselves in a renegotiated REDD

Three stakeholders groups must be involved in establishing a model for jumpstarting AD programming. The central element of the strategy is negotiating a new social contract for reaching agreements at various levels of governance for avoiding deforestation. By definition, this is step 1 in a multistep process creating a framework for social feasibility in AD, under REDD or any denominated program.

The three groups are (a) community level groups across developing countries, (b) national and local developing country governments, and (c) "impact investors": a stakeholder category that has yet to squarely enter into REDD conversations to date.

A fourth group is key to the process – UN agencies that currently have been dominating all aspects of REDD to date, along with ODA donors who can and will jumpstart the process if the logic is clear. The role of UN agencies and donors, however, needs to become one of facilitator, versus decision maker, coordinator, or implementing agency.

Subsidiary to these are technical agencies – BINGOs, consulting firms, etc.

Troubled waters under the bridge

The UNFCCC, UN technical agencies, conservation and development NGOs, private sector technical expertise, and some technical agencies from the Consultative Group on International Agricultural Research (CGIAR) system, have maintained primary leverage over the evolution of strategy and approach to avoiding deforestation. Their domination of processes has not led to a REDD architecture that is functional and enables sustainable growth. This has more to do with negligent, conventional wisdom than intent. Either way, the result is the disempowerment of front line forest stakeholders, central to moving REDD forward.

This arrangement which has produced REDD/REDD+, while enjoying broad conceptual support from thought-leaders, funders, and technical agencies involved in early implementation – a readily understandable situation given the interests involved – is characterized now more by uncertainty and controversy than by confirmed stakeholder buy-in. This has happened because the tail of the issue – carbon finance – has been wagging the dog, AD, since REDD's inception.

To take avoided deforestation, REDD or otherwise, to scale the UNFCCC and UN-REDD will need to remain part of the process at international levels. U.N agencies, together with BINGOs, can both provide options for environmental governance frameworks, or technical assistance to local solutions proposed. No other body is capable of coordinating as the UN.

At national and subnational levels, other entities will be more appropriate to facilitate planning and programming. This will need working out on a country by country basis.

Funding agencies like the World Bank and its FCPF and FIP should be just that – funding agencies for programs that are collaboratively developed by stakeholders, employing technically and socially feasible design criteria. The same should be true for bilateral funding agencies, and other interested funding agencies. All of these should support the process that the national and local stakeholders drive.

Technical agencies like CIFOR, ICRAF, and CIRAD (Centre de coopération internationale en recherche agronomique pour le développement) should ideally be just that – technical support agencies that provide technical assistance and provide information for core decision makers to capitalize upon.

International NGOs such as BINGOs and big development NGOs should identify their niche to provide support as opportunities arise. They should not be in the position to control lands and the destinies of millions of rural peoples *unless invited to do so* by the core group of planners.

Can impact investors play a positive role?

Impact investing is an emergent category of investing in projects with social and environmental benefits. The Monitor Institute (Freireich and Fulton 2009) predicts that this category of investment could reach US$500 billion globally by 2020. There is a spectrum of impact investor types, but all seem to be seeking the theoretical triple win outcomes for the environment, society, and the personal bottom line that REDD has long promoted.

I have discussed why many NGOs remain legitimately skeptical of investors seeking a profit in REDD through carbon trading. The question is, however, whether profit making is forever incompatible with social and environmental outcomes, or whether a balance can be attained. If balance is possible, what are the preconditions for it?

Impact investors are described as follows (Freireich and Fulton 2009):

> [I]mpact investors want to move beyond "socially responsible investment," which focuses primarily on avoiding investments in "harmful" companies or encouraging improved corporate practices related to the environment, social performance, or governance. Instead, they actively seek to place capital in businesses and funds that can provide solutions at a scale that purely philanthropic interventions usually cannot reach. This capital may be in a range of forms including equity, debt, working capital lines of credit, and loan guarantees. Examples in recent decades include many microfinance, community development finance, and clean technology investments.

By involving impact investors who participate in the capitalist system, NGOs with an anti-capitalist ideological viewpoint, for example, will immediately reject the premise. Given the trends and stakes, the issue in AD should not involve rejection on the basis of ideology or academic preferences. Rather, as a global community, it is important to understand whether impact investors would be willing to participate in negotiating the terms of their participation in major challenges like AD programming that must involve government, communities, UN agencies, and other donors to jumpstart the process.

Some impact investors in AD or carbon sequestration could be interested in piecing together the kinds of local, community driven AD "portfolios" that Lou Munden (2011) was suggesting would be appropriate for investors to explore, in lieu of participating in carbon markets. These would need to be for investors where profit is not the principal incentive, and who prioritize the potential opportunity to impact at national, sub-national, and global scales. While there may only be a handful of impact investors who could exert maximal, positive leverage over the direction of AD and its mitigation, the opportunity is clearly there.

Incentives to government

Elected politicians and public administrators in developing countries need to perceive that the best option for their respective countries, and ideally their own careers, will be in facilitating and enabling, positive outcomes in big issues like AD and REDD. Obviously, a transparent and equitable incentive system will need to underpin this such that the perceived opportunity costs via any existing, corrupt BAU channels that are abundant in the forest sector of developing countries, is trumped. For this to occur, civil society will need to demand accountability of public administrators in countries where parliamentarians are democratically elected.

Presently, incentives for government officials are perceived to be linked to the fate of carbon markets, where prospects for generating soft donor funds, along with rents from carbon marketed projects, are the main focus. In many public sectors, these funds become all too fungible. While it has been suggested (Karsenty et al. 2012a) that the level of rents generated from REDD may not be as significant as many anticipate, it is probable that these expectations continue to shape many governments' interests in seeing carbon markets as the central premise in REDD. If alternative incentive structures to enable the emergence of more appropriate and sustainable funding arrangements are to surface, responsibly incentivizing public administrators with preponderant leverage over REDD will be key.

What will the new social contract lead to?

The quest for a new social contract will lead to defining the specific governance arrangements for avoiding deforestation programming at national and sub-national levels. The work at international levels will seek to harmonize approaches for what needs to be achieved nationally, identifying support for working out the processes at that level, and providing concessionary funding to get on with the job. Clearly, there is much from the past seven years of REDD that can be built on. The question is how to capitalize on that to constructively move forward.

Figure 11.1 overviews what needs accomplishing, and by whom, to appropriately move forward. The focus is for the core stakeholders to the negotiation to provide specifics on how an environmental governance framework is to be developed. It needs to address enabling policy, participation, stakeholder capacity building, resource tenure issues, decision making in planning, benefit sharing and management responsibility assignments; reporting responsibilities and standards, and thresholds of accountability, are to be worked out at different scales.

Figure 11.1 Conceptual approach for negotiated environmental governance frameworks

Premises and actions for redeeming REDD

There are a number of premises that I believe now need to be considered in successfully avoiding deforestation and creating a context for redeeming REDD. At present, these are not issues of serious concern to planners and thought-leaders. They should be.

Premises

- REDD design must be social feasibility driven.
  This has been the main topic of this book.
- A new social contract between core stakeholders to avoid deforestation needs negotiation internationally and at national and sub-national levels.
  The logic for this has been presented in prior chapters.

- Communities living with "REDD eligible resources" need to be front and center in planning and decision making to enable feasibility to be sought and obtained.

  This core element in participatory development has been thoroughly neglected in REDD to date. Achieving social feasibility is conditional on effective community participation. Elucidating how this can be achieved at different scales will be essential.

- REDD is a type of conservation activity, and the successes and failures of conservation can guide strategic planning of TMAs that can or cannot work in REDD.

  The fact that biodiversity conservation organizations have assumed thought-leadership in REDD on their own initiative is indicative of REDD being seen as an extension of conservation programming. So too, it represents a business development opportunity linking BINGOs with investment bankers and private sector interests. Yet, conservation programming has few metrics that enable objective assessment of its achievements. This is glaring for social welfare indicators pivotal to achieving, and demonstrating, social feasibility to help justify this leadership. New arrangements for design and implementation of policies and programs on the ground are therefore needed.

- REDD will succeed or fail on the basis of the effectiveness of governance arrangements that bring stakeholders equitably into the process.

  This premise is the basis for the new social contract proposed for underpinning avoided deforestation programming.

- Expert social analysis coupled to IEC strategies are no substitute for effective stakeholder engagement. For too long, conservationists have assumed that providing information on conservation benefits would elicit conservation behaviors. This has never been proven, but plays a central role in REDD as well.

  For REDD to work, policy focus needs to shift from frameworks designed to enable external experts to shape policy and decision making prioritizing MRV. While this is believed to enable carbon transactions, and hence carbon finance, emphasis should be placed on the components and steps required for AD and carbon sequestration to succeed at different scales. This shifts emphasis from an MRV–safeguards framework, to a social feasibility framework.

  The issue is not to provide *outside conservation agencies* with better assessment skills to gauge trade-offs that enable them to make decisions (see McShane et al. 2011). On the contrary, it *should be*, adhering to subsidiarity principles, so as to provide this information to *local* and *national* stakeholders living closest to forest resources, who have ability to oversee, manage, and, to a greater degree, defend these interests. Local people will be most directly responsible for reducing either emissions from deforestation and forest degradation, or overseeing that it is done. They will be tasked with conservation, sustainable management of forests, and enhancement of forest carbon stock responsibilities too.

  While to date, most attention in REDD program and project design has been devoted to safeguards – e.g. assuring that no harm is done, if possible,

– in this new orientation, emphasis would shift by beginning with a simple question: *What needs to be in place for REDD to work for the core stakeholders involved in avoiding deforestation?* Achievement and sustainability of milestones that, at a minimum, are co-established by REDD-impacted peoples, will define success.

- Focusing on formal land titling to reform tenure in the short term in REDD programming may inadvertently undermine constructive steps forward, considering the empirical contexts of overlapping and contentious tenure claims in many nation states.

  In an ideal world, governments would clarify tenure rights in a manner that enabled equitable development to proceed. Yet, the world is far from ideal in most instances. This means that tenure ambiguity constraining REDD will best be approached through negotiation strategies, versus definitive legislative reform, for some years to come.

  Research on less formalized approaches to securing tenure for rights holders from around the world suggests that a more practical approach may be to build on land rights deriving from tradition and local history that provide a de facto basis for managing natural resources, and could serve as a realistic bridge model for moving forward in addressing the complex tenure ambiguities posed in REDD.

  In the same manner that benefit sharing arrangements will require negotiation (Bruce 2012), so too will tenure arrangements. Establishing effective arrangements, versus definitive solutions, should be the guiding principle. Here the examples of Mathieu et al. (2003) and Lavigne Delville (2002) are instructive. So too, the possible model of certificates issued via local land offices, a key institutional support structure used in Madagascar (Burnod et al. 2012) could potentially offer guidance for the piloting of tenure reform activities in REDD eligible countries and landscapes, though the on-the-ground constraints to broad up-scaling are daunting and require capacity building support (see AfricanBrains 2012).

- Expert social analysis coupled to IEC strategies cannot substitute for "good enough" stakeholder participation, analysis, and decision making to enable FPIC to be achieved.

  At present, IEC campaigns are either conflated with, or substitute much of the time, for stakeholder participation in REDD "stakeholder meetings". While meetings sponsored under the aegis of UN-REDD, for example, may have the genuine intention to elicit stakeholder views in national, sub-national, and project planning contexts, the size and duration of the meetings, coupled to the capacity of stakeholders to assimilate and speak knowledgeably on issues, are insufficient. This point has repeatedly been put into question by many local stakeholders involved in the processes, with protests from central American civil society groups on UN-REDD readiness activities most notable (AIDESEP 2010; COONAPIP 2012).

  Systematic data does not exist on how organizations developing REDD projects go about addressing the social dimensions of REDD, beyond attending to prevailing standards for safeguards that promote participation and FPIC to current best practice standards. It is reasonable to conclude that the long-standing tradition of expert social science input, in the form of socioeconomic studies or social analysis, is part and parcel of the planning approach for both development and biodiversity conservation organizations who are currently involved in REDD programming. The role of social scientists in conservation organizations in this regard has been well described (Welch-Devine 2009), and has been clearly inadequate on multiple grounds.

In practice, expert social science analysis can unwittingly come to substitute for local community analysis. By using PRA, conservationists can argue that communities have been fully "consulted", as this is the industry standard that almost all organizations rely upon. Even if social scientists are able to identify valid issues of concern in the design process, through participating in PRAs or implementing other ethnographic methods, this is exclusive of the need for communities to *themselves* process available information to make informed decisions. Just as universalist, technological solutions disengage from social context (see Pretty 2002), PRA and expert social science analysis can also fall prey to reductionism.

Unfortunately, the assumption that this external expertise can somehow represent local understandings and, apparently, local buy-in, is preposterous. And while social science data collected may be relevant to demonstrating minimum understanding of social contexts, as required in project planning documents for VCM REDD projects, the information generated may or may not be relevant to community level analysis needed to establish informed consent and buy-in to projects. Moreover, it hardly can be equated with community empowerment either.

To be credible, analysis that occurs at community levels must extend beyond "needs assessment" competence; for example, communities need to be able to assess costs, benefits, and risks of REDD projects to pronounce FPIC coherently. This capacity gap is currently glaring.

In biodiversity conservation, best practice guidance is for implementors to identify human agency threats to best mitigate these and advance desired outcomes. This needs accomplishing with minimum possible transaction costs, while assigning the minimum package needed to address human welfare issues directly tied to conservation objectives (Conservation Measures Partnership 2012).

REDD will be no different on the topic of minimization of transaction costs to address social dimensions key to project validation (see Richards and Panfil 2011). With cost minimization as an objective, it is essential that efficiency prevails. In this respect, delegating key planning responsibilities to less than fully capable local stakeholders, for example, would be counterproductive for reducing this aspect of transaction costs. However for the longer term, a different transaction cost calculus would be required that accounts for feasibility failures.

The fact that external social analysts may be accurate in assessments will bear little necessary correlation with what *could* be generated through robust community analysis and decision-making processes that external analysts cannot deliver. As we are reminded that *"the social scientist is there not to provide a more thorough understanding of complex political and social contexts, but to make the people 'behave'"* (Welch-Devine 2009), we should recall that there are eventual costs for doing so that need accounting for in REDD.

The assumption that IEC packaged with checklist questionnaires, and a day or two of community reflection can then engender FPIC, is absurd when discussing twenty-five- to thirty-year project commitment periods, implicating the destiny of local communities. Yet in practice, this approach is being perpetuated in REDD planning, and can only lead to infeasible programming. It is unlikely that genuine FPIC can be achieved in these circumstances, no matter how many checklist boxes get ticked.

It is crucial, therefore, that replication of best practice and BAU from the development and conservation industries as pertains to strong IEC programs, and emphasis on expert social science analysis, *not* be a substitute for community analytical and decision-making processes. Data does not exist to demonstrate that IEC and reliance on social science expertise in lieu of community social capital building, analysis, and decision making actually works in complex program situations like REDD.

- COAIT is one tool to consider for adapting to meet many community level planning needs in REDD. It appears to respond to the call from the Forests Dialogue (Gurung et al. 2011) for "a new generation of land-use decision-making" approaches in REDD.

COAIT is a toolkit that was developed to enable community driven land use planning in forest and biodiversity conservation in central African landscapes.[1] It was developed by IRM from 1998–2006 under funding from CARPE to enable local stakeholders to analyze complex conservation and development situations from the standpoint of economic, ecological, and social factors. Costs, benefits, and risks are analyzed leading to identification of options which form the basis of community proposals for land use plans (Brown et al. 2008). These plans can be negotiated with government, NGOs, logging companies, or now, REDD planners. They have the advantage, if well done, of being representative, and reflective of a stronger FPIC process.

COAIT was developed to enable communities to participate in forest management particularly in landscapes, as well as granular scales. For REDD purposes, it would be useful to systematically compare COAIT with other empowering toolkits that potentially get at feasibility issues and transcend PRA – perhaps Appreciative Inquiry and ZOPP, the latter of which is based on logical framework analysis – to determine if feasibility is dealt with similarly.

Linking COAIT with the Merlet–Bastiaensen framework (Merlet and Bastiaensen 2012) could produce a powerful analytical and capacity building tool to facilitate more effective negotiation and outcomes in REDD. Social feasibility in turn would be enhanced, and is an area social scientists could explore.

Actions for planners to consider

- Accept that REDD is not a value-free endeavor that will automatically generate "win–win–win" outcomes for the global climate, people's development, and biodiversity, and make strategic and programmatic corrections accordingly.

  Prevailing technocratic approaches to REDD based on simplistic assumptions that people will readily accept the PR spin for REDD initiatives will fail if and when the same people see they are not feasible, as they define feasibility. Barring miracles, poor, marginalized peoples are more likely to bear undue costs to their livelihood status and development prospects under REDD projects because of untenable assumptions about carbon prices, the questionable viability of benefit-sharing mechanisms in countries where corruption is rife, coupled to weak oversight, analytical, and adaptive management capacities. To change this, a different approach to identifying human welfare benefits in REDD needs to be the product of national and local level negotiation processes.

- Utilize the Merlet–Bastiaensen framework (Merlet and Bastiaensen 2012) for approaching large-scale land acquisitions, of which REDD can be seen as a subset. The framework can be paired with community analysis and decision-making tools to create a toolkit to better enable poor populations to prevent their land rights being encroached upon by more powerful actors in the REDD process.

  The centrality of tenure compared with all other deforestation indicators including budget, staff, and management plans has been recently confirmed (Nolte et al, 2013). The framework proposed here can be used to analyze how the multiplicity of land rights claimants and legitimate rights holders that prevail in REDD contexts are superimposed. These overlays create ambiguous or contradictory normative orders. In the case of REDD, these normative orders constrain local stakeholder participation, the obtaining of FPIC, outside investment, state resolution of tenure conflicts, and preclude social feasibility from being achieved.

  By pairing a framework that identifies points of contradiction between normative order overlays and stakeholders, with a proven toolkit like COAIT (Brown et al. 2008) (or others that exist or may be developed), the "bargaining capacity" of weaker actors can be strengthened. This will be a precondition to sustainability in AD programming. Arguments that avoiding deforestation is so urgent that it cannot wait for capacities to be built, obfuscates the fact that for sustainability, capacities must in fact be built. There is no escaping this.

- Review the logic established by the leading standard setter addressing social issues, the CCBA, that deforestation drivers can be identified in project design through use of PRA and similar participatory tools.

  PRA is very useful for ranking community needs and for general descriptions of situations in village settings. PRA was not designed, however, to address complex, landscape-level environmental governance situations posed by REDD.

If inadequately conducted over short time periods due to the expediency of minimizing transaction costs, PRA can lead to misleading information that can undermine sound project design and lead to conflict. Any FPIC in REDD based on this use of PRA is therefore highly suspect, whether certified by industry-leading auditors or not.

Aside from clear recognition at a policy level that local peoples must be safeguarded in the REDD process, that they should participate in the process, and should provide FPIC, this has not translated into clear consent standards with calibrated thresholds.

• Capitalize upon data empirically substantiating subsidiarity principles showing that community forest management works, and that maximizing community voice and empowerment is well worth shooting for in REDD.

The effectiveness, or lack thereof, of communities in forest management has been a topic of great debate for many years now. This is an extension of the debate about communities' role in biodiversity conservation. Elinor Ostrom, among others (see Blomley et al. 2008; Ellis and Porter-Bolland 2008; Chhartre and Agarwal 2009; Ojha et al. 2009; Nelson and Chomitz 2011; Porter-Bolland et al. 2011), demonstrated that local communities stand as good or a better chance of managing common resources – forests, fisheries, oil fields or grazing lands – than governments or private companies.

Strangely, the implications of Ostrom's work have not, as of yet, been fully capitalized upon in REDD, as the locus of Ostrom's concerns – resource users in communities – remain marginalized. Intergovernmental, state, international NGO and private sector entities continue to coordinate in REDD planning.

• Understand where capacity needs strengthening to achieve genuine participation that can lead to credible FPIC, and implement accordingly.

For communities to take on a central role in REDD, investment in capacity building must be comprehensive and robust enough to enable situational analysis, take decisions about consent or otherwise, and negotiate concrete steps in design and implementation with policy makers and other decision makers.

IP groups like AIDESEP or COONAPIP, and civil society organizations like La Via Campesina, are examples of organizations with demonstrated, strong rights-based advocacy skills in addressing REDD issues. The stronger their analytical and negotiation skill sets become, the more capable they will be at participating in decision-making fora, demand-driving appropriate programming, and enhancing advocacy for initiatives that are appropriate and correspond to local sustainable development visions.

• Private capital can be encouraged to support REDD, but should be matched with activities that reflect feasibility and low risk, with investor portfolios aggregated as assets only when clear social feasibility thresholds have been met.

The Munden project has called the current market-based model for REDD "hopeless" (Munden 2011), not because it was irrationally exuberant, but because the model is dysfunctional (The Munden Project 2011). While the findings and conclusion have been criticized by the Climate Markets and Investors Association (2011), the points were reinforced by Lou Munden in October 2011.[2]

As a substitute to carbon markets, Munden (2011a) proposed a broad performance-oriented framework that tries to take a rights-based approach tied to performance measurements. He argues that, from this, determination could be made as to whether there is indeed some kind of scalable activity that could be aggregated out of the small scale, local solutions that could potentially be cobbled together. The objective would be to create portfolios that could, presumably, be invested in or even themselves potentially traded, with the caveat being that they are more locally defined.

To Munden, "the money is there", and investors are simply looking for a solid opportunity to invest in a new type of asset class. Perhaps this could be the basis for a revised REDD offering. Here too, negotiating a new social contract guided by social feasibility principles cannot be avoided.

Not throwing the baby out with the bathwater

From the beginning there have been Pollyannas[3] and Cassandras as to whether REDD can be a key component to global climate change solutions, or simply a hot-air charade that will more likely contribute to international criminality, the abuse of human rights,[4] and exacerbated poverty among many of the world's poorest peoples.

That said, if deforestation could be reduced, and it could be done without exacerbating poverty or violating human rights, would governments, BINGOs, and those stakeholders currently critical of REDD *not* support the initiative, whether or not biodiversity conservation was a co-benefit? If there is a will to correct REDD by focusing realistically on the conditions required for making it work, not simply in theory, then the REDD baby should not be thrown out with the current dirty bathwater. More to the point – tropical deforestation should not be abandoned as an issue, as it certainly is a legitimate one for many reasons.

While many are at odds with the means that REDD proponents have set forth to accomplish its objectives, many also remain favorable on *the intent* of the original concept – reducing deforestation and forest degradation is important to the international community as well as to local communities. The problem is figuring out how to do it in a timely manner so that processes, results, and impacts are sustainable, and that conflict is not created in the process.

Reports of shootings, jailings, and forceful evictions associated with REDD programming have come from the DRC, Indonesia, and countries in central America. There is considerable scope for demanding accountability of the planning processes, standards, and prevailing operating procedures employed, particularly where taxpayer funds have been used to promote standards and best practice.

Realistically though, is there the clarity of vision and political will to do what it takes to get REDD right? To move constructively in this direction, will those now driving programming reconsider what it will take to make REDD work? Will governments negotiate security-of-tenure agreements that provide incentives for communities to invest and not feel instrumentalized as the *campesinos* in the Chiapas appear to do, based on their widely publicized resistance to GCF-supported activities?[5]

Before REDD is taken to scale, several questions merit answering: will there be open negotiation between states and claimants to indigeneity in regions targeted for REDD programming, who have explicitly not provided FPIC? Can penny-pinching on expenditures for stakeholder capacity building be reversed so that genuine FPIC can be achieved? Can metrics be established that show that AD can become attractive and feasible for local stakeholders with most leverage over deforestation, when sufficient capacity building has been provided? Most importantly, can a forum and seating arrangement be established to enable key stakeholders to AD to bargain out a governance framework that will be a cornerstone for a new social contract that realistically approaches REDD's many challenges?

At present, trends in REDD design and implementation do not provide reasoned grounds for optimism. Even among constructive critics who see the allure of a REDD program gotten right, there is arguably too much reliance on moral imperatives and overly general recommendations. This approach will not lead to stronger plans for postioning REDD to work. On an operational level, the current REDD Readiness Phase does not appear to be achieving what is needed to methodologically inform REDD, beyond in a nominal and normative sense. To work, nuanced understandings of deforestation drivers and feasible solutions are needed. These transcend normative ideals and nominal acknowledgement of the issues. From there, negotiated outcomes are unavoidable if REDD is to work.

Even while an NGO critic like Global Witness[6] is correct in arguing for "well designed" programs and an emphasis on building governance capacity, the lack of specificity as to *the parameters* for what constitutes "well designed" REDD projects perpetuates ambiguity in its own right. Specifically: what degree of capacity building, targeting who exactly, is needed? What level of civil society participation in fighting and overseeing corruption is needed, and what actions would this realistically comprise? With too much leeway currently for broad interpretation and compliance with standards for project developers, even REDD critics are not providing specifics for exactly what needs correcting and how to go about doing it, to improve probability of success.[7]

Contrary to the contorted, technocratic project that REDD has become, its ultimate efficacy will be determined by the social relationships that various groups of people succed in establishing. This includes local communities and IPs local and national government agencies, international donor agencies, national and international investors, local and international NGOs, standard setters and auditors of carbon. For at the end of the day, conservation is a social and political process (Brechin et al. 2002), and AD will require a socially collaborative and politically coherent process to succeed. In 2013, we still have miles to go before we can sleep soundly with REDD.

If the REDD narrative continues to be dominated by the convergent interests of (a) big international conservation organizations and (b) shareholder corporations seeking to capture major funding flows from (c) CSR-sensitive investment banks looking to speculate on a major new class of assets coming into new markets, who in turn require the imprimatur of (d) national governments anticipating rents from on investments made within their national jurisdictions, and (e) concessionary bilateral funders along with the framers of international negotiations,the facilitators of best practice, and convenors of national processes ,then REDD will be a colossal failure. The informed voice of REDD-impacted peoples needs to be integrated into decision making too. As noted throughout this book, many social justice NGOs, activists, and researchers have predicted REDD's failure. With emphasis on the roll-out of the so-called green economy in the Rio+20 conference on sustainable development, in which REDD foreshadows PES' markets for diverse natural capital assets including biodiversity and water, tying profit through derivatives instruments (see Friends of the Earth 2009) to common pool resources is, as The Munden Project (2011) noted, "manifestly dysfunctional".

As I have argued throughout this book, there still is time to right the ship and enable REDD to be far more than what it currently is shaped up to be. For this, a combination of objective thinking coupled to strong political will is needed to move beyond conventional wisdom. People-driven solutions that may once have been deemed radical need to be tested, evaluated, adapted, and reformulated again. Regretably more piloting is inevitable in REDD, to fill remaining gaps. Ignoring them will not make them go away..

I began the book by suggesting that local forest peoples with customary rights to forest resources who are being offered REDD incentives in 2013 are, for all intents and purposes, disempowered and unaware of REDD costs and potential benefits. I hypothesized that for REDD to work, they must become aware of the full cost, benefit, and risk implications of REDD as a first step in feasibility and sustainability, satisfying FPIC requirements in turn. This is but one practical justification for establishing a new social contract between REDD stakeholders that can frame the establishment of plans and environmental governance arrangements at national and sub-national levels.

Putting local communities first in driving alliances and partnerships where international finance and MRV stakeholders, big NGO or otherwise, collaborate in technical partnership with communities would not seem to be a radical proposition. Considering the predominance of bottom-up rhetoric in development over the past thirty years, this should already be the norm. Yet, operationalizing the rhetoric has, as discussed, proven to be disappointing overall. If there ever was an arena where the justification for transformational change in approach, for putting people first, had a compelling logic, REDD is clearly it.

The days when big conservation NGOs in collaboration with governments, and backed by strong international foundation and donor support, drives REDD as it has biodiversity conservation must change if deforestation is to be mitigated. Nor can shareholder companies or investment bank-supported private sector interests able to cut deals directly with agencies in nation states where transparency and accountability is weak, be relied upon to solve deforestation. Another institutional model for project design and implementation putting people first is needed in practice. Getting past the rhetoric that REDD is learning-by-doing, and is on the road to succeeding, when all know full well that it is not, is an essential first step.

Is this unlikely to occur? Most probably it is unlikely.

Is it possible to achieve? Perhaps it could be, though major shifts in stakeholder attitudes and behaviors are required.

It boils down to one question: Is solving deforestation and its contribution to climate change the issue for those decision makers and thought-leaders, or is it about maintaining power?

If it's about avoiding deforestation, there is still hope for REDD to be redeemed.

Notes

Introduction

1 See IEG (2012) for a reasonably frank internal review of the The World Bank's REDD programming.
2 These refer to biodiversity conservation and development NGOs.
3 See http://sonencapital.com/news-events.php for these illustrative figures.
4 See Paula Goldman's Stanford Social Innovation Review blog at http://www.ssireview.org/blog/entry/the_distortion_risk_in_impact_investing.

1 Grounds for pessimism and optimism

1 See http://cpi.transparency.org/cpi2011/.
2 Lauriston Sharpe's 1952 classic "Steel axes for stone age Australians" appropriately frames the type of situation we are witnessing at the interface between those offering payment schemes over the coming decades for sustained avoidance of deforestation, and those potentially accepting those schemes. As was the case with steel axes, the implications for poor, remote peoples from a single new input into their production and resource management systems, can have a cascade of unforeseen impacts, not always for the better.
3 *The National*, August 16, 2012. http://www.tokstret.com.
4 See Ugo Ribet's July 24, 2012 posting "On the need for REDD+ safeguards as carbon credit dealers move into forests and threaten indigenous rights" (http://www.blog.clientearth.org/on-the-need-for-redd-safeguards-as-carbon-credit-dealers-move-into-forests-and-threaten-indigenous-rights/).
5 See Winn (2012) for an exposé on how millions of hectares of customary forest lands have been "stolen" from clans in the past ten years in Papua New Guinea.
6 A parody that sums up the core argument of opponents at the 2011 UNFCCC Durban COP 17 can be found at http://climate-connections.org/2011/12/06/durban-world-business-summit-corporate-clowns-press-statement-video/. See Patrick Bond on the politics of climate in the Durban COP 17 – "green fascism for a green capitalist economy..." at http://www.redd-monitor.org/2011/12/11/redd-texts-from-the-conference-of-polluters-durban-cop-17-11-december-2011/?utm_source=feedburner&utm_medium=email&utm_campaign=Feed%3A+Redd-monitor+%28REDD-Monitor%29.
7 See http://theamericanscene.com/2010/04/26/re-reply-to-jim-manzi. A similar debate of costs and benefits in the REDD component has not emerged.
8 "BINGO" has become entrenched in the conservation and development lexicon to refer mainly to big "mainstream conservation" (Brockington et al. 2008) organizations. See Chapin (2004) and Dowie (2009). R. Schutt (2009) broadens this to include all big

international NGOs. While originally perceived by some as pejorative, I believe it has largely become a default term for big international NGOs, particularly those working in conservation.

9   See http://bclc.uschamber.com/blog/2011-09-29/share-value-or-shared-values.
10 Stakeholder engagement provisions for the UNFCCC's Technology Executive Committee include the following: "Engaging stakeholders through other models that the TEC may consider establishing, such as consultative groups, stakeholder forums and technical task forces." See UNFCCC (2011b). If it chose to, the TEC has the means to be as innovative as it chooses to be in environmental governance of REDD.
11 For Miliband's position on personal carbon trading, See http://www.publications. parliament.uk/pa/cm200708/cmselect/cmenvaud/565/56505.htm.
12 See Laurance et al. 2012.
13 See Nielsen 2011.
14 Quoted in Agence France-Presse,2011.
15 For example, see Arturo Escobar (2001), William Easterly (2006), Paul Collier (2007), Dambisa Moyo (2009), and Dan Brockington and R. Duffy (2010) for a selection of critiques of both development and conservation.
16 Brockington et al. (2008) argue for analysis of winners and losers to be front and center in analyses of important topics, of which REDD is arguably one.
17 For a sampling of some of the best of this critique see Karsenty (2009, 2010, 2011), Lang (2010), Global Witness (2011), FERN (2012), and Böhm et al. (2012).
18 See http://news.blogs.cnn.com/2012/01/13/report-california-slips-to-worlds-9th-largest-economy/.
19 Available at: http://www.gcftaskforce.org/documents/GCF_annual_meeting2012_agenda.pdf.

2   Theses and theory of change

1   Reducing emissions from deforestation and forest degradation (REDD) was first adopted at the 13th Session of the Conference of the Parties (COP 13) to the United Nations Framework Convention on Climate Change (UNFCCC) as a global mechanism to mitigate climate change. REDD+ refers to the international mechanism which was agreed to at the UN climate change conference in Cancun, Mexico, in December 2010, the objective of which is to reduce emissions from deforestation and forest degradation, and address the role of conservation, sustainable management of forests, and enhancement of forest carbon stocks. REDD+ thus embraces a broader spectrum of activities from REDD alone. See UN documents: Decision 2/CP.13, "Reducing emissions from deforestation in developing countries: approaches to stimulate action". Also see UNFCCC (2008) and (2009).
2   Bloomberg New Energy Finance July 26, 2012.
3   Sustainable development is a term that is even more complex to define than "forest" or "participation". For example, Columbia University's Masters Degree program in Public Administration and Development Practice notes the following: "Sustainable Development is broadly defined as 'meeting the needs of the present without compromising the ability of future generations to meet their own needs'" (Brundlandt Commission 1987). The program description goes on to note further that "Nearly half of the world's population lives on less than two dollars a day, resulting in unnecessary human suffering caused by debilitating hunger, poor health, environmental degradation and lack of basic infrastructure." See http://www.sipa.columbia.edu/academics/degree_programs/mpa-dp/. Arguably, Columbia probably means that basic needs are not being met reliably, a far cry from what the Brundtland Commission suggests, which likely includes education, health, and other forms of social welfare perceived as "needed".

4 Big, mainstream conservation NGOs are, again euphemistically, known as "BINGOs". Through repeated use over the past ten to fifteen years, it has come to be a catch-all label to probably refer to a very small group of NGOs, with their main base of operation in the US, who have enjoyed inordinate leverage over the programming and ultimate receipt of over US$4 billion in annual conservation funding. Indicative of the strength of the industry, among the leaders who are at the forefront of REDD, are groups like TNC which reported US$5.1 billion in total assets and US$997 million in total revenues in its US Internal Revenue Service (IRS) 990 form for 2011. Another leader, CI, reported US$248.7 million in total assets in 2010 and US$140.7 million in total revenues in 2010. WWF reported US$400.5 million in total assets and US$182 million in total revenues in 2010 for the US affiliate alone. Others like the WCS reported US$792.8 million in total assets and US$206.1 million in total revenue, and the World Resources Institute reported total assets of US$65 million and total revenues of US$50 million in 2010. In contrast, small US NGOs with the same 501(c)3 IRS status as larger known NGOs, but who are also involved in REDD conservation concession and carbon marketing, are more modest. For example, Bonobo Conservation Initiative (BCI) reported US$929,000 in total revenue in 2010, with total assets of US$238,532 (See http://www.guidestar.org/PartnerReport.aspx?partner=networkforgood&source=donate_nfg&ein=52-2146443, downloaded September 22, 2012). The disparity between big and small conservation NGOs working in REDD is thus tremendous.

5 A carbon credit is a certificate or permit which represents the right to emit one tonne of CO_2, and they can be traded for money. There are two categories of carbon credits: voluntary emission reductions (VERs) and certified emission reductions (CERs).

6 Barbara Ehrenreich (2009) has written on the phenomenon of bright-sidedness, and how it has become a debilitating cultural trait in America.

7 Frankfurt (2005) and Macintosh (2003, 2006) have taken the term "bullshit" into different academic disciplines for the sake of adding to rather than detracting from clarity. As defined by the Oxford English Dictionary, bullshit is 1: Nonsense, rubbish (noun). Definition 2: Trivial or insincere talk or writing (verb, *bullshitted, bullshitting*).

8 http://www.ucsusa.org/global_warming/solutions/forest_solutions/   (downloaded October 13, 2012).

9 One thinks of UN-REDD, based as it is in New York, and the big conservation NGOs supporting REDD including WWF/US, CI, WCS, TNC, and African Wildlife Fund among others, who are all US-based organizations.

10 This personal communication from a well placed senior manager at one of the BINGOs will remain anonymous, as it was sent as a private email and may well not have been expressed otherwise.

11 The fact that Kareiva et al.'s (2012) argument inevitably leads them to argue for closer direct collaboration with big corporations is not being examined here. What is at issue is their statement that "protecting biodiversity for its own sake has not worked". Why then would the global community employ TMAs from biodiversity conservation in REDD, *even if* economic incentives are being introduced into the equation?

12 The USAID-funded PVO-NGO/NRMS project organized this workshop. The author coordinated and facilitated the event.

13 Barack Obama quoted in *New York Times*, March 22, 2013.

14 See http://pdf.usaid.gov/pdf_docs/PDACL817.pdf;   http://rmportal.net/library/content/nric/1718.pdf; (Brown et al. 2008).

3 REDD's path to date

1 Anthropogenic climate change (ACC) and anthropogenic global warming (AGW) are used interchangeably, with usage varying by author and citation.

2 In mid-2012.

3 Near daily critiques have been posted over the past several years on blog sites such as www.REDD-Monitor.org, www.climate-justice-now.org, globaljusticeecology.org, and others.

4 For full, recent display of what for lack of a more accurate expression has become an exhibition of white collar, professorial, vitriolic "mud wrestling" between climate scientists and climate change deniers, see the Editorial Page of the *Wall Street Journal*, January 27, 2012, "No need to panic about global warming: There's no compelling scientific argument for drastic action to 'decarbonize' the world's economy", available at: http://online.wsj.com/article/SB10001424052970204301404577171531838421366.html?mod=WSJ_Opinion_LEADTop, and Peter Gleick's retort, "Remarkable editorial bias on climate science at the *Wall Street Journal*," available at:http://www.forbes.com/sites/petergleick/2012/01/27/remarkable-editorial-bias-on-climate-science-at-the-wall-street-journal/.

5 See http://www.nma.org/pdf/c_production_state_rank.pdf.

6 Quoted in *The West Australian*, "Consumption driving unprecedented damage: UN", June 7, 2012. Available at http://au.news.yahoo.com/thewest/a/-/world/13889171/consumption-driving-unprecedented-environment-damage-un/.

7 Adapted from http://bassinducongo.reddspot.org/web/en/47-introduction-au-mecanisme-redd.php.

8 Copied from Sukhdev et al. 2012, "REDD+ and a Green Economy", UN-REDD Programme Policy Brief. The green economy serves as one of two principal themes at the 2012 Rio+20 United Nations Conference on Sustainable Development (See http://www.uncsd2012.org/rio20/about.html).

9 A blog by the ETC Group, "Rio+20 or Silent Spring −50?" presents the viewpoint of those disappointed by the Rio+20 process, who are also opposed to the green economy. See http://climate-connections.org/2012/06/24/rio20-or-silent-spring-50/.

10 Questions and Answers on the Commission's proposal to revise the EU Emissions Trading System, MEMO/08/35, Brussels, January 23, 2008.

11 REDD activists note that any offset programs are premised on perpetuating business as usual (BAU) emissions' practices, thus avoiding the root of the problem. Where emissions directly impact communities, as in Richmond, California, it is also argued that some developed country peoples directly bear health-related externalities from perpetuation of BAU emissions practices. See http://grassrootsclimatesolutions.net/node/777. Dr Henry Clark, of the West County Toxics Coalition of Richmond California, states in a video called *A Darker Shade of Green* (Global Forest Coalition and the Global Justice Ecology Project 2012) that "from an environmental justice community perspective, we want polluting companies like Chevron here and others in our community to not to produce the pollution in the first place and reduce it." See http://globalforestcoalition.org or http://www.youtube.com/watch?v=FPFPUhsWMaQ&feature=relmfu.

12 The National Research Council (2004) provides a reference on valuation of ecosystem services, particularly regarding methodologies and mitigating for uncertainty and the thorny issue of "framing" which influences methodologies employed and results generated.

13 "The precautionary principle or precautionary approach states that if an action or policy has a suspected risk of causing harm to the public or to the environment, in the absence of scientific consensus that the action or policy is harmful, the burden of proof that it is *not* harmful falls on those taking the action. The principle implies that there is a social responsibility to protect the public from exposure to harm, when scientific investigation has found a plausible risk. These protections can be relaxed only if further scientific findings emerge that provide sound evidence that no harm will result." Quoted in Wikipedia, available at: http://en.wikipedia.org/wiki/Precautionary_principle.

14 Hansen (2012) suggests that the 2°C rise is at this point overly optimistic, and the probability for both greater temperature rises and thus economic costs is higher.

15 A "dispassionate" argument that the costs of mitigating AGW outweigh the economic benefits based on the IPCC's science and cost estimates, has been most clearly argued by Jim Manzi. According to the IPCC's own estimates, Manzi points out that a temperature rise of 4°C will reduce global GDP by about 3 percent in 2100. Conversely, the IPCC estimates the cost of maintaining carbon concentrations in the atmosphere below a "safe" level of 450 parts per million at around 6 percent of GDP. Manzi concludes that mitigation probably isn't worth it based on the math. See http:// theamericanscene.com/2010/04/26/re-reply-to-jim-manzi. A similar debate of costs and benefits in the REDD component has not emerged. Most debate has focused on the plausibility, or not, of markets or moral imperatives in REDD. Cost–benefit analysis per se has not been a central feature of the REDD debate to date, as the arguments have been sweeping, and broad brush stroke in nature.

16 The IPCC is the leading international body for the assessment of climate change. It reviews and assesses the most recent scientific, technical, and socioeconomic information produced worldwide relevant to the understanding of climate change. See www.ipcc.ch.

17 The estimates for deforestation are highly variable, with a range of 12–25 percent; 17 percent appears anecdotally to be most frequently cited.

18 There have been lessons learned from conservation that are applicable to REDD. Their value, however, has been overstated to justify the term "best practice". Learning relates as much or more to what *to avoid* doing in program and project design.

19 The comparative investment in environmental and social impact aspects of projects appears minor in comparison with MRV if the Central African Republic (CAR) experience is any indication. The CAR has budgeted US$43,000 for work on environmental and social impacts in its REDD Readiness Preparation Proposal (R-PP) which is approximately 0.6 percent of a total budget of US$6.7 million in comparison to approximately 20 percent budgeted for MRV of carbon (The Rainforest Foundation UK 2012).

20 The information on the FCPF was downloaded on October 16, 2012, from the World Bank's FCPF website: http://www.climatefinanceoptions.org/cfo/node/57.

21 http://www.redd-monitor.org/wordpress/wp-content/uploads/2008/11/papua_ new_guinea_tap_consolidated_pin_review.pdf provides a critique submitted to the FCPF for the PNG R-PIN.

22 See PwC 2012 for recommendations made to the World Bank Program on Forests project.

23 Public domain data are scarce on what conservation is actually spent on. While Balmford et al. (2002) claim high benefits from conservation funding in developing countries, there are few specifics as to how monies are spent, or the sustainability of results and impacts. This leaves interpretation open to anecdote and conjecture on conservation investment based on reading of the literature and extrapolation from site visits on the ground. Chapin's (2004) critical review of the evolution of conservation is as timely in 2013 as it was in 2004. His recommendation that "public discussion that can lead towards the creation of conservation programs that are responsive to the needs of both biological and human diversity worldwide" be pursued remains equally valid as well (for the questionnaire, See http://www.conservationmeasures.org/wp-content/ uploads/2010/05/Performance-Measurement-Survey-Foundations.pdf).

24 See Tan et al. (2010) for an example of a good capacity-building tool for promoting FPIC.

25 While ICDPs have largely been categorized as failures and disavowed in favor of landscape conservation approaches favoring policy levers, they arguably were never properly designed and implemented. The basic imperative for ICDPs – development as a rationale or compensation for local conservation – have yet to be successfully addressed in biodiversity conservation programs. The same issues will plague REDD.

26 "Me" refers to Michael I. Brown.

4 What do Pygmies circa Mobutu's Zaire have to do with REDD?

1 International development in its early years was premised on the success of the Marshall Plan. Its focus was on improving the economic and social welfare of countries and poor people. Work covering its shortcomings over the years include Easterly (2006), Collier (2007), and Moyo (2009).
2 The World Bank (2009) notes that the DRC does not concede a special status to Pygmies, that they are Congolese citizens with the same rights as other nationals. In practice, however, Pygmies "cannot access these rights due to discrimination and marginalization", and furthermore, Pygmies "lack secure rights to their ancestral lands, as their rights are not recognised in statutory or customary law" and often lack identity cards to support claims to citizenship and basic entitlements (see IRIN 2006).

5 Science and policy

1 Parts per million.
2 Anthropogenic warming of the climate system is widespread and can be detected in temperature observations taken at the surface, in the free atmosphere, and in the oceans.
3 Bello and Heydarian (2012) provide critical perspective on Doha, COP 18.
4 See http://bassinducongo.reddspot.org/web/en/47-introduction-au-mecanisme-redd.php along with http://bassinducongo.reddspot.org/web/en/48-de-montreal-a-durban.php for the French Development Agency's view on REDD's evolution.
5 Cited by Worldwatch Institute, September 27, 2012, available at: http://www.worldwatch.org/node/6318.
6 Lenton (2011) notes that crucial climate events are ultimately driven by changes in energy fluxes. The one metric that unites them, radiative forcing, has largely been absent from most discussions of dangerous climate change.
7 Annex 1 countries are those with binding GHG ER targets.
8 What this means is that government has ownership rights over land and carbon while communities have "*rights*" to manage. Cynics may argue this does not sound like a good deal. Why would local people agree to assume management responsibility in the absence of any authority over revenues, or negotiation over revenue allocations?
9 Validation and Verification Approved - CCB Standards Second Edition Gold Level (Dec 6, 2012) See http://www.climate-standards.org/2012/09/03/mai-ndombe-redd-project/.
10 On October 6, 2012, the author received an email from a DRC agroforester working at the time for a major international NGO in the DRC entitled: "Subject: Tr : Fwd: incidents sur site d'ERA (Bandundu/Basengele/Bongo)". The email sender had forwarded correspondence from a thread of messages from key actors in the DRC REDD network on the incident. S/he noted this was "extremely confidential". The source of the email forwarding is anonymous here. The situation was further confirmed in an email on April 7, 2013, with indications that the community relations in Basengele and Bongo remain problematic in terms of a range of what may be termed "social feasibility issues". The example is presented to consider how the prevailing theory of change guiding the REDD project accounts, or does not, for situations like the one portrayed.
11 Time will tell whether this statement is valid or not. In the meantime, in the case of SIA, it will continue to be marketed as "good enough" to meet REDD design and implementation needs. This will enable REDD programs to be rolled out at scales which, in due course, may prove to be insufficient in the manner approached.
12 http://web.worldbank.org/WBSITE/EXTERNAL/TOPICS/ENVIRONMENT/EXTBIODIVERSITY/0,,contentMDK:23264854~pagePK:210058~piPK:210062~theSitePK:400953,00.html.

13 The Kyoto Protocol, negotiated in 1997, is the main treaty to date designed to reduce GHG emissions. Neither REDD nor REDD+ fall under Kyoto. Newell and Paterson (2010) provide a history of the evolution of Kyoto, from the role of the US and EU, to private market actors and banks, along with the key role that emergent stakeholder associations such as the International Emissions Trading Association, the Emissions Marketing Association, and Climate Markets and Investors Association played in holding politicians' feet to the fire on emissions trading after the US withdrawal from Kyoto in 2001.

14 Copied on September 28, 2012, from http://www.climate-standards.org/standards/index.html.

15 See https://s3.amazonaws.com/CCBA/Projects/April_Salumei_Sustainable_Forest_Management_Project/April+Salumei+PDD+Validated+-+June+2011.pdf.

16 One critique of the scheme is available at http://forestryanddevelopment.com/site/2010/07/12/redd-rears-its-head/. "The PNG Government has publicly distanced itself from a voluntary carbon trading scheme being developed for East Sepik Province in PNG. The Rainforest Management Alliance (RMA) is proposing to develop a carbon trading scheme for the April Salumei FMA in accordance with standards set up by the Climate Community and Biodiversity Alliance. According to a public notice published by the RMA, one of the 'project partners' involved in the scheme is PNG Government's Office of Climate Change and Development (OCCD). However the PNG Government has been keen to deny any such involvement. Dr Wari Iamo, Executive Director of the OCCD, published an emphatic statement refuting the Office had offered any such support to the project. A similar carbon trading project has been developed for the Kamula Doso FMA by Nupan Trading Corporation. Nupan is run by former disqualified Australian horse trainer and Philippine cock-fighting syndicate operator, Kirk Roberts. Roberts has been accused by Papua New Guinea's former Forest Minister Belden Namah of running a carbon 'cargo cult' in the country. The PNG Government was also put under close media scrutiny late last year after inconsistencies in the development of its own carbon conservation projects. New carbon scandals would be unwelcome for the government, and more hollow promises of 'sky money' would be unwelcome in forest communities. Both these schemes make grand commitments to conserve forests, whilst delivering economic and social outcomes, yet there have been no REDD schemes in Papua New Guinea that have delivered clear economic benefits to local communities."

17 See Matt Leggett (2010), Payments for Environmental Services/REDD Coordinator, WWF Western Melanesia Programme's "Comments on the 'April-Salomei Sustainable Forest Management Project' proposal submission for validation under the Climate, Community and Biodiversity Alliance standards", where he expresses "some concerns about the quality of the design of the April-Salomei [sic] project that has recently been submitted for validation under the Climate, Community and Biodiversity Alliance (CCBA) standards", clarifying that he believes "in certain key areas, the current project design fails to meet the CCBA standards, and that some critical areas have been left unaddressed." He states that "WWF WMPO is concerned that if this project design is validated and ultimately receives approval, the communities within the project region may not benefit from the project equitably. WWF WMPO is therefore concerned that this may endanger both the perception of the CCBA standards internationally, and the future development of voluntary carbon projects in PNG." See https://s3.amazonaws.com/CCBA/Projects/April_Salumei_Sustainable_Forest_Management_Project/April_Salumei_Comments_Final.pdf.

18 "Papua New Guinea and Australia: Near neighbors, worlds apart". *The Economist*, August 8, 2011.

19 See https://s3.amazonaws.com/CCBA/Projects/Rimba_Raya_Project/CCB_InfiniteEarth_ValidationReport_Final_101811.pdf.

20 See: https://s3.amazonaws.com/CCBA/Projects/Rimba_Raya_Project/CCBA_PDD_2011_05.15_Final percent5B1 percent5D.pdf.

21 Nested REDD programming refers to the accounting of emissions reductions by project-level activities which is then reconciled with "jurisdictional" (regional or national level) accounting of deforestation ERs. Practically speaking, the US State of California, for example, has adopted regulations for its cap and trade program that allow for the acceptance of jurisdictional level, so-called "sector-based offset credits" from reducing deforestation emissions in other countries. The California regulation notes explicitly that a nested approach to sector-based REDD that includes project-level activities can potentially be reconciled against jurisdictional accounts.

22 Hajek et al. 2011 present results from Andean Peru that point to the importance of hybrid institutional logics, and the key role played by highly networked individuals in pushing project-level REDD forward. Understanding the construction of the REDD+ credit value chain as the fundamental innovation taking place, the development of standards, technologies, and other norms are complementary to the basic task of defining and reconfiguring roles on this chain. Arguably, if decision makers can encourage the "bottom-up" construction of REDD as a strategy to encourage innovation and flexibility, REDD will have a greater chance of succeeding.

23 Sikor and Stahl (2011) present a strong argument for attention being paid to forest peoples' rights in nested REDD+ governance systems. They fail, however, to go beyond a normative framework to discuss what will be required in practice for the nested governance systems to work. The impression that the limiting factor is simply participation – a seat at "the table" at different REDD+ governance scales – is, I believe, misleading and ultimately counterproductive. While necessary, if those seated at the "nested" tables lack the analytical skills to participate effectively and contribute to adequate decision making, their participation is likely to reflect co-optation versus "genuine" participation. Genuine participation can only be achieved if skills are such to enable analysis and optimal decision making.

24 Anderson (2012) provides convincing logic for the long-term illogic of carbon offsets as a solution for climate change mitigation, while Bachram (2004) introduces an early analysis of how offset trading is "tenable or desirable" from an environmental justice standpoint.

6   Stakeholders and REDD

1 These NGOs are also termed environmental NGOs, or ENGOs, and more recently, "mega-environmental NGOs", and include CI, TNC, WWF, WCS, and the African Wildlife Foundation (see Sullivan 2010).

2 The G8 unambiguously stated at its 2008 gathering in L'Aquila, Italy: "We remain engaged in seeking the reduction of emissions from deforestation and forest degradation and in further promoting sustainable forest management globally... We will cooperate to identify innovative instruments in this respect, including through initiatives such as [the] UN Programme on Reducing Emissions from Deforestation and Forest Degradation [UN-REDD Programme], [the] Forest Carbon Partnership Facility (FCPF) and the Informal Working Group on Interim Finance for Reducing Emissions from Deforestation and Forest Degradation (IWG-IFR)." See http://www.un-redd.org/NewsCentre/Newsletterhome/1News1/tabid/1592/language/en-US/Default.aspx.

3 http://www.un-redd.org/AboutUNREDDProgramme/tabid/583/Default.aspx.

4 An excellent guide to the evolution of REDD negotiations at the UNFCCC level, particularly as regards whether the agreement will be legally binding on parties to the UNFCCC or not, from 2013 onward is the Foundation for International Environmental Law and Development (FIELD) "Guide for REDD-plus negotiators: February 2013". FIELD. London. Available at: http://www.field.org.uk/files/field_redd-plus_guide_feb_2013_en.pdf.

5 Contributions and grants as reported on US IRS Form 990 for EDF fell from $126,116,250 in 2008 to $52,480,737 in 2009, a possible sign of the impact of the US

recession on a New York based not-for-profit organization. The figure has however jumped back to $94,076,678 in 2010 and $110,051,773 in 2011.

6 Cited in http://v-c-s.org/news-events/news/vcs-ccb-join-forces-streamline-project-approval-and-credit-issuance.

7 While ethnic groups may have been homogeneous and "remote enough" during the colonial era to have warranted studies portraying them in a static manner, suggesting "authenticity" and cultural endurance, the late twentieth century and early twenty-first century has wrought enormous change in all respects. Remoteness for most IPs is far from a given, leading to novel forms of contact between ethnic groups and a variety of other stakeholders who often are interested in the natural resources found on lands occupied under what is referred to as customary law.

8 Oxford University's Professor Douglas Gollin (2012) notes that even in Ghana with its remarkable urban center growth, "rural areas remain poor, and the economic changes there have been far less visible and dramatic than the changes in urban areas", leading him to an obvious question: "why has change come so slowly in rural areas"?

9 This figure is a tally of Freudenberger's (2010) Madagascar Environmental Interventions Time Line Table.

10 SRCs (*cahiers des charge* in francophone Africa) are widespread, and are the source of years of lingering conflict between dissatisfied communities, the private sector, and governments with whom logging companies have signed 50–100-year concession agreements. REDD is inserting itself into contexts where lingering conflict over unfulfilled SRCs predominates and discolors any new forest sector initiatives (Brown 2010a).

11 See "REDD+ and Indigenous Peoples". IWGIA. 1–2/09, at http://www.iwgia.org/graphics/Synkron-Library/Documents/publications/Downloadpublications/IndigenousAffairs/IA percent201-2_09/IA percent201-2009.pdf

12 See http://climate-connections.org/2012/08/31/coonapip-panamas-indigenous-peoples-coordinating-body-denounces-un-redd/.

13 See http://www.fenamad.org.pe/noticias2.htm for a detailed account and analysis of Peruvian IPs' communities' objections to all aspects of the REDD planning process.

14 See "We have to fight for truth", redd-monitor.org, November 2, 2012.

15 See http://www.survivalinternational.org/news/8710

16 See http://www.ohchr.org/en/hrbodies/upr/pages/uprmain.aspx.

17 See http://www.wri.org/press/2012/09/release-indonesian-president-yudhoyono-honored-valuing-nature-award-nyc.

7 Social feasibility and its components

1 See http://www.fastcompany.com/1596661/aveda-and-yawanawa-csr-chief-chief.

2 Matt Somerville (2011) states: "Investment in REDD+ performance in situations where tenure is unclear is a waste of resources. Therefore, further investment in pilot projects or readiness should be contingent on the development and implementation of sound land policies and progress in achieving broader tenure security for affected populations." This builds on an earlier IIED paper by Lorenzo Cotula and James Mayers (2009) where the point was made: "It appears evident that many countries are ill-equipped in practice to ensure that REDD schemes benefit local people. Improvements in tenure alone will not achieve this."

3 See www.satyadi.com.

4 See Richards and Panfil (2011; Richards 2011).

5 Hochschild (1998) cites Charles Gréban de St. Germain, a colonial magistrate in the Belgian Congo in 1905, as attributing porterage, in addition to sleeping sickness and malnutrition, to the woes of the country, stating "I've seen nowhere in the Congo as sad a spectacle as that along the road from Kasongo to Kabambare [in the Mai Ndombe], with few people, sick women and children and little food." The situation was not materially

much different in 2002–2007 when I worked in Bandundu province, and arguably is part of the backdrop for any REDD projects hoping to succeed in the Mai Ndombe today.

6 Social capital has no consensus definition. It is used here to refer to an attribute of groups sharing common interests, which facilitates both individual and collective action, is generated by networks of relationships, reciprocity, and social norms. Portes (1998) provides a good overview.

7 Human capital is "an aggregate economic view of the human being acting within economies, which is an attempt to capture the social, biological, cultural and psychological complexity as they interact in explicit and/or economic transactions" (Wikipedia, downloaded October 2, 2012).

8 Social contracts are the product of citizens' assent to honor the rights of others in return for assurances that their own rights will be protected. When these cease to function, this leads to conflict (White 2007).

9 Financing issues

1 See http://climate-connections.org/2012/05/18/european-parliament-absent-in-sustainability-summit/ for an opinion piece on why carbon markets are not a solution to tackling the climate crisis or moving towards a low-carbon, sustainable economy.

2 See http://unfccc.int/resource/docs/2011/cop17/eng/09a01.pdf.

3 See http://foreignassistance.gov/Initiative_GCC_2012.aspx?FY=2012.

10 Risks related to REDD

1 See http://rmportal.net/news/usaid-rm-portal-events/social-dimensions-of-redd-current-practice-and-challenges-open-forum.

2 In 1988 I presented a paper at the American Anthropological Association meetings on why development anthropologists need to assume roles as project coordinators, managers, etc. if they are to have an impact and not have their input systematically marginalized. The same point holds true twenty-five years later.

3 An objective assessment of social risk would include quantitative measures and some probabilistic projections. While this type of projecting is attempted for the MRV elements central to verification and validation, the SIA tools employed by auditors and standard setters demand far less precision.

4 See Chris Lang, June 12, 2012, available at: http://www.redd-monitor.org/2012/06/12/the-story-of-redd-is-just-a-lie-says-ulu-masen-villager/?utm_source=feedburner&utm_medium=email&utm_campaign=Feed%3A+Redd-monitor+%28REDD-Monitor%29.

5 Commonwealth Of Australia, Proof Committee Hansard, Senate, Foreign Affairs, Defence And Trade Legislation Committee Estimates (Public), Thursday, 31 May 2012, Canberra, By Authority Of The Senate. Available at: http://www.redd-monitor.org/wordpress/wp-content/uploads/2012/06/Foreign-Affairs-Defence-and-Trade-Legislation-Committee_2012_05_31_1071.pdf.

6 See Petermann interview in http://www.democracynow.org/2010/12/9/is_redd_the_new_green_injustices for the video interview: "Is REDD the New Green?", December 9, 2010.

7 See Soros' interview in http://www.democracynow.org/2010/12/9/is_redd_the_new_green_injustices for the video interview: "Is REDD the New Green?", December 9, 2010.

8 See http://www.uncsd2012.org/rio20/about.html for an overview of the 2012 UN Conference on Sustainable Development, or Rio+20, in which the green economy is highlighted as one of two principal themes.

9 These are all PAs I have personally visited under professional circumstances over the past twenty years.
10 See www.redd-monitor.com.
11 See http://www.terraglobalcapital.com/press/Environmental%20Finance%20Awards %202012.pdf (downloaded October 6, 2012).
12 The assumption is that this document accurately reflects on what OPIC ultimately provided "political insurance" for.
13 No date is provided in the report, though it appears to have been drafted in 2009.

11 A new social contract for moving forward

1 The author together with Zephirin Mogba and Jean Martial Bonis-Charancle took the lead, and were assisted at the time by colleagues from CIFOR/Cameroon and CIRAD including Guillaume Lescuyer, Anne Marie Tiani, and Chimère Diaw.
2 See http://www.youtube.com/watch?v=o-AWu4LkCVw for video.
3 The Union of Concerned Scientists was one of the earliest groups to come out strongly in support of the rationale and prospects for REDD. See http://www.ucsusa. org/assets/documents/clean_energy/Briefing-1-REDD-costs.pdf for an analysis of why it believes that REDD "remains a very cost effective way to address global warming".
4 See http://www.globalwitness.org/sites/default/files/library/Forest%20Carbon,%20 Cash%20and%20Crime.pdf.
5 A contentious encounter between REDD sub-national promoters and the GCF occurred in the Chiapas, Mexico, on September 26, 2012. See http://climate-connections.org/2012/10/08/anti-redd-statement-read-to-governors-climate-change-task-force-in-chiapas-mexico/. Protesters stated: "We have come before you today to denounce the programs and projects that threaten to dispossess us of our territories and our resources; programs which bad governments have attempted to impose for a long time; now they have a new pretext: climate change and the project they call REDD+."
6 See page 2, "Forest Carbon, Cash, and Crime", http://www.globalwitness.org/sites/ default/files/library/Forest%20Carbon,%20Cash%20and%20Crime.pdf.
7 For example, recommending that capacities be built "at all levels, including among civil society, within government institutions and in forest law enforcement to ensure all stakeholders can engage effectively in REDD+ design and implementation. For law enforcement agencies this includes improving international coordination, with neighbouring and regional countries as well as with timber importing countries" (Global Witness 2011b), is simply *too vague* to elicit anything more than knee jerk compliance. Should the focus be on grand corruption and national policy, something that is standard amongst agencies and has led to scant measurable results to date? Or should the focus be on local anti-corruption committees that can progressively generate demand-driven policy change and compliance, as the USAID-funded "Relance Economique" project implemented by IRM in the DRC succeeded in demonstrating as feasible from 2002–2006? I would personally argue for the latter. See http://s4rsa.wikispaces.com/file/view/Relance+Economique+Brochure.pdf for a description of IRM's Relance Economique's approach to combating administrative corruption in the DRC, or http://www.google.com/search?q=Relance+economiqu e+IRM&ie=utf-8&oe=utf-8&aq=t&rls=org.mozilla:en-US:official&client=firefox-a for a PowerPoint presentation.

References

4CMR. (2012). "Overcoming Legal Barriers to REDD+ Implementation", Project Description, University of Cambridge. Available at: http://www.4cmr.group.cam.ac.uk/land-use-projects.

Achebe, C. (1977). An Image of Africa: Racism in Conrad's *Heart of Darkness*, Massachusetts Review. 18. Rpt. in Kimbrough, R. (ed), *Heart of Darkness: An Authoritative Text, Background and Sources Criticism*. 1961. 3rd ed. W. W. Norton and Co., London, 1988: 251–261.

Activist San Diego, Amazon Watch, Asian Pacific Environmental Network, Battle Creek Alliance-Bus Riders Union, California Environmental Justice Alliance, Cascade Action Now!, Causa Justa: Just Cause, Center For Biological Diversity, Center On Race, Poverty And The Environment, Communities For A Better Environment, Filipino/American Coalition For Environmental Solidarity (Faces), Forests Forever, Friends Of Lassen Forest, Friends Of The Earth, Global Exchange, Global Justice Ecology Project, Grassroots Global Justice Alliance, Greenpeace Us, Indigenous Environmental Network, International Accountability Project, International Forum On Globalization, International Indian Treaty Council, International Rivers, Just Transition Alliance, Justice In Nigeria Now (Jinn), Movement Generation: Justice And Ecology Project, National Network For Immigrant And Refugee Rights, Pachamama Alliance, Priority Africa Network, Rainforest Action Network, Seventh Generation Fund For Indian Development, Urban Communities & Environment, Antioch University Los Angeles (2012). Re: Climate change. *Letter to Governor Jerry Brown, State of California*. Available at: http://libcloud.s3.amazonaws.com/93/ca/b/2271/Letter_to_Governor_and_ARB_re_CA_REDD_final.pdf.

Adams, J. and McShane, T. (1992). *The Myth of Wild Africa: Conservation Without Illusion*. W. W. Norton & Co., New York.

Adger, W. N. and Barnett, J. (2009). Four reasons for concern about adaptation to climate change. *Environment and Planning* A 41(12): 2800–2805.

AfricanBrains. (2012). *Madagascar: Small steps toward land reform*. Available at: http://africanbrains.net/2012/04/14/madagascar-small-steps-towards-land-reform/.

Agence France-Presse. (2011). *World deforestation rate "accelerating": U.N.*, Agence France-Presse, November 30.

Agrawal, A. and Angelsen, A. (2009). Using community forest management to achieve REDD+ goals. In A. Angelsen et al. (eds), *Realising REDD+: National Strategy and Policy Options*. CIFOR, Bogor, Indonesia.

Agrawal, A. and Redford, K. H. (2006). *Poverty, development, and biodiversity conservation: Shooting in the dark?* Working Paper No. 26. Wildlife Conservation Society, Bronx, New York.

Aguilar, Y., Erazo, M. and Soto, F. (2012). REDD-plus schemes in El Salvador: Low profile, friendly fancy dresses and commodification of ecosystems and territories. Available at http://svmrevproxy01.unfccc.int/files/methods_science/redd/submissions/application/pdf/redd_20120823_equipo.pdf.

AIDESEP. (2010). *Pronunciamiento: Sin Territorios, Derechos y Consulta Indígenas no puede haber concesiones REDD, forestales, petroleras y de servicios ambientales.* Lima, Peru: 1. Available at: http://servindi.org/actualidad/34447.

Akwah, G. (2013). Personal communication.

Alcorn, J. (1990). Indigenous agroforestry strategies meeting farmers' needs. In Anderson, A. B. (ed.), *Alternatives to Deforestation: Steps Toward Sustainable Use of the Amazon Rain Forest.* Columbia University Press, New York.

Alcorn, J. (2010). *Getting REDD Right: Best practices that protect indigenous peoples' rights and enhance livelihoods.* CIFOR. Available at: http://www.slideshare.net/mobile/CIFOR/jalcorn-reddbpip-sept2010.

Alcorn, J. (2011). USAID issue brief tenure and indigenous peoples the importance of self determination, territory, and rights to land and other natural resources property rights and resource governance briefing paper #13. Available at: http://usaidlandtenure.net/sites/default/files/USAID_Land_Tenure_2012_Washington_Course_Module_3_Land_Tenure_Indigenous_Peoples_Issue_Brief.pdf.

Alden Wily, L. (2012). The global land grab: The new enclosures. In Bollier, D. and Helfrich, H. (eds), *The Wealth of the Commons: A World Beyond Market and State.* Levellers Press, Amherst, MA. Available at: http://www.wealthofthecommons.org/essay/global-land-grab-new-enclosures.

Almeida, F., Hatcher, J., White, A., Corriveau-Bourque, A. and Hoffman, Z. (2012). *What Rights?: A Comparative Analysis of Developing Countries' National Legislation on Community and Indigenous Peoples' Forest Tenure Rights.* Rights and Resources Initiative, Washington, DC.

American Electric Power, Conservation International, Duke Energy, El Paso Corporation, Environmental Defense Fund, Marriott International, Mercy Corps, National Wildlife Federation, Natural Resources Defense Council, Pacific Gas & Electric, Rainforest Alliance, Republicans for Environmental Protection Sierra Club, Starbucks Coffee Company, The Green Belt Movement, The Nature Conservancy, The Walt Disney Company, Union of Concerned Scientists, Wildlife Conservation Society, Woods Hole Research Center. (2009). *Tropical forest-climate unity agreement: Consensus principles on international forests for U.S. climate legislation.* Facilitated by AD Partners on behalf of the endorsing groups. Available at: http://adpartners.org/pdf/ADP%20Forest-Climate%20Unity%20Agreement-%205-18-09.pdf.

Anderegg, W., Prall, J., Harold, J. and Schneider, S. (2010). Expert credibility in climate change. *PNAS*, 21 June. doi: 10.1073/pnas.1003187107.

Anderson, D. and Grove, R. (eds) (1987). *Conservation in Africa: Peoples, Policies and Practices.* Cambridge University Press, Cambridge, UK.

Anderson, K. (2012). The inconvenient truth of carbon offsets. *Nature*, 484, 7, doi:10.1038/484007a. Available at: http://www.nature.com/news/the-inconvenient-truth-of-carbon-offsets-1.10373.

Anderson, K. and Bows A. (2011). Beyond "dangerous" climate change: Emission scenarios for a new world. *Phil. Trans. R. Soc. A*, 369, 20–44, doi:10.1098/rsta.2010.0290.

Anderson, P. (2011). *Free, Prior, and Informed Consent: Principles and Approaches for Policy and Project Development.* Bangkok, RECOFTC and GIZ.

Angelsen, A. (ed.) (2008). *Moving Ahead With REDD: Issues, Options, and Implications.* CIFOR. Bogor, Indonesia.

Angelsen, A. and Atmadja S. (2008). *What is this Book About? Moving Ahead with REDD: Issues, Options and Implications.* Bogor, Indonesia, CIFOR: 156.

Angelsen, A., Brockhaus, M., Sunderlin, W. and Verchot, L. (eds.) (2012). *Analysing REDD: Challenges and Choices.* CIFOR, Bogor, Indonesia.

Angelsen, A. with Brockhaus, M., Kanninen, M., Sills, E., Sunderlin, W. D. and Wertz-Kanounnikoff, S. (eds.) (2009). *Realising REDD+: National Strategy and Policy Options*. CIFOR, Bogor, Indonesia.

Argus Media (2013). Argus European Emissions Markets 2013. Available at http://www.argusmedia.com/Events/Argus-Events/Europe/Argus-Euro-Emissions/Sponsors

Asen A., Boscolo, M., Carrillo, R., van Dijk, K., Nordheim-Larsen, C., Oystese, S., Savenije, H., Thunberg, J. and Zapata, J. (2012). *Unlocking National Opportunities: New Insights on Financing Sustainable Forest and Land Management*. FAO, Tropenbos International, the Global Mechanism (GM), the National Forest Programme (NFP) Facility, and ITTO, Rome.

Ashoka. (2008). Tashka Yawanawá. Ashoka Innovators for the Public. Available at: http://www.ashoka.org/fellow/tashka-yawanaw%C3%A1.

Aurora, L. (2011). *REDD+ biggest success in climate change talks, Norway says*. Forests Blog, Center for International Forestry Research. Available at: http://blog.cifor.org/6264/redd-biggest-success-in-climate-change-talks-norway-says/.

Babon, A., McIntyre, D. and Sofe, R. (2012). *REDD+ politics in the media: A case study from Papua New Guinea*. Working Paper 97. CIFOR. Bogor, Indonesia.

Bachram, H. (2004). Climate fraud and carbon colonialism: The new trade in greenhouse gases. *Capitalism, Nature, Socialism* 15(4), 10–12.

Bahuchet, S. (1994). *Situation des populations indigènes des forêts denses humides*. Laboratoire des langues et civilisations à tradition orale (France), Université libre de Bruxelles. Centre d'anthropologie culturelle, Commission of the European Communities, Brussels.

Bailey, R. (2008). *Another inconvenient truth: How biofuel policies are deepening poverty and accelerating climate change*. #114 Oxfam briefing paper series. Oxfam, Oxford. Available at: http://www.oxfam.org/policy/another-inconvenient-truth.

Balmford, A. et al. (2002). Economic reasons for conserving wild nature. *Science, New Series* 297(5583): 950–953 Published by: American Association for the Advancement of Science Stable URL: http://www.jstor.org/stable/3832034.

Banarjee, A. and Duflo, E. (2011). *Poor Economics: A Radical Rethinking of the Way to Fight Global Poverty*. Public Affairs, New York.

Bank Information Center (BIC) (2010). *SESA, Safeguards and the FCPF: A Guide for Civil Society*. BIC, Washington, DC.

Bank Information Center (BIC) (2011). *Introduction to the FCPF Readiness Package (R-Package) and the Carbon Fund (CF) Operational*. October 13, 2011. Available at: http://www.bicusa.org/updates/introduction-fcpf-r-package-and-carbon-fund/.

Barnett-Hart, A. (2009). *The story of the CDO market meltdown: An empirical analysis*. Bachelor of Arts thesis. Harvard University. Available at: http://www.hks.harvard.edu/m-rcbg/students/dunlop/2009-CDOmeltdown.pdf.

Barofsky, N. (2012). *Bailout: An Inside Account of How Washington Abandoned Main Street While Rescuing Wall Street*. Free Press, New York.

Barr, C. (2011). Governance risks for REDD+: How weak forest carbon accounting can create opportunities for corruption and fraud. In Transparency International, *Global Corruption Report: Climate Change*. Earthscan, London and Washington, DC.

Barry, D. and Taylor, P. L. (2008). *An Ear to the Ground: Tenure Changes and Challenges for Forest Communities in Latin America*. Rights and Resources Initiative, Washington, DC.

Batterbury, S. and Fernando, J. (2006). Rescaling governance and the impacts of political and environmental decentralization: An introduction. *World Development*, 34, 1851–1863.

Bayon, R., Hawn, A. and Hamilton, K. (eds) (2007). *Voluntary Carbon Markets*. Earthscan, Sterling, VA.

Bello, W. and Heydarian, R. (2012). *Towards a grand compromise in the climate negotiations*. Global Justice Ecology Project. Available at: http://climate-connections.org/2012/11/27/towards-a-grand-compromise-in-the-climate-negotiations/.

Benneker, C. (2012). *Forest governance in DRC: Artisinal logging.* Tropenbos. ETFRN News 53, April. Available at: http://www.etfrn.org/etfrn/newsletter/news53/Articles/1.4.C.Benneker.pdf.

Berger, L. (2010). *Why Maurice Bloch's work on "religion" is nothing special but is central. Religion and Society: Advances in Research,* 1, 14–18, doi:10.3167/arrs.2010.010103. Available at: http://api.ning.com/files/64cG790Afy1rqDQbBaMaGc5hdqSDC4AHbV0Ld8THPcGU*0W3FCOIRYljmpr4DJqltwPQa3vOb89ryNHqB9QWqR2uZMYXCzTr/Berger2010MauriceBlochsWorkonReligion.pdf.

Berkes, F. (2004). Rethinking community-based conservation. *Conservation Biology* 18(3): 621–630, June 2004.

Berkes, F. and Folke, C. (eds) (1998). *Linking Social and Ecological Systems: Management Practices and Social Mechanisms for Building Resilience.* Cambridge University Press, New York.

Berkes, F., Colding, J. and Folke, C. (2000). Rediscovery of traditional ecological knowledge as adaptive management. *Ecological Applications* 10: 1251–1262.

Beymer-Farris, B. and Bassett, T. (2012). The REDD menace: Resurgent protectionism in Tanzania's mangrove forests. *Global Environmental Change,* 22(2), 332–341.

Bloch, M. (1975). Property and the end of affinity. In Bloch, M. (ed.) *Marxist Analyses and Social Anthropology,* A.S.A. Studies. Malaby Press, London.

Blom, B., Sunderland, T. and Murdiyarso, D. (2010). Getting REDD to work locally: Lessons learned from integrated conservation and development projects. *Environ. Sci. Policy,* doi:10.1016/j.envsci.2010.01.002.

Blomley, T. et al. (2008). Seeing wood for the trees: An assessment of the impact of participatory forest management on forest condition in Tanzania. *Oryx,* 42(3), 380–391.

Blomley, T. and Iddi, S. (2009). *Participatory Forest Management in Tanzania: Lessons Learned and Experiences To Date.* For Forestry and Beekeeping Division, Ministry of Natural Resources and Tourism, United Republic of Tanzania.

Böhm, S. and Siddartha, D. (eds) (2009). *Upsetting the Offset: The Political Economy of Carbon Markets.* MayFly Press, UK. Available at: www.mayflybooks.org.

Böhm, S., Murtola, A.-M. and Spoelstra, S. (eds) (2012). *The atmosphere business. Ephemera.* Available at: http://www.ephemeraweb.org/journal/12-1/12-1ephemera-may12.pdf.

Bonis Charancle, J., Tiani, A., Brown, M., Akwah, G., Mogba, Z., Lescuyer, G., Warne, R. and Greenberg, B. (2009.) Strengthening local analytical capability: Community options analysis and investment. In Diaw, M., Aseh, T. and Prabhu, R. (eds), *In Search of Common Ground: Adaptive Collaborative Management in Cameroon.* Center for International Forestry Research (CIFOR), Bogor, Indonesia.

Booker, C. (2012). *How climate change has got Worldwide Fund for Nature bamboozled. The Telegraph,* May 5. Available at: http://www.telegraph.co.uk/earth/environment/climatechange/9246853/How-climate-change-has-got-Worldwide-Fund-for-Nature-bamboozled.html.

Borenstein, S. (2011). *Biggest jump ever seen in global warming gases.* Associated Press. Available at: http://news.mongabay.com/2011/1102-hance_extremeweather.html#ixzz1d1xGzZ9k.

Borrini-Feyerabend, G. (2004). *Governance of protected areas – innovation in the air.* Available at: http://www.earthlore.ca/clients/WPC/English/grfx/sessions/PDFs/session_1/Borrini_Feyerabend.pdf.

Boucher, D. (2008). *Out of the Woods: A Realistic Role for Tropical Forests in Curbing Climate Change.* Union of Concerned Scientists. Cambridge, MA. Available at: www.ucsusa.org/REDD.

Boudewijn, I. (2012). *Inclusion and benefit sharing in REDD+, the case of Oddar Meanchey.* Danida Research Portal. Available at: http://drp.dfcentre.com/project/inclusion-and-benefit-sharing-redd-case-oddar-meanchey.

Bowyer, C. (2010). *Anticipated indirect land use change associated with expanded use of biofuels and bioliquids in the EU – an analysis of the national renewable energy action plans.* Institute in European Environmental Policy. Available at: www.iiep.eu.

Boyd, E. and Folke, C. (eds) (2012). *Adapting Institutions: Governance, Complexity and Social–Ecological Resilience.* Cambridge University Press, Cambridge, UK.

Boyd, E., Boykoff, M. and Newell, P. (2011). The "new" carbon economy: What's new? *Antipode*, ISSN 0066-4812, pp. 1–11, doi: 10.1111/j.1467-8330.2011.00882.x.

Boyle, D. and Titthara, M. (2012). *Blind eye to forest's plight. Phnom Penh Post*, March 26. Available at: http://www.phnompenhpost.com/index.php/2012032655247/National-news/blind-eye-to-forests-plight.html.

Boyle, J. (2012). *Building REDD+ Policy Capacity for Developing Country Negotiators and Land Managers: Lessons Learned.* Task Force Meeting Report. International Institute for Sustainable Development, Winnipeg, Manitoba; ASB Partnership for the Tropical Forest Margins, World Agroforestry Center, Nairobi, Kenya.

Bray, D. B. (2008). Collective Action, Common Property Forests, Communities and Markets. *The Commons Digest.* 6: 1–4.

Bray, D., Duran, E., Romas, V., Mas, J.-F., Velazquez, A., McNab, R., Barry, B. and Radachowsky, J. (2008). Tropical deforestation, community forests, and protected areas in the Maya Forest. *Ecology and Society*, 13, 56.

Brechin, S., Wilshusen, P., Fortwangler, C. and West, P. (2002). Beyond the square wheel: Toward a more comprehensive understanding of biodiversity conservation as social and political process. *Society & Natural Resources: An International Journal*, 15(1), doi:10.1080/089419202317174011.

Brockington, D. and Duffy, R. (2010). Capitalism and conservation: The production and reproduction of biodiversity conservation. *Antipode* 42 (3): 469–484.

Brockington, D., Duffy, R. and Igoe, J. (2008). *Nature Unbound: Conservation, Capitalism and the Future of Protected Areas.* Earthscan, London and Sterling, VA.

Brown, K. (2004). Trade-off analysis for integrated conservation and development. In McShane, T. O. and Wells, M. P. (eds), *Getting Biodiversity Projects to Work.* Columbia University Press, New York: 232–255.

Brown, M. (2001a). *Community management of forest resources.* CARPE Briefing Sheet #17. For the Congo Basin Information Series. Biodiversity Support Program, Washington, DC. Available at: http://www.worldwildlife.org/bsp/publications/africa/127/congo_17.html.

Brown, M. (2001b). *Mobilizing communities to conserve forest resources.* CARPE Briefing sheet #20. For the Congo Basin Information Series. Biodiversity Support Program, Washington, DC. Available at: http://www.worldwildlife.org/bsp/publications/africa/127/congo_20.html.

Brown, M. (2007). *A demand driven model for fighting corruption in the Democratic Republic of Congo.* Presentation at Chatham House, London. Innovative Resources Management. http://www.chathamhouse.org.uk/publications/papers/view/-/id/505/.

Brown, M. (2009). *Capitalizing on a proven model for combating corruption in Africa from an unlikely source: Enhancing business participation in a multi-stakeholder Congo Basin initiative to broaden and sustain impacts.* World Bank Institute Anti Corruption Portal. Available at: http://info.worldbank.org/etools/antic/docs/Case%20Studies/Anti-Corruption%20Winners/Practitioners/Satya%20Development%20International%20LLC.pdf.

Brown, M. (2010). *Conservation Monitoring Platform for Better Public Information: Addressing Prototype Development Design Issues.* Proposal solicited for submission to the Gordon and Betty Moore Foundation. Satya Development International LLC, Washington, DC.

Brown, M. (2011). *Social feasibility in REDD+: A brief concept note.* Contribution to the Forest Trends Convened Workshop held in Goma, DRC, 20–21 September. Submitted to Forest

Trends and the USAID TRANSLINKS Project. Satya Development International LLC, Washington, DC.

Brown, M. and Holzknecht, H. (1993). An assessment of institutional and social conservation issues in Papua New Guinea. In Alcorn, J. and Beehler, B. (eds), *Papua New Guinea Conservation Needs Assessment*, Volumes 1 & 2. Department of Environment and Conservation, Government of Papua New Guinea, Waigani, and Biodiversity Support Program, Washington, DC.

Brown, M. and Mogba, Z. (2008). *Le consentement libre et informé: Principes et Critères 2 & 3 de FSC: Rapport de consultation auprès de la Congolaise Industrielle des Bois. République du Congo.* Confidential. Satya Development International LLC, Washington, DC.

Brown, M. and Mogba, Z. (2010). *REDD+ Perspectives in Central Africa: Can the Katoomba Process Add Value in a Crowded Institutional Landscape, and If So, How, Where and Why?* Forest Trends. For the USAID-funded TransLinks Project. Satya Development International LLC, Washington, DC.

Brown, M. and Wyckoff-Baird, B. (1992). *Designing integrated conservation and development projects.* The Biodiversity Support Program. World Wildlife Fund. Washington, DC. Available at: http://www.worldwildlife.org/bsp/publications/bsp/designing_eng/icdp-latest.pdf.

Brown, M., Ngwala, P., Songo, A. and Wande, L. (2004). *Combating low-level corruption on waterways in the Democratic Republic of Congo: Approaches from Bandundu and Equateur Provinces.* GWU Law School Public Law Research Paper No. 116. Available at: http://papers.ssrn.com/sol3/papers.cfm?abstract_id=627684.

Brown, M., Bonis-Charancle, J. M., Mogba, Z., Sundararajan, R. and Warne, R. (2008). Linking the Community Options, Assessment and Investment Tool (COAIT), Consen*sys*® and Payment for Environmental Services (PES): A model to promote sustainability in African gorilla conservation. In Stoinski, T. S. et al. (eds), *Conservation in the 21st Century: Gorillas as a Case Study.* Springer, New York.

Brown, M. with Bagabo, S., Dakouo, J., Dembele, E., Kazoora, C., Koopman, J., Maiga, S., Mukyala Makiika, R., Moyini, Y., Mubbala, S., Ndeso-Atanga, A., Nuwanyakpa, M., Rajaofara, H., Ramaroharinosy, W., Rizika, J., Sebukeera, C., Sidibe, I., French Smith, M., Soumare, M., Tanjong, E. and Taryatunga, F. (1996). *Non-governmental organizations and natural resources management: Synthesis assessment of capacity building issues in Africa.* PVO-NGO/NRMS Project, Funded by the US Agency for International Development. World Learning/CARE/World Wildlife Fund. Washington, DC. Available at: http://pdf.usaid.gov/pdf_docs/PNABZ603.pdf.

Brown, M. L. (2010). Limiting corrupt incentives in a global REDD+ regime. *Ecology Law Quarterly*, 37: 237–267.

Bruce, J. (2012). *Identifying and Working with Beneficiaries When Rights Are Unclear: Insights for REDD+ Initiatives.* The World Bank, Program on Forests (PROFOR), Washington, DC.

Brundtland Commission (1987). *Our Common Future: Report of the World Commission on Environment and Development.* United Nations, New York.

Buchanan, G. M., Donald, P. F. and Butchart, S. H. M. (2011). Identifying priority areas for conservation: A global assessment for forest dependent birds. *PLoS ONE* 6(12), e29080.

Bucki, M., Cuypers, D., Mayaux, P., Achard, F., Estreguil, C. and Grassi, G. (2012). Assessing REDD+ performance of countries with low monitoring capacities: The matrix approach. *Environmental Research Letters*, 7, doi:10.1088/1748-9326/7/1/014031.

Bulkan, J. (2012). *The rule of law? – Not in the forest sector of Guyana.* Illegal-logging.info. Available at: http://www.illegal-logging.info/item_single.php?it_id=6240&it=news.

Bunting, M. (2011). *How land grabs in Africa could herald a new dystopian age of hunger: Africa is up for sale by the acre to the highest bidder. But how can rice exports from Ethiopia to Saudi Arabia be justified? The Guardian.* Available at: http://www.guardian.co.uk/global-development/poverty-matters/2011/jan/28/africa-land-grabs-food-security.

Burkart, K. (2010). *Unholy alliance: Green NGO's buddy up with big corporations. Nature Conservancy, Conservation International, WWF, Sierra Club – are the biggest environmental NGO's losing credibility by receiving funds from major corporations?* Available at: http://www.mnn.com/green-tech/research-innovations/blogs/unholy-alliance-green-ngos-buddy-up-with-big-corporations.

Burnod, P., Andrianirina, N., Boue, C., Gubert, F., Rakoto-Tiana, N., Vaillant, J., Rabeantoandro, R. and Ratovoarinony, R. (2012). *Land reform and certification in Madagascar: Does perception of tenure security matter and change?* Draft version. Paper prepared for presentation at the Annual World Bank conference and Land and Poverty, The World Bank, Washington, DC, April 23–26.

Burnod, P., Gingembre, M., Andrianirina-Ratsialonana, R. and Ratovoarinony, R. (2011). *From international land deals to local informal agreements: Regulations of and local reactions to agricultural investments in Madagascar.* Paper presented at the International Conference on Global Land Grabbing, April 6–8. LDPI, Sussex University. Available at: http://www.iss.nl/content/download/24187/227376/version/3/file/35+Perrine_Mathilde_Rivo+and+Raphael.pdf.

Burren, C. (2012). *The Makira REDD Pilot Project and REDD Revenues.* Wildlife Conservation Society. Available at: http://www.forestcarbonpartnership.org/fcp/sites/forestcarbonpartnership.org/files/Documents/PDF/May2012/Makira%20Revenue%20Sharing.pdf.

Business Green. (2012). *Investment in UN's carbon scheme to 'dry up' as prices plunge: Thomson Reuters Point Carbon predicts value of CERs will drop to €0.50 by the end of the decade as oversupply bites.* Available at http://www.businessgreen.com/bg/news/2216163/investment-in-uns-carbon-scheme-to-dry-up-as-prices-plunge.

Butler, R. (2009a). *Changing drivers of deforestation provide new opportunities for conservation.* Available at: http://news.mongabay.com/2009/1208-drivers_of_deforestation.html.

Butler, R. (2009b). *Brazilian tribe owns carbon rights to Amazon rainforest land.* Available at: http://news.mongabay.com/2009/1208-surui_carbon.html#JCX2zOlzmR4KABk6.99.

Butler, R. and Laurance, W. (2008). New strategies for conserving tropical forests. *Trends in Ecology & Evolution*, 23(9): 469–472, 1 September 2008. doi:10.1016/j.tree.2008.05.006

Cadman, T. (ed.) (2013). *Climate Change and Global Policy Regimes Towards Institutional Legitimacy.* Palgrave Macmillan, Basingstoke.

Campese, J., Borrini-Feyerabend, G., de Cordova, M., Guigner, A. and Oviedo, G. (2007). "Just" conservation? What can human rights do for conservation…and vice versa?! *Policy Matters: Conservation and Human Rights*, Issue 15, July. IUCN Commission on Environmental, Economic, and Social Policy.

Carrere, R. (2006). *GREENWASH: Critical Analysis of FSC Certification of Industrial Tree Monocultures in Uruguay.* World Rainforest Movement, Montevideo.

Carrington, D. and Valentino, S. (2011). Biofuels boom in Africa as British firms lead rush on land for plantations: Controversial fuel crops linked to rising food prices and hunger, as well as increased greenhouse gas emissions. *The Guardian*, May 31.

Carter, N., Shrestha, B., Karki, J., Pradhan, N. and Liu, J. (2012). *Coexistence between wildlife and humans at fine spatial scales.* Proceedings of the National Academy of Sciences USA 109: 15360–65.

Castree, N. (2008). Neoliberalising nature: The logics of deregulation and reregulation. *Environment and Planning A*, 40, 131–152.

Castree, N. (2011). Neoliberalism and the biophysical environment 3: Putting theory into practice. *Geography Compass.* 5(1), 35–49. doi: 10.1111/j.1749-8198.2010.00406.x.

Catley, A., Burns, J., Abebe, D. and Suji, O. (2008). *Participatory Impact Assessment: A Guide for Practitioners.* Feinstein International Center, Tufts University, Medford, MA. Available at: http://reliefweb.int/sites/reliefweb.int/files/resources/DF552E1C45A24AFCC12574F600509CDA-Tufts_Oct2008.pdf.

Cavanagh, C. (2012). *Unready for REDD+: Lessons for corruption from Uganda conservation areas.* U4 Brief June 2012: 3. Anti-corruption Resource Center, Norwegian Institute for Nature Research, Chr. Michelsen Institute, Bergen, Norway.

CCBA. (2012). *Rules for the use of the climate, community, and biodiversity standards.* The Climate, Community, and Biodiversity Alliance. Available at: http://www.climate-standards. org/pdf/CCB_Standards_Rules_Version_June_21_2010.pdf.

CCBA and CARE. (2010). *REDD+ SES, 2010.* Fact Sheet. Version 1. Available at: http://www.redd-standards.org/files/pdf/lang/english/FactSheet-logo_En.pdf.

CDM Watch. (2012). *An insider's view: Fraud, corruption and environmental integrity of the CDM. CDM Watch Newsletter*, issue 19. Available at: www.cdm-watch.org.

Chagas, F., Streck, C., O'Sullivan, R., Olander, J. and Seifert-Granzin, J. (2011). *Nested Approaches to REDD+: An Overview of Issues and Options.* Forest Trends, Climate Focus, Washington, DC.

Chambers, R. (1983). *Rural Development: Putting the Last First.* Longman Scientific & Technical, University of Michigan, Ann Arbor, MI.

Chambers, R. and Belshaw, D. (1973). "Managing Rural Development: Lessons and Methods from Eastern Africa", IDS Discussion Paper no. 15, IDS Sussex, 1973.

Chapin, M. (2004). A challenge to conservationists. *World Watch*, 17(6), November/ December.

Chatterton, P. (2012). *REDD+ at scale: Mai Ndombe.* PowerPoint presentation. Douala, Cameroon. WWF Forest & Climate Initiative. Available at: http://www.google.com/url?sa=t&rct=j&q=&esrc=s&source=web&cd=3&ved=0CDcQFjAC&url=http%3A%2F%2Fccr-rac.pfbc-cbfp.org%2Fdocumentation-de-la-10ieme-reunion-des-partenaires-du-pfbc.html%3Ffile%3Ddocs%2FRdP2012%2Fresultats%2FRdP%2Fsession4%2FM2-P3-marche%2520de%2520information-WWF%2520Forest%2520%2526%2520Climate%2520Initiative-Paul%2520Chatterton-WWF.pdf&ei=B5RoUPX2N-nd0QHL0oGwBw&usg=AFQjCNF5MWp6f2MbSA__mYOhQJTIbCw-CQ&sig2=ASon5T4fw0uQKApduJftaQ.

Chen, S. and Ravallion, M. (2008). *The developing world is poorer than we thought, but no less successful in the fight against poverty.* Policy Research Working Paper 4703. World Bank Development Research Group, Washington DC. August.

Cherfas, J. (2012). *"Unjustly condemned" slash and burn agriculture found essential for forest regeneration.* CIFOR Blog, 12 February. Available at: http://blog.cifor.org/7394/unjustly-condemned-slash-and-burn-agriculture-found-essential-for-forest-regeneration/#.T6NBB1K3jzw.

Chhartre, A. and Agarwal, A. (2009). Trade-offs and synergies between carbon storage and livelihood benefits from forest commons. *PNAS* 106: 17667–17670.

Childs, M (2012). *Privatising the atmosphere: A solution or dangerous con? Ephemera*, 12(1/2), 12–18. Available at: http://www.ephemeraweb.org/journal/12-1/12-1ephemera-may12.pdf.

Chomitz, K., with Buys, P., De Luca, G., Thomas, T. and Wertz-Kanounnikoff, S. (2007). *At Loggerheads?: Agricultural Expansion, Poverty Reduction, and Environment in the Tropical Forests.* The International Bank for Reconstruction and Development/The World Bank. Washington, DC.

CI. (2010). *Hotspots science: Conservation response.* http://www.biodiversityhotspots.org/xp/hotspots/hotspotsscience/conservation_responses/Pages/protected_area_coverage.aspx.

CI Japan. (2012). *New mechanism feasibility study for REDD+ in Prey Long Area, Cambodia.* Available at: http://gec.jp/gec/en/Activities/fs_newmex/2011/2011newmex23_eCIJ_Cambodia_rep.pdf.

CIFOR. (2012). *Global database of REDD+ and other forest carbon projects: Interactive map.* Available at: http://www.forestsclimatechange.org/redd-map/.

Clark, P. and Blas, J. (2012). *Carbon prices tumble to record low.* ft.com, 2 April. Available at: http://www.ft.com/intl/cms/s/0/b36fa102-7cc3-11e1-9d8f-00144feab49a. html#axzz1rYMerXq6.

Cleary, D. (2005). *The questionable effectiveness of science spending by international conservation organizations in the tropics. Conservation Biology.* Available at: http://www.aseanbiodiversity. info/Abstract/51008744.pdf.

ClimateChangeCorp. (2009). *Is carbon trading the most cost-effective way to reduce emissions?* Available at: http://www.climatechangecorp.com/content.asp?ContentID=6064.

Climate Markets and Investors Association. (2011). *CMIA response to the Munden Report.* Available at: http://www.tropical-forestry.org/2011/08/carbon-markets-and-investors-association-response-to-the-munden-report/.

Coase, R. H. (1960). The problem of social cost. *Journal of Law and Economics*, 3 (Oct., 1960): 1–44.

Cohen, J. and Easterly, W. (2009). *What Works in Development: Thinking Big and Thinking Small.* The Brookings Institution. Washington, DC.

Colchester, M. (2007). *Beyond Tenure: Rights-based approaches to peoples and forests. Some lessons from the Forest Peoples Programme.* Moreton-in-Marsh, UK.

Colchester, M. (2010). *Free, Prior and Informed Consent: Making FPIC Work for Forests and Peoples.* Research Paper. Number 11. The Forests Dialogue. School of Forestry and Environmental Studies. Yale University, New Haven, CT.

Colchester, M. (2011). *Letter to Madagascar Environmental Justice Network.* April 18. Available at: http://madagascarenvironmentaljustice.ning.com/forum/topics/explaining-redd-to-local?commentId=5844155%3AComment%3A11455&xg_source=msg_mes_network.

Collier, P. (2007). *The Bottom Billion: Why the Poorest countries are Failing and What Can Be Done About It.* Oxford University Press, New York.

Comision Ejuctiva Nacional et al. (2011). *Declaration of Patihuitz.* Chiapas, Mexico. http://www.redd-monitor.org/2011/04/07/redd-alert-in-chiapas-mexico/?utm_source=feedburner&utm_medium=email&utm_campaign=Feed%3A+Redd-monitor+%28REDD-Monitor%29.

Comunidades de la Región Amador Hernández, Chiapas, México, Grupos comunitarios de Marqués de Comillas, Chiapas, México, Grupos comunitarios de Benemérito de las Américas, Chiapas, México, Consejo de Médicos y Parteras Indígenas Tradicionales de Chiapas (Compitch), Chiapas, México, Sociedad Civil Las Abejas, Comité de DH de Base "Digna Ochoa", Chiapas, México, Koman Ilel, Chiapas, México, Kinal Antzetik, Chiapas, México, Otros Mundos, A. C./Amigos de la Tierra México, Movimiento Mexicano de Afectados por las Represas y en Defensa de los Ríos (MAPDER), Red Mexicana de Afectados por la Minería (REMA), Amigos de la Tierra América Latina y El Caribe (ATALC), Amigos de la Tierra Estados Unidos, COECO-CEIBA/ Amigos de la Tierra Costa Rica, CEIBA/Amigos de la Tierra Guatemala, NAT/ Amigos de la Tierra Brasil, Red Latinoamericana contra los Monocultivos de Árboles (RECOMA), Movimiento Mesoamericano contra el Modelo Extractivo Minero (M4), Movimiento Mexicano de Alternativas a las Afectaciones Ambientales y al Cambio Climático (MOVIAC)/Capítulo Chiapas, Movimiento de Víctimas y Afectados por el Cambio Climático (MOVIAC) El Salvador, Movimiento de Víctimas y Afectados por el Cambio Climático (MOVIAC) Guatemala, Alianza Mundial de los Pueblos Indígenas y Comunidades Locales sobre Cambio Climático en contra de REDD y por la Vida, Movimiento Mundial por los Bosques Tropicales (WRM) Consejo de Pueblos de Occidente (CPO), Guatemala, Frente Petenero contra las Represas, Consejo Cívico de Organizaciones Populares de Honduras (COPINH), Organización Fraternal Negra Hondureña (OFRANEH), Jubileo Sur, Tsunel Bej. (2012). *Declaration of Chiapas in REDDellion: Enough of REDD+ and the green economy*. Available at: http://climate-connections.org/2012/09/24/declaration-of-chiapas-in-reddellion-enough-of-redd-and-the-green-economy/.

Connor, S. (2011). *Russian research team astonished after finding "fountains" of methane bubbling to surface*. Available at: http://ascendingstarseed.wordpress.com/tag/siberian-sea-arctic-methane-release/.

Conservation Measures Partnership. (2010). *Integrating social strategies and human welfare targets into the Open Standards for the Practice of Conservation*. Available at: http://www.google.com/url?sa=t&rct=j&q=&esrc=s&source=web&cd=1&ved=0CCUQFjAA&url=http%3A%2F%2Fwww.conservationmeasures.org%2Fwp-content%2Fuploads%2F2010%2F10%2FCMP-Social-Strategies-Initiative-2010-10-12.doc&ei=yzmDT7LENqPh0gGqheGcCA&usg=AFQjCNHBr9rRI-jxNJe8KO05VDYRNI9oeA&sig2=_AjM5j-prgxzaBEYTORKNw.

Conservation Measures Partnership. (2012). *Addressing social results and human wellbeing targets in conservation projects*. Draft guidance, June 27. Available at: http://www.fosonline.org/wordpress/wp-content/uploads/2012/09/DRAFT-Guidance-on-HWT-and-Social-Results-in-Conservation-Projects-v2012-06-27.pdf.

Cooke, B. and Kothari, U. (eds) (2001). *Participation: The New Tyranny?* Zed Books, London and New York.

COONAPIP. (2012). *Letter to ANAM/Panama and United Nations/Panama*. REDD-Monitor, 30 August. Available at: http://www.redd-monitor.org/2012/08/30/coonapip-panamas-indigenous-peoples-coordinating-body-denounces-un-redd/#cl.

Corbera, E. and Schroeder, H. (2011). Governing and implementing REDD+. *Environmental Science & Policy*, 14 (2): 89–99.

Corson, C. (2011). Territorialization, enclosure and neoliberalism: Non-state influence in struggles over Madagascar's forests. *Journal of Peasant Studies*, 38(4), 703–726.

Costanza, R. (2012). *Response to George Monbiot: The valuation of nature and ecosystem services is not privatization*. RTCC. Available at: http://www.rtcc.org/policy/response-to-monbiot-valuation-is-not-privatization/.

Cotula, L. and Mayers, J. (2009). *Tenure in REDD: Start Point or Afterthought?* International Institute for Environment and Development, UK.

Council on Foreign Relations (2012). *The global climate change regime*. Available at: http://www.cfr.org/climate-change/global-climate-change-regime/p21831.

Cracknell, R. (2012). *From Project-based to Nested REDD+: Monitoring, Reporting and Verifying (MRV) Standards for Carbon Accounting*. WWF, Gland, Switzerland. Forest and Climate Initiative.

Creed, A. and Nakhooda, S. (2011). *REDD+ Finance Delivery: Lessons from Early Experience*. Overseas Development Institute, London and Heinrich Boll Stiftung, North America, Washington DC.

Cronkleton, P., Pulhin, J. M. and Saigal, S. (2012). Co-management in community forestry: How the partial devolution of management rights creates challenges for forest communities. *Conservation and Society*, 10, 91–102.

Czebiniak, R. P. (2012). "Commentary: Protecting the people, not the polluters, says Greenpeace". Guest commentary by Roman Paul, Greenpeace Senior Policy Advisor on Climate and Forests, special to mongabay.com September 27, 2012. Available at: http://news.mongabay.com/2012/0926-greenpeace-vs-nepstad.html.

Dalton, G. (1971). Theoretical issues in economic anthropology. In Dalton, G. (ed.), *Economic Development and Social Change*. American Museum of Natural history, Washington, DC.

Danielsen, F., Beukema,H., Burgess, N., Parish, F., Bruhl, C., Donald, P., Murdiyarso, D., Phalan, B., Reijnders, L., Struebig, M. and Fitzherbert, E. (2008). Biofuel plantations on forested lands: Double jeopardy for biodiversity and climate. *Conservation Biology*, 23(2), 348–358.

DARA and The Climate Vulnerable Forum (2012). *The climate vulnerability monitor: A guide to the cold calculus of a hot planet.* Available at: http://www.daraint.org/wp-content/uploads/2012/09/CVM2ndEd-FrontMatter.pdf.

Daviet, F., Davis, C., Goers, L. and Nakhooda, S. (2009). *Ready or not?: A review of the World Bank Forest Carbon Partnership R-Plans and the UN REDD Joint Program Documents.* World Resources Institute, Washington, DC. Available at: http://pdf.wri.org/working_papers/ready_world_bank_redd.pdf.

Davis, et al. (2009). *A review of 25 readiness idea plan notes from the World Bank Forest Carbon Partnership Facility.* WRI Working Paper. World Resources Institute. Available at: http://www.wri.org/gfi.

Diaw, C., M., Tiani, A.-M., Jum, C., Milol, A. and Wandji, D. (2003). *Assessing Long-term Management Options for the Villages in the Korup National Park: An Evaluation of All Options.* CIFOR, Yaounde, Cameroon.

Diaz, D., Hamilton, K. and Johnson, E. (2011). *State of the Forest Carbon Markets.* Ecosystem Marketplace and Forest Trends, Washington, DC.

Dietz, T., Ostrom, E. and Stern, P. (2003). The struggle to govern the commons. Available at: http://stephenschneider.stanford.edu/Publications/PDF_Papers/DietzOstromStern.pdf.

Di Gregorio, S. and Davidson, J. (2008). *Qualitative Research Design for Software Users.* McGraw-Hill Open University Press, Maidenhead.

Di Gregorio, M., Brockhaus, M., Cronin, T. and Muharrom, E. (2012). Politics and power in national REDD+ policy processes. In Angelsen, A., Brockhaus, M., Sunderlin, W. and Verchot, L. (eds) *Analysing REDD+: Challenges and Choices.* CIFOR, Bogor, Indonesia: 69–90.

Dooley, K. (2010). Quoted in *The Ecologist*, UK. *Fears of corruption as REDD forest-protection schemes begin.* Available at: http://www.illegal-logging.info/item_single.php?it_id=4851&it=news.

Dooley, K, and Horner, K. (2011). *FW Special Report – Durban aimed to save the market, not the climate.* December. Available at: www.fern.org.

Dooley, K., Martone, F. and Ozinga, S. (2011). *Smoke and Mirrors: A Critical Assessment of the Forest Carbon Partnership Facility.* FERN and Forest Peoples Programme. Moreton-in-Marsh, UK.

Dooley, K., Griffiths, T., Leake, H. and Ozinga, S. (2008). *Cutting corners: World Bank's forest and carbon fund fails forests and peoples.* FERN, Forest Peoples Programme. Moreton-in-Marsh, UK. Available at: http://www.redd-monitor.org/wordpress/wp-content/uploads/2008/12/document_4312_4313.pdf.

Dove, M. (1993). *A revisionist view of tropical deforestation and development.* East-West Center, Honolulu, Hawaii. Available at: http://www2.eastwestcenter.org/environment/CBFM/1_Dove_1993.pdf.

Dowie, M. (2009). *Conservation Refugees: The Hundred Year Conflict between Global Conservation and Native Peoples.* The MIT Press, Cambridge, MA.

Duffy, R. (2006a) Global governance and environmental management: the politics of transfrontier conservation areas in Southern Africa, *Political Geography*, 25(1): 89–112.

Duffy, R. (2006b). NGOs and governance states: the impact of transnational environmental management networks in Madagascar, *Environmental Politics*, 15(5): 731–749.

Duffy, R. (2010). *Nature Crime: How We're Getting Conservation Wrong.* Yale University Press, New Haven, CT.

Duflo, E. (2010). *Social Experiments to Fight Poverty.* Ted Talks. http://www.ted.com/talks/lang/eng/esther_duflo_social_experiments_to_fight_poverty.html.

Dutta, D. (2009). Elite Capture and Corruption: Concepts and Definitions. National Council of Applied Economic Research. New Delhi, India. Available at: http://www.ruralgov-ncaer.org/images/product/doc/3_1345011280_EliteCaptureandCorruption1.pdf.

Dyer, N. and Counsell, S. (2010). *McREDD+: How McKinsey "cost-curves" are distorting REDD+*. Rainforest Foundation UK – Climate and Forests Policy Brief. November. Available at: http://www.rainforestfoundationuk.org/files/McREDD+%20English.pdf.

Easterly, W. (2006). *The White Man's Burden: Why the West's Efforts to Aid the Rest Have Done So Much Ill and So Little Good*. The Penguin Press, New York.

EcoNexus. (2012). *Why we should continue to oppose the inclusion of agriculture in the climate negotiations*. February. Available at: http://econexus.info/publication/why-we-should-continue-oppose-inclusion-agriculture-climate-negotiations.

EcoSecurities. (2010). *The Forest Carbon Offsetting Report (2010)*. EcoSecurities, Conservation International, the Climate, Community and Biodiversity Alliance, ClimateBiz and Norton Rose.

Ehrenreich, B. (2009). *Bright-sided: How the Relentless Promotion of Positive Thinking Has Undermined America*. Picador, New York.

EIA (The U.S. Energy Information Administration) (2009). *U.S. Carbon Dioxide Emissions in 2009: A Retrospective Review*. Available at: http://www.eia.gov/oiaf/environment/emissions/carbon/.

Eilpirin, J. (2011). "World on track for nearly 11-degree temperature rise, energy expert says." November 28, 2011. Available at: http://articles.washingtonpost.com/2011-11-28/national/35282761_1_celsius-climate-talks-energy-expert.

Eklöf, G. (2013). REDD Plus or REDD "Light"? – Biodiversity, communities, and forest certification. Swedish Society for Nature Conservation. Stockholm, Sweden. Available at: http://www.naturskyddsforeningen.se/sites/default/files/dokument-media/REDD%20Plus%20or%20REDD%20Light.pdf.

Eliasch, J. (2008). *The Eliasch report*. http://www.official-documents.gov.uk/document/other/9780108507632/9780108507632.pdf.

Ellis, E. and Porter-Bolland, L. (2008). Is community-based forest management more effective than protected areas? A comparison of land use/land cover change in two neighboring study areas of the Central Yucatan Peninsula, Mexico. *Forest Ecology and Management*, 256, 1971–1983.

Enright (2012). *Exploring pro-poor options for distributing REDD+ benefits in Vietnam*. SNV. Available at: http://www.snvredd.com/index.php?option=com_content&view=category&layout=blog&id=25&Itemid=46.

ERA. (2011). ERA Ecosystem Restoration Associates Inc. / Mai Ndombe, DRC. Available at: http://www.coderedd.com/redd-project-devs/era-ecosystem-restoration-associates-inc-mai-ndombe/.

ERA. (2012). *Mai Ndombe REDD+ Project*. Available at: http://www.forestcarbongroup.de/Projects-of-the-Forest-Carbon-Group-Mai-Ndombe-REDD-and-Project/494.

ERA and Wildlife Works (2012a). *Mai Ndombe REDD+*. A Joint Project of ERA, Wildlife Works. Available at: https://s3.amazonaws.com/CCBA/Projects/Mai_Ndombe_REDD_Project/Mai_Ndombe_REDD_CCB_PDD_Aug_30%5B1%5D.pdf.

ERA and Wildlife Works (2012b). *Mai Ndombe REDD: Project monitoring plan for climate, community, and biodiversity benefits*. September 18. Available at: https://s3.amazonaws.com/CCBA/Projects/Mai_Ndombe_REDD_Project/v2.0_Mai_Ndombe_CCB_Monitoring_Plan_Sept_18%5B1%5D.pdf.

Escobar, A. (1995). *Encountering Development: The Making and Unmaking of the Third World*. Princeton University Press, Princeton, NJ

European Commission. (2011). *Mobilising $100 billion per year by 2020 for climate actions in developing countries "challenging but feasible", says Commission*. April 8. http://www.egovmonitor.com/node/41637.

Fairhead, J. and Leach, M. (1996). *Misreading the African Landscape: Society and Ecology in a Forest-savanna Mosaic*. Cambridge University Press, Cambridge, UK.

Fairhead, J. and Leach, M. (2000). Reproducing locality: A critical exploration of the relationship between natural science, social science, and policy in West African ecological problems. In Broch-Due, V. and Schroeder, R. (eds), *Producing Nature and Poverty in Africa*. Nordiska Afrikainstitutet, Stockholm, Sweden.

Fairhead, J., Leach, M. and Scoones, I. (eds) (2012). Green grabbing: A new appropriation of nature? *Journal of Peasant Studies*, 39(2), special issue, March.

FAO, UNDP, UNEP. (2008). UN Collaborative Program on Reducing Emissions from Deforestation and Forest Degradation in Developing Countries (UN-REDD). Rome: FAO, UNDP, UNEP Framework Document.

Farley, J. and Costanza, R. (2010). *Payments for ecosystem services: From local to global. Ecological Economics*, 69 (2010), 2060–2068. Available at: https://www.pdx.edu/sites/www.pdx.edu.sustainability/files/Farley%20and%20Costanza%202010.pdf.

FCPF. (2011). *Update on the Carbon Fund: Proposal for a methodological approach*. Available at: http://www.forestcarbonpartnership.org/fcp/sites/forestcarbonpartnership.org/files/Documents/PDF/Sep2011/FCPF%20CF%20Methods%2009-01-11%20final.pdf.

FCPF and UN-REDD. (2012). R-PP Template Version 6, for Country Use (April 4, 2012) (To replace R-PP draft v. 5, Dec. 22, 2010). The World Bank and UN-REDD, Geneva.

Feeny, D., Hanna, S. and McEvoy, A. F. (1996). Questioning the assumptions of the tragedy of the commons model of fisheries. *Land Economics* 72: 187–205.

Feeny, D., Berkes, F., McCay, B. and Acheson, J. (1999). The tragedy of the commons: Twenty-two years later. In Opschoor, J. B., Button, K. and Nijkamp, P. (eds), *Environmental Economics and Development*. Elgar, Cheltenham, UK: Chapter 8.

FENAMAD. (2010). *Propuesta de la Federacion Nativa del Rio Madre de Dios y Afluentes (FENAMAD) a los hermanos indigenas de Peru y la Amazonia – REDD Indigena*. Puerto Maldonado. Available at: http://www.fenamad.org.pe/noticias2.htm.

FERN. (2012). *Suffering here to help them over there*. 12 minute video. Available at: http://www.fern.org/sufferinghere.

Ferraro, P. J. and Pattanayak, S. K. (2006). Money for nothing? A call for empirical evaluation of biodiversity conservation investments. *PLoS Biol* 4(4), e105. doi:10.1371/journal.pbio.0040105.

Figuieres, C., Leplay, S., Midler, E. and Thoyer, S. (2010). *The REDD Scheme to Curb Deforestation: An Ill-Defined System of Incentives?* World Congress of Environmental and Resource Economists, 2010/06/28-2010/07/02. Montreal, Canada.

Financial Services Authority. (2012). *Carbon credit trading*. Available at: http://www.fsa.gov.uk/Pages/consumerinformation/scamsandswindles/investment_scams/carbon_credit/index.shtml.

Forest Carbon Portal. (2012). *Code REDD aims for scale and quality*. June 20. Available at: http://www.forestcarbonportal.com/content/code-redd-aims-for-scale-and-quality.

Forest Peoples Programme. (2007). *Making FPIC – free, prior and informed consent – work: Challenges and prospects for indigenous people*. Available at: http://www.forestpeoples.org/topics/civil-political-rights/publication/2010/making-fpic-free-prior-and-informed-consent-work-chal.

Forest Peoples Programme. (2012a). *Carbon concessions in the Democratic Republic of Congo (DRC) neglect communities*. Available at: http://www.forestpeoples.org/topics/redd-and-related-initiatives/news/2012/04/carbon-concessions-democratic-republic-congo-drc-ne.

Forest Peoples Programme. (2012b). *Civil society groups in DRC suspend engagement with National REDD coordination process*. Available at: http://www.forestpeoples.org/topics/redd-and-related-initiatives/news/2012/07/civil-society-groups-drc-suspend-engagement-nationa.

Forum Barcelona (2004). "Dialogue: beyond models?" Available at http://www.barcelona2004. org/www.barcelona2004.org/eng/banco_del_conocimiento/documentos/ficha18c3. html?IdDoc=599.

Frankfurt, H. (2005). *On Bullshit*. Princeton University Press, Princeton, NJ and Oxford, UK.

Freireich, J. and Fulton, K. (2009). *Investing for social and environmental impact: A design for catalyzing an emerging industry*. The Monitor Institute. Available at: http://monitorinstitute.com/ downloads/what-we-think/impact-investing/Impact_Investing.pdf.

Freudenberger, K. (2010). *Paradise lost?: Lessons from 25 years of USAID environment programs in Madagascar*. International Resources Group. Washington, DC. Available at: http:// transition.usaid.gov/locations/sub-saharan_africa/countries/madagascar/paradise_ lost_25years_env_programs.pdf.

Friends of the Earth. (2009). *Sub-Prime Carbon: Re-thinking the World's Largest New Derivatives Market*. Friends of the Earth, Washington, DC.

Galbraith, J. (2012). *Inequality and Instability: A Study of the World Economy Just Before the Great Crisis*. Oxford University Press, New York.

Garnett, S. T., Sayer, J. and Du Toit, J. (2007). *Improving the effectiveness of interventions to balance conservation and development: A conceptual framework*. Ecology and Society 12(1): 2. Available at: http://www.ecologyandsociety.org/vol12/iss1/art2/.

GCF Task Force (2012). Annual Meeting, September 25–27, 2012. Available at http:// www.gcftaskforce.org/documents/GCF_Annual_Meeting_Agenda_Final_EN.pdf.

Geisler, C. and de Sousa, R. (2001). From refuge to refugee: The African case. *Public Administration and Development*, 21, 159–170.

Gibson, C., Ostrom, E. and Ahn, T.-K. (2000). The concept of scale and the human dimensions of global change: a survey. *Ecological Economics* 32: 217–239.

Giddens, A. (2009). *The Politics of Climate Change*. Polity Press, Cambridge, UK.

Gillam, C. (2012). *TIAA-CREF forms global farmland investing company*. Food Crisis and the Global Land Grab. Available at: http://farmlandgrab.org/post/view/20494.

Gledhill, R., Streck, C., Maginnis, S. and Brown, S. (2011). *Funding for forests: UK government support for REDD+*. PricewaterhouseCoopers LLP. Available at: http://www.decc.gov. uk/assets/decc/internationalclimatechange/1832-funding-for-forests-uk-government-support-for-red.pdf.

Gleick, P. (2012). *Climate change, disbelief, and the collision between human and geologic time*. 16 January. Available at: http://www.forbes.com/sites/petergleick/2012/01/16/climate-change-disbelief-and-the-collision-between-human-and-geologic-time/.

Global Forest Coalition and the Global Justice Ecology Project. (2012). *A darker shade of green: REDD alert and the future of forests*. Video. Available at: http://www. redd-monitor.org/2012/01/25/new-video-a-darker-shade-of-green-redd-alert-and-the-future-of-forests/?utm_source=feedburner&utm_medium=email&utm_ campaign=Feed%3A+Redd-monitor+%28REDD-Monitor%29.

Global Justice Ecology Project. (2012). *Farmers demand the World Bank and Wall Street stop grabbing their lands at opening of Bank's annual conference in Washington, DC*. Available at: http://climate-connections.org/2012/04/23/farmers-demand-the-world-bank-and-wall-street-stop-grabbing-their-lands-at-opening-of-the-banks-annual-conference-in-washington-dc/.

Global Witness (2006). *The Sinews of War: Eliminating the Trade in Conflict Resources*. London. Available at: http://www.globalwitness.org/sites/default/files/import/the_sinews_of_ war.pdf.

Global Witness (2010). *Principles for Independent Monitoring of REDD+ (IM-REDD+)*. June. Global Witness, London.

Global Witness (2011a). *Congo's Minerals Trade in the Balance*. Available at: http://www. globalwitness.org/sites/default/files/library/Congo%27s%20minerals%20trade%20 in%20the%20balance%20low%20res.pdf.

Global Witness (2011b). *Forest Carbon Cash and Crime: The Risk of Criminal Engagement in REDD+*. Global Witness, London.

Global Witness (2012). *Safeguarding REDD+ Finance: Ensuring Transparent and Accountable International Financial Flows*. Global Witness, London. Available at: http://www.globalwitness.org/sites/default/files/library/Safeguarding%20REDD+%20Finance.pdf.

Gollin, D. (2012). *The rural imperative: Targeting agriculture for transformation*. Available at: http://acetforafrica.org/publications/post/the-rural-imperative-targeting-agriculture-for-transformation/.

Goodland, R. (2004). Free, prior and informed consent and the World Bank Group. *Sustainable Development Law and Policy*, Summer, 66–74.

Go-REDD+ (2012) "REDD+ and markets: any lessons to be learned from voluntary carbon markets?" UN-REDD Programme.

Gorenflo, L., Romaine, S., Mittermeier, R. and Walker-Painemilla, K. (2012). *Co-occurrence of linguistic and biological diversity in biodiversity hotspots and high biodiversity wilderness areas*. *PNAS Early Edition*. Available at: www.pnas.org/cgi/doi/10.1073/pnas.1117511109.

Gough, I. and Wood, G. D. (eds.) (2004). *Insecurity and Welfare Regimes in Asia, Africa and Latin America: Social Policy in Developing Countries*. Cambridge University Press, Cambridge, UK.

Government of Nepal (2010). Ministry of Forests and Soil Conservation. Nepal's Readiness Preparation Proposal: REDD. 2010–2013. Khatmandu, Nepal. Available at: http://www.theredddesk.org/sites/default/files/r-pp_nepal_revised_october.pdf.

Greenpeace. (2007). *Carving up the Congo*. Volumes 1–4. Greenpeace, Amsterdam.

Greenpeace. (2010). *Turning REDD into green in the DRC: Can a national REDD plan in the Democratic Republic of Congo set a new course for the protection of forests, people and global climate?* Available at: http://www.greenpeace.org/africa/Global/africa/publications/Turning%20REDD%20into%20Green%20in%20the%20DRC.pdf.

Greenpeace. (2012). *Outsourcing hot air*. Available at: http://www.greenpeace.org/international/Global/international/publications/forests/2012/REDD/OutsourcingHotAir.pdf.

Gregersen, H., El Lakany, H., Bailey, L. and White, A. (2011). *The Greener Side of REDD+: Lessons from Countries where Forest Area Is Increasing*. Rights and Resources Institute, Washington, DC.

Gregersen, H., El Lakany, H., Karsenty, A. and White, A. (2010). *Does the Opportunity Cost Approach Indicate the Real Cost of REDD+?: Rights and Realities in Paying for REDD+*. Rights and Resources Initiative, Washington, DC.

Groves, C., Jensen, D., Valutis, L., Redford, K., Shafffer, M., Scott, M., Baumgartner, J., Higgins, J., Beck, M. and Anderson, M. (2002). Planning for biodiversity conservation: Putting conservation science into practice. *Bioscience*. 52(6), 499.

Grubb, M. (2011). Interview in *"Carbon Markets: Trading With Our Future"*. Occupy UNFCCC COP 17. Available at: http://vimeo.com/32995647.

Gudeman, S. (2008). *Economy's Tension: The Dialectics of Community and Market*. Berghahn Books, Oxford, UK.

Gurung, C., Maskey, T., Poudel, N., Lama, Y., Wagley, M., Manandhar, A., Khaling, S., Thapa, G., Thapa, S. and Wikramanayake, E. (2006). The sacred Himalayan landscape: Conceptualizing, visioning and planning for conservation of biodiversity, cultures and livelihoods. In McNeely, J. A., McCarthy, T. M., A. Smith, Olsvig-Whittaker, L. and Wikramanayake, E. D. (eds), *The Eastern Himalayas. Conservation Biology in Asia*. Society for Conservation Biology Asia Section/Resources Himalaya, Kathmandu, Nepal.

Gurung, J., Krasny, S., Pircher, H. and Taylor, R. (2011). *Scoping Dialogue on Changing Outlooks on Food, Fuel, Fiber and Forests 1-3 June, 2011*. The Forests Dialogue, Washington DC, United States Co-Chairs' Summary Report.

Hajek, F., Ventresca, M. J., Scriven, J. and Castro, A. (2011). Regime-building for REDD+: Evidence from a cluster of local initiatives in south-eastern Peru. *Environmental Science & Policy*, 14 (2), 201–215.

Hall, A. (2012). *Forests and Climate Change: The Social Dimensions of REDD in Latin America*. Edward Elgar, Cheltenham, UK.

Hansen, J. (2012). *Climate change is here – and worse than we thought. Washington Post*. August 3. Available at: http://www.washingtonpost.com/opinions/climate-change-is-here--and-worse-than-we-thought/2012/08/03/6ae604c2-dd90-11e1-8e43-4a3c4375504a_story.html.

Hardin, G. (1968). Tragedy of the commons. *Science*, 162, 1243–1248.

Hart, K., Laville, J.-L. and Cattani, A. (2010). Building the human economy together. In Hart, K., Laville, J.-L. and Cattani, A. (eds), *The Human Economy*. Polity Press, Cambridge, UK and Malden, MA, USA.

Harvey, C. and Dickson, B. (2010). Tools and measures for ensuring REDD+ provides biodiversity benefits. *ITTO Tropical Forest Update*, 20(1): 13–15.

Hecht, J., Gibson, D. and App, B. (2008). *Protecting Hard-won Ground: USAID Experience and Prospects for Biodiversity Conservation in Africa*. Biodiversity Assessment and Technical Support Program (BATS). Chemonics International, Washington, DC.

Helm, D. (2012). *The Carbon Crunch: How we're Getting Climate Change Wrong*. Yale University Press, New Haven, CT and London.

High Level Panel on the CDM Dialogue. (2012). *Climate Change, Carbon Markets, and the CDM: A Call to Action*. UNFCCC, Bonn.

Hitchcock, R. (1996). *Bushmen and the Politics of the Environment in Southern Africa*. IWGIA, Copenhagen.

HLPE. (2011). *Land Tenure and International Investments in Agriculture*. A report by the High Level Panel of Experts on Food Security and Nutrition of the Committee on World Food Security, Rome.

Hochschild, A. (1998). *King Leopold's Ghost: A Story of Greed, Terror, and Heroism in Colonial Africa*. Mariner Books, Boston, MA.

Hockley, N. and Andriamarovololona, M. (2007). *The Economics of Community Forest Management in Madagascar: Is There a Free Lunch? An Analysis of Transfert de Gestion*. Development Alternatives Inc. for USAID/University of Wales, Bangor.

Hockley, N. and Razafindralambo, R. (2006). *A social cost-benefit analysis of conserving the Ranomafana Andringitra-Pic d'Ivohibe Corridor in Madagascar*. University of Wales, Bangor, UK & Conservation International, Madagascar. USAID Cooperative Agreement No. 687-A-00-04-00090-00. Available at: http://pdf.usaid.gov/pdf_docs/PNADI193.pdf.

Holmgren, P. and Marklund, L. (2007). National Forest Monitoring Systems – purposes, options and status. CABI Publishing. In Freer-Smith, P. H. and Broadmeadow, M. (eds). *Forestry and Climate Change*. CABI Publishing, Wallingford, UK.

Homewood, K. and Rodgers, W. A. (1991). *Maasailand Ecology: Pastoralist Development and Wildlife Conservation in Ngorongoro, Tanzania*. Cambridge University Press, Cambridge, UK.

Horton, R. (1993). *Patterns of Thought in Africa and the West*. Cambridge University Press, Cambridge, UK.

Hughes, R. and Flinton, F. (2001). *Integrating Conservation and Development Experience: A Review and Bibliography of the ICDP Literature*. IIED, London.

Hulme, M. (2009). *Why I Disagree about Climate Change: Understanding Controversy, Inaction and Opportunity*. Cambridge University Press, Cambridge, UK.

Hurley, M. (2010). Personal communication. Bonobo Conservation Initiative. Washington, DC.

Hurley, M. (2012). Personal communication. Bonobo Conservation Initiative. Washington, DC.

IFC (2011). *Update of IFC's Policy and Performance Standards on Environmental and Social Sustainability, and Access to Information Policy*, May 12. Available at: http://www.treasury.gov/resource-center/international/development-banks/Documents/IFC%20policy%20review%20-%20final%20policy%20May%2012%202011%20-%20US%20position%20to%20post.pdf.

Igoe, J. and Brockington, D. (2007). Neoliberal conservation: A brief introduction. *Conservation and Society*, 5(4), 432–449.

Independent Evaluation Group (IEG). (2011). The Forest Carbon Partnership Facility. *Global Program Review*, 6(3). The World Bank Group, Washington, DC.

Independent Evaluation Group (IEG). (2012). Global Program Review Forest Carbon Partnership Facility. August 27, 2012. Country Corporate and Global Evaluations. Available at: http://ieg.worldbankgroup.org/content/dam/ieg/grpp/fcpf_gpr.pdf.

Indradi, Y. (2012). Interview with Bustar Maitar and Yuyun Indradi, Greenpeace: *"REDD is not answering the real problems of deforestation, yet"*. Available at: http://www.redd-monitor.org/2012/04/10/interview-with-bustar-maitar-and-yuyun-indradi-greenpeace-redd-is-not-answering-the-real-problems-of-deforestation-yet/#more-11858.

Innovative Resources Management (IRM). (2005). *Community Options Analysis and Investment Toolkit (COAIT): A manual, v1.0*. Washington, DC. Available at: http://www.satyadi.com.

Innovative Resources Management (IRM). (2006). *Congo Livelihood Improvement and Food Security Project*. Cooperative Agreement No. 623-A-00-03-00068-00. Available at: http://pdf.usaid.gov/pdf_docs/PDACH746.pdf.

International Institute for Sustainable Development. (2012). *Engaging the private sector in REDD+: Challenges and opportunities*. Discussion paper. Winnipeg, Manitoba. Available at: http://www.theredddesk.org/sites/default/files/resources/pdf/2012/redd_engaging_private_sector.pdf.

International Rivers. (2008). *Rip-offsets: The failure of the Kyoto Protocol's Clean Development Mechanism*. Available at: http://www.internationalrivers.org/resources/rip-offsets-the-failure-of-the-kyoto-protocol-s-clean-development-mechanism-2649.

International Work Group for Indigenous Affairs (IWGIA) and Asia Indigenous Peoples Pact (AIPP). (2011). *Understanding community-based REDD+: A manual for indigenous community trainers*. Available at: http://www.aippnet.org/home/images/stories/CB-REDD-Trainers_small-20120117172426.pdf.

IPCC. (2007). *Climate Change 2007: The Physical Science Basis*. Contribution of Working Group I to the Fourth Assessment Report of the Intergovernmental Panel on Climate Change. Solomon, S., Qin, D., Manning, M., Chen, Z., Marquis, M., Averyt, K. B., Tignor, M. and Miller, H. L. (eds). Cambridge University Press, Cambridge, UK and New York: 996 pp.

IPCC. (2011). Summary for policymakers. In Field, C. B., Barros, V., Stocker, T. F., Qin, D., Dokken, D., Ebi, K. L., Mastrandrea, M. D., Mach, K. J., Plattner, G.-K., Allen, S., Tignor, M. and P. M. Midgley (eds), *Intergovernmental Panel on Climate Change Special Report on Managing the Risks of Extreme Events and Disasters to Advance Climate Change Adaptation*. Cambridge University Press, Cambridge, UK and New York.

IRIN (2006). *In-depth Minorities Under Siege – Pygmies today in Africa*. Available at: http://www.irinnews.org/InDepthMain.aspx?InDepthId=9&ReportId=58627.

Isenberg, J. and Potvin, C. (2010). *Financing REDD in developing countries: A supply and demand analysis*. Available at: http://biology.mcgill.ca/faculty/potvin/articles/Potvin_08CP604_suppl.pdf.

IUCN. (2009). *REDD-plus scope and options for the role of forests in climate change mitigation strategies*. International Union for conservation of Nature. November. Washington, DC. Available at: http://cmsdata.iucn.org/downloads/redd_scope_english.pdf.

IUCN. (2012). *Wanted: The pro-poor REDD+ principles.* World Conservation Congress. Available at: http://portals.iucn.org/2012forum/?q=node/2688.

IUFRO. (2012). *REDD+ May Cut Both Ways – Potential Trade-offs between Climate Change Mitigation and Biodiversity Conservation Require Well Thought Out Measures.* Global Forest Expert Panels. CBD Briefing Note, October 15, Hyderabad, India.

Johns, T., Nepstad, D., Merry, F., Laporte, N. and Goetz, S. (2008). A three-fund approach to incorporating government, public and private forest stewards into a REDD funding mechanism. *International Forestry Review*, 10(3), 458–464.

Josserand, H. (2001). *Community Based Natural Resource Management in Madagascar.* PTE Programme, USAID Madagascar.

Kahneman, D. (2011). *Thinking, Fast and Slow.* Farrar, Straus and Giroux, New York.

Kanak, D. and Henderson, I. (2012). Closing the REDD+ Gap: the Global Forest Finance Facility. Available at: http://www.theredddesk.org/sites/default/files/resources/pdf/2012/redd_discussion_paper-final-pdf.pdf.

Kanninen, M., Murdiyarso, D., Seymour, F., Angelsen, A., Wunder, S. and German, L. (2007). *Do trees grow on money? The implications of deforestation research for policies to promote REDD.* Center for International Forestry Research (CIFOR), Bogor, Indonesia. Available at: http://www.cifor.org/publications/pdf_files/Books/BKanninen0701.pdf.

Kareiva, P., Lalasz, R. and Marvier, M. (2012). *Conservation in the Anthropocene. Breakthrough Journal.* Available at: http://breakthroughjournal.org/content//2012/02/conservation_in_the_anthropoce-print.html.

Karsenty, A. (2008). The architecture of proposed REDD schemes after Bali: Facing critical choices. *International Forestry Review*, 10(3).

Karsenty, A. (2009). What the (carbon) market cannot do. *Perspective: Forests/Climate Change.* CIRAD, Montpellier, France. No 1, November.

Karsenty, A. (2010). *La lutte contre la deforestation dans les "Etats fragile": une vision renouvellee a l'aide au developpement.* Centre d'analyse strategique. La Note de Veille. No 180. Paris.

Karsenty, A. (2011). Combining conservation incentives with investment. *Perspective: Forests/Climate Change.* CIRAD, Montpellier, France. No. 7, January.

Karsenty, A. (2013). Personal communication.

Karsenty, A. and Ongolo, S. (2012). Les terres agricoles et les forêts dans la mondialisation: de la tentation de l'accaparement à la diversification des modèles? *Cahiers Demeter.* Available at http://agents.cirad.fr/pjjimg/alain.karsenty@cirad.fr/DEMETER_Accaparement_des_terres_et_mondialisation.pdf.

Karsenty, A., Vogel, A. and Castell, F. (2012a). "Carbon rights", REDD+ and payments for environmental services. *Environmental Science & Policy*, in press.

Karsenty, A. with Tulyasuwan, N., Global Witness and Ezzine de Blas, D. (2012b). *Financing options to support REDD+ activities.* CIRAD, Montpellier, France.

Katoomba Group. (2009). *Indigenous engagement in REDD: Developing a Project with the Suruí in the Southwest Amazon of Brazil.* Forest Trends, Washington, DC. Available at: http://www.overbrook.org/newsletter/03_09/pdfs/env/Katoomba_Group.pdf.

Kaufman, F. (2011). *How Goldman Sachs created the food crisis: Don't blame American appetites, rising oil prices, or genetically modified crops for rising food prices. Wall Street's at fault for the spiraling cost of food.* Foreign Policy. Available at: http://www.foreignpolicy.com/articles/2011/04/27/how_goldman_sachs_created_the_food_crisis?page=0,1.

Killeen, T., Schroth, G., Turner, W., Harvey, C., Steininger, M., Dragisic, C. and Mittermeier, R. (2011). *Stabilizing the agricultural frontier: Leveraging REDD with biofuels for sustainable development.* PowerPoint. Science and Knowledge (CABS), Center for Environmental Leadership in Business (CELB), Conservation International. Available at: http://www.nwf.org/~/media/PDFs/Global-Warming/Policy-Solutions/Stopping-Deforestation/REDD-Workshop/KilleenREDDbiofuels%20Sept%207.ashx.

Kissinger, G., Herold, M., and De Sy, V. (2012). *Drivers of Deforestation and Forest Degradation: A Synthesis Report for REDD+ Policymakers*. Lexeme Consulting, Vancouver Canada, August.

Knorringa, P. and Helmsing, A. (2008). Beyond an enemy perception: Unpacking and engaging the private sector. *Development and Change*, 39, 1053–1062.

Knox, A., Caron, C., Goldstein, A. and Miner, J. (2010). "The Interface of Land and Natural Resource Tenure and Climate Change Mitigation Strategies: Challenges and Options". Paper prepared for the Expert Meeting on Land Tenure Issues for Implementing Climate Change Mitigation Policies in the AFOLU sectors. Sponsored by the U.N. Rome: FAO.

Knutti, R. and Plattner, G.-K. (2012). Comments on "Why Hasn't Earth Warmed as Much as Expected?" *J. Climate*, 25, 2192–2199. doi: http://dx.doi.org/10.1175/2011JCLI4038.1.

Kothari, A. (2004). *Displacement fears. Frontline*. Available at: http://www.frontlineonnet.com/fl2126/stories/20041231000108500.htm.

Koyunen, C. and Yilmaz, R. (2009). *The impact of corruption on deforestation: A cross-country evidence. The Journal of Developing Ideas*. Spring. Available at: http://pod-208.positive-internet.com/uploads/TheImpactofCorruptiononDeforestation.pdf.

Kremen, C., Niles, J. O., Dalton, M. G., Daily, G. C., Ehrlich, P. R., Fay, J. P., Grewal, D. and Guillery, R. P. (2000). Economic incentives for rain forest conservation across scales. *Science* ,288, 1828–1832.

Krugman, P. (2010). *Building a green economy. The New York Times*. 7 April. Available at: http://www.nytimes.com/2010/04/11/magazine/11Economy-t.html.

Lalande, J. (2012). *"Cash Investigation": WWF poursuit France 2 en justice*. Available at: http://www.ozap.com/actu/-cash-investigation-quand-wwf-veut-faire-censurer-une-emission-de-france-2/440702.

Lang, C. (2010). *Forests, carbon markets and hot air: Why the carbon stored in forests should not be traded*. REDD-Monitor, 11 January. Available at: http://www.redd-monitor.org/2010/01/11/forests-carbon-markets-and-hot-air-why-the-carbon-stored-in-forests-should-not-be-traded/.

Lang, C. (2011a). *REDD+ and carbon markets: Ten myths exploded*. Available at: http://www.redd-monitor.org/2011/08/23/redd-and-carbon-markets-ten-myths-exploded/.

Lang, C. (2011b). *Increasing deforestation in Guyana gives Norway a headache*. Available at: http://www.redd-monitor.org/2011/01/27/increasing-deforestation-in-guyana-gives-norway-a-headache/.

Lang, C. (2011c). *Ecosystem Restoration Associates project in DR Congo: Plenty of REDD-hot air?* Available at: http://www.redd-monitor.org/2011/08/24/ecosystem-restoration-associates-project-in-dr-congo-plenty-of-redd-hot-air/.

Lang, C. (2011d). *WWF scandal (part 3): Embezzlement and evictions in Tanzania*. Available at: http://www.redd-monitor.org/2012/05/09/wwf-scandal-part-3-corruption-and-evictions-in-tanzania/.

Lang, C. (2012a). *Conservation International, illegal logging and corruption in the Cardamoms, Cambodia*. REDD-Monitor, April 25. Available at: http://www.redd-monitor.org/2012/04/25/conservation-international-illegal-logging-and-corruption-in-the-cardamoms-cambodia/?utm_source=feedburner&utm_medium=email&utm_campaign=Feed%3A+Redd-monitor+%28REDD-Monitor%29.

Lang, C. (2012b). *Environmental activist Chut Wutty shot dead in Cambodia*. REDD-Monitor, April 28. Available at: http://www.redd-monitor.org/2012/04/27/environmental-activist-chut-wutty-shot-dead-in-cambodia/?utm_source=feedburner&utm_medium=email&utm_campaign=Feed%3A+Redd-monitor+%28REDD-Monitor%29.

Lang, C. (2012c). *Up in flames. Tripa peatswamp forest and Indonesia's moratorium*. Available at: http://www.redd-monitor.org/2012/03/30/up-in-flames-tripa-peatswamp-forest-and-indonesias-moratorium/.

Lang, C. (2012d). *WWF scandal (part 3): Embezzlement and evictions in Tanzania.* Redd-Monitor, May 9. Available at: http://www.redd-monitor.org/2012/05/09/wwf-scandal-part-3-corruption-and-evictions-in-tanzania/?utm_source=feedburner&utm_medium=email&utm_campaign=Feed%3A+Redd-monitor+%28REDD-Monitor%29.

Lang, C. (2012e). *Victory: Indonesian court revokes oil palm concession in Tripa peat swamp.* REDD-Monitor, 7 September. Available at: http://climate-connections.org/2012/09/09/victory-indonesian-court-revokes-oil-palm-concession-in-tripa-peat-swamp/.

Lang, C. (2012f). *OPIC political risk insurance for the Oddar Meanchey REDD project in Cambodia: Who benefits?* Available at: http://www.redd-monitor.org/2012/06/26/opic-political-risk-insurance-for-the-oddar-meanchey-redd-project-in-cambodia-who-benefits/.

Lang, C. (2012g). *No REDD+! in RIO +20: A declaration to decolonize the earth and the sky.* Available at: http://www.redd-monitor.org/2012/06/19/no-redd-in-rio-20-a-declaration-to-decolonize-the-earth-and-the-sky/.

Lang, C. (2013). *COONAPIP, Panama's Indigenous Peoples Coordinating Body, withdraws from UN-REDD.* 6 March 2013. Available at: http://www.redd-monitor.org/2013/03/06/coonapip-panamas-indigenous-peoples-coordinating-body-withdraws-from-un-redd/.

LaPiere, R. (1934). Attitudes vs. actions. *Social Forces,* 13(2), 230–237.

Larson, A. P. and Ribot, J. C. (2008). *Democratic Decentralisation through a Natural Resource Lens.* Routledge, London.

Larson, A., Corbera, E., Cronkleton, P., van Dam, C., Bray, D., Estrada, M., May, P., Medina, G., Navarro, G. and Pacheco, P. (2011). *Rights to forests and carbon under REDD+ initiatives in Latin America.* Available at: http://www.theredddesk.org/sites/default/files/resources/pdf/2011/rights_to_forest_brief.pdf.

Laurance, W., Useche, D., Rendeiro, J., Kalka, M., Bradshaw, J., Sloan, S., Laurance, S., Campbell, M., Abernethy, K., Alvarez, P., Arroyo-Rodriguez, V., Ashton, P., Benitez-Malvido, J., Blom, A., Bobo, K., Cannon, C., Cao, M., Carroll, R., Chapman, C., Coates, R., Cords, M., Danielsen, F., De Djin, B., Dinerstein, E., Donnelly, M., et al. (2012). Averting biodiversity collapse in tropical forest protected areas. *Nature,* doi:10.1038/nature11318.

La Via Campesina. (2011). *Our carbon is not for sale!* REDD-Monitor. http://www.redd-monitor.org/2011/09/17/our-carbon-is-not-for-sale-via-campesina-rejects-redd-again/#more-9606.

La Via Campesina. (2012a). *Rio+20: La Via Campesina rejects REDD.* 28 February. Available at: http://www.carbontradewatch.org/take-action/rio-20-la-via-campesina-rejects-redd.html.

La Via Campesina. (2012b). *Rio+20: International Campaign of Struggles: Peoples of the World against the Commodification of Nature.* Quoted in Global Justice Ecology Project, May 8. Available at: http://climate-connections.org/2012/05/08/rio20-international-campaign-of-struggles-peoples-of-the-world-against-the-commodification-of-nature/.

La Via Campesina (2012c). *UN climate negotiations move towards burning the planet.* September 18. Available at: http://viacampesina.org/en/index.php/actions-and-events-mainmenu-26/-climate-change-and-agrofuels-mainmenu-75/1296-bangkok-un-climate-negotiations-move-towards-burning-the-planet.

Lavigne Delville, P. (2002). When farmers use pieces of paper to record their transactions in francophone rural Africa: Insights into the dynamics of institutional innovation. *European Journal of Development Research,* 14(2), 89–108.

Lazarus, R. (2009). Super wicked problems and climate change: Restraining the present to liberate the future. *Cornell Law Review,* 94, 1153.

LeClair, E. and Schneider, H. (1968). *Economic Anthropology: Readings in Theory and Analysis.* Holt, Rhinehart, and Winston, New York.

Lederer, M. (2011). From CDM to REDD+ – What do I know for setting up effective and legitimate carbon governance? *Ecological Economics*, doi:10.1016/j.ecolecon.2011.02.003.

Leighton Schwartz, M. and Notini, J. (1994). *Desertification and Migration: Mexico and the United States*. Natural Heritage Institute, San Francisco.

Leiserowitz, A., Maibach, E., Roser-Renouf, C. and Hmielowski, J. (2012). "Global Warming's Six Americas", March 2012 & Nov. 2011. Yale University and George Mason University. Yale Project on Climate Change Communication, New Haven, CT. Available at: http://environment.yale.edu/climate/files/Six-Americas-March-2012.pdf.

Leisher, C., Sanjayan, M., Blockhus, J., Kontoleon, A. and Larsen, S. N. (2010). *Does Conserving Biodiversity Work to Reduce Poverty? A State of Knowledge Review*. The Nature Conservancy, University of Cambridge, IIED and Conservation Development North South.

Lenton, T. (2011). *2°C or not 2 °C? That is the climate question*. Available at: http://www.nature.com/news/2011/110504/full/473007a.html.

Lewis, D. and Mosse, D. (2006). Encountering order and disjuncture: Contemporary anthropological perspectives on the organization of development. *Oxford Development Studies*, 34(1), March. Routledge, Taylor & Francis Group.

Llanos, R. E. and Feather, C. (2011). *The reality of REDD+ in Peru: Between theory and practice*. AIDESEP, FENAMAD, CARE and FPP. Available at: http://www.forestpeoples.org/sites/fpp/files/publication/2011/11/reality-redd-peru-between-theory-and-practice-website-english-low-res.pdf.

Lohmann, L. (2006). *Carbon Trading: A Critical Conversation on Climate Change, Privatization, and Power*. Development dialogue. No. 48, September. Dag Hammarskjöld Foundation. Uppsala, Sweden.

Lohmann, L. (2008). Carbon trading, climate justice and the production of ignorance: Ten examples. *Development*, 51, 359–365, doi:10.1057/dev.2008.27.

Lohmann, L. (2009). *Regulation as corruption in the carbon offset markets: Cowboys and choirboys united*. Available at: http://www.thecornerhouse.org.uk/sites/thecornerhouse.org.uk/files/Athens%2010.pdf.

Lohmann, L. (2011a). Capital and climate change. *Development and Change*, 42(2), 649–668, doi: 10.1111/j.1467-7660.2011.01700.x.

Lohmann, L. (2011b). The endless algebra of carbon markets. *Capitalism Nature Socialism*, 22(4), December.

Lopez de Victoria, S. (2012). *Media manipulation of the masses: How the media psychologically manipulates*. *Psych Central*. Available at: http://psychcentral.com/blog/archives/2012/02/06/media-manipulation-of-the-masses-how-the-media-psychologically-manipulates/.

Lowrey, K. (2008). Incommensurability and new economic strategies among indigenous and traditional peoples. *Journal of Political Ecology*, 15: 61–74.

Luttrell, C., Loft, L., Gebara, M. F. and Kweka, D. (2012). Who should benefit and why: Discourses on REDD+ benefit sharing. In Angelsen, A., Brockhaus, M., Sunderlin, W. and Verchot, L. (eds), *Analysing REDD: Challenges and Choices*. CIFOR, Bogor, Indonesia.

MacDonald, C. (2008). *Green Inc.: An Environmental Insider Reveals How a Good Cause Has Gone Bad*. The Lyons Press, Guilford, CT.

Macintosh, N. (2003). *From Rationality to Hyperreality: Paradigm Poker*. Research Paper No. 04-09. Queen's University School of Business. Kingston, Ontario, Canada.

Macintosh, N. (2006). Commentary: The FASB and accounting for economic reality. Accounting—truth, lies, or "bullshit"? A Philosophical Investigation. *Accounting and the Public Interest*, 6(1): 22–36.

MacKinnon, K. (2011). Are we really getting conservation so badly wrong? *PLoS Biology*, January, 9(1): e1001010. Available at: 10.1371/journal.pbio.1001010.

Mansuri, G. and Rao, V. (2004). Community based and driven development: A critical review. *The World Bank Research Observer*, 19(1). The International Bank for Reconstruction and Development/The World Bank. Washington, DC.

Mansuri, G. and Rao, V. (2013). *Localizing Development: Does Participation Work?* A World Bank Policy Research Report. Washington, DC. Available at: http://siteresources.worldbank.org/INTRES/Resources/469232-1321568702932/8273725-1352313091329/PRR_Localizing_Development_full.pdf.

Manzi, J. (2008). *Weitzman: Formalism run amok*. The American Scene. Available at: http://theamericanscene.com/2008/01/04/weitzman-formalism-run-amok.

Manzi, J. (2010a). *Re: Reply to Jim Manzi*. The American Scene. http://theamericanscene.com/2010/04/26/re-reply-to-jim-manzi.

Manzi, J. (2010b). *Why the decision to tackle climate change is not as simple as Al Gore says. The New Republic*, 22 June. Available at: http://www.tnr.com/blog/critics/75757/why-the-decision-tackle-climate-change-isn%E2%80%99t-simple-al-gore-says?page=0%2C1.

Marshall, G. (2007). *Nesting, Subsidiarity, and Community-based Environmental Governance Beyond the Local Level*. Occasional Paper 2007/01, Institute for Rural Futures, University of New England, Armidale, Australia. Available at: http://www.ruralfutures.une.edu.au/publications/2.php?nav=Occasional%20Paper.

Mathieu, P., Lavigne Delville, P., Ouédraogo, H., Zongo, M. and Paré, L. with Baud, J., Bologo, E., Koné, N. and Triollet, K. (2003). *Making Land Transactions More Secure in the West of Burkina Faso*. Issue Paper No. 117, IIED, London.

Mayers, D. (2012). *Infographic: Mobile phone growth transforms Africa*. SmartPlanet. Available at: http://www.smartplanet.com/blog/global-observer/infographic-mobile-phone-growth-transforms-africa/7006.

McAfee, K. (2012a). The contradictory logic of global ecosystem services markets. *Development and Change*, 43(1), 105–131. doi: 10.1111/j.1467-7660.2011.01745.xC_2012. International Institute of Social Studies. Blackwell Publishing, Oxford, UK and Malden, MA.

McAfee, K. (2012b) Nature in the market-world: ecosystem services and inequality. *Development* 55(1): 25–33.

McKinsey & Company (2011). *Pathways to a Low Carbon Economy*. Available at: http://www.mckinsey.com/en/Client_Service/Sustainability/Latest_thinking/Pathways_to_a_low_carbon_economy.aspx.

McShane, T. and Wells, M. (eds) (2004). *Getting Biodiversity Projects to Work: Towards More Effective Conservation and Development*. Columbia University Press, New York.

McShane, T., O'Connor, S., Kinzig, A., Pulgar-Vidal, M., Monteferri, B., Dammert, J. L., Hirsch, P., Van Thang, H. and Brosius, P. (2011). *Advancing Conservation in a Social Context: Working in a World of Trade-offs*. Final report. Arizona State University. Tempe, AZ.

Meckling, J. (2011). *Carbon Coalitions: Business, Climate Politics, and the Rise of Emissions Trading*. The MIT Press, Cambridge, MA and London.

Meinshausen, M., Meinshausen, N., Hare, W., Raper, S. C. B, Frieler, K., Knutti, R., Frame, D. J. and Allen, M. R. (2009). Greenhouse-gas emission targets for limiting global warming to 2 °C. *Nature* 458, 1158–1162.

Meridian House. (2011). *Guidelines for REDD+ and Reference Levels: Principles and Recommendations*. Prepared for the government of Norway by Arild Angelsen, Doug Boucher, Sandra Brown, Velerie Merckx, Charlotte Streck and Daniel Zarin. Available at: www.REDD-OAR.org.

Merlet, P. and Bastiaensen, J. (2012). *Struggles over Property Rights in the Context of Large-scale Transnational Land Acquisitions. Using Legal Pluralism to Re-politicize the Debate. Illustrated with Case Studies from Madagascar and Ghana*. AGTER. IOB Discussion Paper 2012-02, May. Available at: http://www.agter.asso.fr/IMG/pdf/2012-02_dp_merlet_bastiaensen_legal-pluralism.pdf.

Miliband, D. (2006). *The Great Stink: Towards an Environmental Contract.* Audit Commission Annual Lecture, London, 19 July.

Millennium Ecosystem Assessment (MEA). (2005). *Ecosystems and Human Ill-being: Biodiversity Synthesis.* World Resources Institute, Washington, DC.

Miller, A. (2012). *The Carbon Rush.* Video. Available at: http://www.thecarbonrush.net/.

Milne, S. (2012). Chut Wutty: Tragic casualty of Cambodia's dirty war to save forests. *New Mandala*, 30 April. Available at: http://asiapacific.anu.edu.au/newmandala/2012/04/30/chut-wutty-tragic-casualty-of-cambodia%E2%80%99s-dirty-war-to-save-forests/.

Milne, S. and Adams, B. (2012). Market Masquerades: Uncovering the politics of community-level payments for environmental services in Cambodia. *Development and Change*, 43, 133–158. doi:10.1111/j.1467-7660.2011.01748.x. Blackwell Publishing Ltd, Oxford, UK.

Mitchell, T. (2002). *Rule of Experts: Egypt, Techno-politics, Modernity.* University of California Press, Berkeley, CA.

Molnar, A., Scherr, S. and Khare, A. (2004). *Who Conserves the World's Forests?* Forest Trends, Washington, DC.

mongabay.com. (2012). *Drivers of deforestation.* Available at: http://rainforests.mongabay.com/deforestation_drivers.html.

mongabay.com. (2013). Experts outline how REDD+ credits could fit into California's cap-and-trade program. January 27, 2013. Available at http://news.mongabay.com/2013/0127-ca-redd-panel.html#1CTdLQoLKMWAzag4.99.

Mosse, D. and Lewis, D. (2004). *The Aid Effect.* Pluto Press. London and Ann Arbor, MI.

Moyo, D. (2009). *Dead Aid: Why Aid is Not Working and How There is a Better Way for Africa.* Farrar, Strauss and Giroux, New York.

Munden, L. (2011). Presentation at Eleventh RRI Dialogue on Forests, Governance and Climate Change. London. Available at: http://www.redd-monitor.org/2011/12/15/the-munden-project-investing-in-communities-is-the-most-effective-way-of-reducing-deforestation/.

Munden Project. (2011). *REDD AND FOREST CARBON: Market-based Critique and Recommendations.* Available at: http://www.mundenproject.com/forestcarbonreport2.pdf.

Myers, N. A., Mittermeier, R. A., Mittermeier, G. C., da Fonseca, G. A. B. and Kent, J. (2002). Biodiversity hotspots for conservation priorities. *Nature*, 403, 853–858.

National Audobon Society, EETAP, US Fish and Wildlife Service and TogetherGreen. (2011). *Tools of Engagement: A Toolkit for Engaging People in Conservation.* Available at: http://web4.audubon.org/educate/toolkit/pdf/the-toolkit.pdf.

National Council of Nonprofits. (2013). What is Capacity Building and Why is it Needed? Available at: http://www.councilofnonprofits.org/capacity-building/what-capacity-building.

National Research Council. (2004). *Valuing Ecosystem Services: Toward Better Environmental Decision-Making.* The National Academies Press, Washington, DC.

National Snow and Ice Data Center. (2013). *Arctic Sea Ice News and Analysis.* Available at: http://nsidc.org/arcticseaicenews/ (downloaded February 19, 2013).

Nellemann, C. (ed.) (2012). *Green Carbon, Black Trade: Illegal logging, Tax Fraud and Laundering in the World's Tropical Forests.* UNEP and INTERPOL, Nairobi, Kenya.

Nelson, A. and Chomitz, K. (2011). Effectiveness of strict vs. multiple use protected areas in reducing tropical forest fires: A global analysis using matching methods. *PLoS ONE* 6(8): e22722. doi:10.1371/journal.pone.0022722.

Nepstad, D. (2012). *Commentary: Greenpeace report threatens climate change mitigation and tropical forests.* Special to mongabay.com. September 25. Available at: http://news.mongabay.com/2012/0925-nepstad-greenpeace-report.html#2i4o4eCKsRhIXlRC.99.

Nepstad, D., Schwartzman, S., Bamberger, B., Santilli, M., Ray, D., Schlesinger, P., Lefebvre, P., Alencar, A., Prinz, E., Fiske, G. and Rolla, A. (2006). Inhibition of Amazon deforestation and fire by parks and indigenous lands. *Conservation Biology*, 20, 65–73.

Neumann, R. P. (1992). Political ecology of wildlife conservation in the Mt. Meru area of Northeast Tanzania. *Land Degradation and Rehabilitation*, 3: 85–98.

Newell, P. and Paterson, M. (2010). *Climate Capitalism: Global Warming and the Transformation of the Global Economy*. Cambridge University Press, Cambridge, UK.

Newell, P., Boykoff, M. and Boyd, E. (eds) (2012). *The New Carbon Economy: Constitution, Governance, and Contestation*. John Wiley & Sons, Chichester, UK.

Nielsen, S. (2011). *Amazon deforestation rates double as farmers anticipate pardon*. Bloomberg, July 1, 2011.

Niezen, R. (2003). *The Origins of Indigenism: Human Rights and the Politics of Identity*. University of California Press, Berkeley.

Niles, J. (2011). *What is the current status of REDD+?* Interview on mongabay.com. March 23. Available at: http://print.news.mongabay.com/2011/0323-niles_REDD+_interview.html.

Nolte, C., Agrawal, A., and Barreto, P. (2013). Setting priorities to avoid deforestation in Amazon protected areas: are we choosing the right indicators? Available at http://iopscience.iop.org/1748-9326/8/1/015039/article;

NPR. (2010). *Energy Policy Explored as Cap-and-Trade Dies*. November 5. Available at: http://www.npr.org/templates/story/story.php?storyId=131104674;

Obama, B. (2010). Quoted in NPR. *Energy Policy Explored as Cap-and-Trade Dies*. 5 November. Available at: http://www.npr.org/templates/story/story.php?storyId=131104674.

O'Brien, K., Hayward, B. and Berkes, F. (2009). Rethinking social contracts: Building resilience in a changing climate. *Ecology and Society*, 14(2), 12. Available at: http://www.ecologyandsociety.org/vol14/iss2/art12/.

OECD. (2005). *The Paris Declaration on Aid Effectiveness*. OECD, Paris. Available at: http://www.oecd.org/development/aideffectiveness/34428351.pdf.

Ojha, H., Persha, L. and Chhartre, A. (2009). *Community Forestry in Nepal: A Policy Innovation for Local Livelihoods*. International Food Policy Research Institute (IFPRI), Washington, DC.

Olbrei, E. and Howes, S. (2012). *A Very Real and Practical Contribution? Lessons from the Kalimantan Forests and Climate Partnership*. Australian National University. Discussion Paper 16. Available at: http://devpolicy.org/a-very-real-and-practical-contribution-lessons-from-the-kalimantan-forests-and-climate-partnership/.

Olivier, J., Janssens-Maenhout, G. and Peters, J. (2012). *Trends in global CO_2 emissions: 2012 Report*. PBL Netherlands Environmental Assessment Agency, The Hague/Bilthoven.

O'Neill, E. and Muir, M. (2010). *Performance Measurement in the Conservation Community*. PowerPoint Presentation at the Measuring Conservation Success Summit. May 5–6. Gordon and Betty Moore Foundation, Palo Alto, CA.

Onta, N. (2012). *When Pigs Fly: Why is Including Women in Managing Forests Still So Unusual?* IIED News and Blogs. Available at: http://www.iied.org/when-pigs-fly-why-including-women-managing-forests-still-so-unusual.

Or, A. (2012). *TIAA-CREF raises $2 bln for agriculture investment*. Market Watch. Available at: http://farmlandgrab.org/post/view/20494.

Ostrom, E. (1990). *Governing the Commons: The Evolution of Institutions for Collective Action*. Cambridge University Press, Cambridge, UK.

Pacific Environment, FERN, Focus on the Global South. (2012). *Precedent-setting insurance for REDD project in Cambodia raises concerns*. Available at: http://www.redd-monitor.org/wordpress/wp-content/uploads/2012/06/OPIC-Risk-Insurance-REDD-Cambodia.pdf.

Pagiola, S. and Bosquet, B. (2009). *Estimating the costs of REDD at the country level*. Forest Carbon Partnership Facility, World Bank Version 2.2 – September 22, 2009. Available at: http://mpra.ub.uni-muenchen.de/18062/1/REDD-Costs-22.pdf.

Painter, M. (2004). Quoted in J. Roach, *Unique Bolivia Park Begun by Indigenous Peoples. National Geographic News*. January 13. Available at: http://news.nationalgeographic.com/news/2004/01/0113_040113_chacopark.html.

Parekh, P. (2011). *Durban package lacks ambition and equity.* Available at: http://www.climate-consulting.org/2011/12/12/durban-package-lacks-ambition-and-equity/.

Paz-Rivera, C. (2011). *REDD+ negotiations and key milestones from Cancun to Durban.* PowerPoint. UN-REDD Secretariat, Geneva.

Pearce, F. (2010). *Forest carbon stores may be massively overestimated. New Scientist.* September 6. Available at: http://www.newscientist.com/article/dn19408-forest-carbon-stores-may-be-massively-overestimated.html.

Pearce, F. (2012). *Turning Point: What future for forests peoples and resources in the emerging world order?* Rights and Resources Group. Washington, DC. Available at: http://www.rightsandresources.org/documents/files/doc_4701.pdf.

Pelletier, J., Ramankutty, N. and Potvin, C. (2011). Diagnosing the uncertainty and detectability of emission reductions for REDD+ under current capabilities: an example from Panama, *Environmental Research Letters,* 6024005.

Peluso, N. (1992). *Rich Forests, Poor People: Resource Control and Resistance in Java.* University of California Press, Berkeley, CA.

Pereira, J. (2012). *Cashing in on climate change? Assessing whether private funds can be leveraged to help the poorest countries respond to climate challenges.* European Network on Debt and Development (EURODAD). Available at: http://eurodad.org/wp-content/uploads/2012/04/CF-report_final_web.pdf.

Persha, L., Agrawal, A. and Chhatre, A. (2011). Social and ecological synergy: Local rulemaking, forest livelihoods, and biodiversity conservation. *Science,* 331, 1606 (2011). DOI: 10.1126/science.1199343.

Peskett, L. and Brodnig, G. (2011). "Carbon rights in REDD+: exploring the implications for poor and vulnerable people." World Bank and REDD-net.

Peters, G. (2008). *Who should pay for Norway's greenhouse gas emissions? Klima* (5). Available at: http://www.cicero.uio.no/fulltext/index.aspx?id=6865&lang=NO.

Peters, G. P., Andrew, R. M., Boden, T., Canadell, J. G., Ciais, P., Le Quéré, C., Marland, G., Raupach, M. R. and Wilson, C. (2012). The challenge to keep global warming below 2°C. *Nature Climate Change,* 3, 4–6 (2013) doi:10.1038/nclimate1783.

Peters-Stanley, M. (2012). *Bringing it Home: Taking Stock of Government Engagement with the Voluntary Carbon Market.* Forest Trend's Ecosystem Marketplace. Washington, DC.

Peters-Stanley, M., Hamilton, K., Marcello, T. and Sjardin, M. (2011). *Back to the Future: State of the Voluntary Carbon Markets (2011).* A Report by Ecosystem Marketplace & Bloomberg New Energy Finance. Washington, DC and New York.

Peters-Stanley, M., Hamilton, K., Yin, D., Castillo, S. and Norman, M. (2012). *Leveraging the Landscape: State of the Forest Carbon Markets (2012).* Ecosystem Marketplace, Washington, DC.

Phelps, J., Webb, E. and Adams, W. (2012). Biodiversity co-benefits of policies to reduce forest-carbon emissions. *Nature Climate Change,* 2, 497–503, doi:10.1038/nclimate1462.

Pirard, R. and Belna, K. (2012). Agriculture and deforestation: Is REDD+ rooted in evidence?, *Forest Policy and Economics,* doi:10.1016/j.forpol.2012.01.012.

Poffenberger, M. (2009). "Sharing REDD Benefits with Forest Dependent Communities". Powerpoint. Community forestry International. Available at: http://www.unredd.net/index.php?option=com_docman&task=doc_details&gid=1241&Itemid=53.

Poffenberger, M., De Gryze, S. and Durschinger, L. (2009). *Designing Collaborative REDD Projects: A Case Study from Oddar Meanchey Province, Cambodia.* Available at: http://www.terraglobalcapital.com/press/ReddProjects.pdf.

Polycarp, C. and Brown, L. (2012). *Priorities for the Green Climate Fund in 2012.* Available at: http://insights.wri.org/news/2012/06/priorities-green-climate-fund-2012.

Polycarp, C. and Patel, M. (2012). *What's the future of the climate investment funds?* Available at: http://insights.wri.org/news/2012/10/whats-future-climate-investment-funds.

Porter-Bolland, L., Ellis, E., Guariguata, M., Ruiz-Mallén, I., Negrete-Yankelevich, S. and Reyes-García, V. (2011). Community managed forests and forest protected areas: An assessment of their conservation effectiveness across the tropics. *Forest Ecology Management*, doi:10.1016/j.foreco.2011.05.034.

Portes, A. (1998). Social capital: Its origins and applications in modern sociology. *Annual Review of Sociology*, 24: 1.24.

Potsdam Institute for Climate Impact Research and Climate Analytics (2012). *Turn Down the Heat: Why a 4°C Warmer World must be Avoided*. The World Bank, Washington, DC.

Poynton, S. (2012). Quoted in C. Lang, Interview with Scott Poynton, TFT: *"We help companies clean up their supply chains – that is the way we need to go to protect the world's forests".* REDD-Monitor, 23 August. Available at: http://www.redd-monitor.org/2012/08/23/interview-with-scott-poynton-tft/#more-12729.

Pretty, J. N. (1995). Participatory learning for sustainable agriculture. *World Development*, 23(8), 1247–1260.

Pretty, J. N. (2002). People, livelihoods and collective action in biodiversity management. Available at: http://userpage.fu-berlin.de/deltongo/osi-biodiversity/downloads/CUP_4.pdf.

Pretty, J. N., Guijt, I., Thompson, J. and Scoones, I. (1995). *Participatory Learning and Action: A Trainer's Guide*. IIED, London.

Prizibisczki, C. (2010). *Indigenous Leaders Taking REDD Into Their Own Hands. Ecosystem Marketplace.* Available at: http://www.ecosystemmarketplace.com/pages/dynamic/article.page.php?page_id=7611&section=home.

PwC. (2012). *Assessing Options for Effective Mechanisms to Share Benefits: Insights for REDD+ Initiatives.* Program on Forests (PROFOR), The World Bank. Washington, DC.

Quarles van Ufford, P. and Giri, A. (eds) (2003). *A Moral Critique of Development: In Search of Global Responsibilities.* Routledge, Taylor and Francis Group, London and New York.

Rayner, J., Humphreys, D., Perron Welch, F., Prabhu, R. and Verkooijen, P. (2010). Introduction. In *Embracing Complexity: Meeting the Challenges of International Forest Governance. A Global Assessment Report.* Prepared by the Global Forest Expert Panel on the International Forest Regime. IUFRO World Series Volume 28. Vienna. 172pp.

Reardon, S. and Hooper, R. (2012). *How the Mafia is destroying the rainforests. New Scientist,* 1 October. Available at: http://www.newscientist.com/article/dn22321-how-the-mafia-is-destroying-the-rainforests.html.

Redford, K. and Robinson, J. (1985). Hunting by indigenous peoples and conservation of game species. *Cultural Survival Quarterly,* 9(1): 41–44.

RED por la paz Chiapas (2012). *De la tierra al asfalto: Informe de la mission civil de observación de la RED por la paz Chiapas y CAIK al programa ciudades rurales sustentables.* Available at: http://chiapaspaz.files.wordpress.com/2012/05/de-la-tierra-al-asfalto-informe-red-por-la-paz-2012.pdf.

REDD-Monitor. (2008). *Risk – the fatal flaw in carbon trading.* http://www.redd-monitor.org/2008/12/05/risk-the-fatal-flaw-in-forest-carbon-trading.

REDD-Monitor. (2012). Interview with Bustar Maitar and Yuyun Indradi, Greenpeace: *"REDD is not answering the real problems of deforestation, yet".* Available at: http://www.redd-monitor.org/2012/04/10/interview-with-bustar-maitar-and-yuyun-indradi-greenpeace-redd-is-not-answering-the-real-problems-of-deforestation-yet/#more-11858.

REDD-Net. (2010). *Monitoring, reporting and verification of social and development issues.* Overseas Development Institute. Available at: http://redd-net.org/files/MRV%20of%20social%20development.pdf.

Responding to Climate Change. (2012). *CBD COP11: REDD+ needs careful management to protect biodiversity.* Available at: http://www.rtcc.org/climate-change-tv/cbd-cop11-not-simply-redd/.

Revkin, A. (2012). *Another round: Conservation on a human-shaped planet.* The Opinion Pages, *New York Times*, April 11. Available at: http://dotearth.blogs.nytimes.com/2012/04/11/another-round-conservation-on-a-human-shaped-planet/#postComment.

Rew, A. and Khan, S. (2006). The moral setting for governance in Keonjhar: The cultural framing of public episodes and development processes in Northern Orissa, India. *Oxford Development Studies*, 34:1, 99–115.

Ribot, J. (2010). *Seeing REDD for Local Democracy: A Call for Democracy Standards.* Invited essay, Beckmann Institute. University of Illinois-Champagne-Urbana. Available at: http://sdep.beckman.illinois.edu/files/Ribot_Redd_CV3.pdf.

Ribot, J. and Larson, A. (2012). Reducing REDD risks: Affirmative policy on an uneven playing field. *International Journal of the Commons*, 6(2), 233–254. Available at: http://www.thecommonsjournal.org.

Ribot, J., Treue, T. and Lund, J. (2010). Democratic decentralization in Sub-Saharan Africa: Its contribution to forest management, livelihoods, and enfranchisement. *Environmental Conservation* 37, 35–44.

Richards, M. (2011). *Social and Biodiversity Impact Assessment (SBIA) Manual for REDD+ Projects: Part 2 – Social Impact Assessment Toolbox.* Climate, Community & Biodiversity Alliance and Forest Trends with Rainforest Alliance and Fauna & Flora International. Washington, DC.

Richards, M. and Panfil, S. N. (2011). *Social and Biodiversity Impact Assessment (SBIA) Manual for REDD+ Projects: Part 1 – Core Guidance for Project Proponents.* Climate, Community & Biodiversity Alliance, Forest Trends, Fauna & Flora International, and Rainforest Alliance. Washington, DC.

Riddell, R. (2009). *Is aid working? Is this the right question to be asking?* Open Democracy. Available at: http://www.opendemocracy.net/roger-c-riddell/is-aid-working-is-this-right-question-to-be-asking.

Rights and Resources Initiative. (2008). *Seeing People through the Trees: Scaling Up Efforts to Advance Rights and Address Poverty, Conflict and Climate Change.* Rights and Resources Initiative, Washington DC.

Rights and Resources Initiative. (2011). *Pushback: Local Power, Global Realignment.* Rights and Resources Initiative, Washington, DC.

Rights and Resources Initiative. (2012). *What Rights? A Comparative Analysis of Developing Countries' National Legislation on Community and Indigenous Peoples' Forest Tenure Rights.* Rights and Resources Initiative, Washington DC.

Robertson, A. F. (1987). *The Dynamics of Productive Relationships: African Share Contracts in Comparative Perspective.* Cambridge University Press, Cambridge.

Robinson, J. G. (2012). Common and conflicting interests in the engagements between conservation organizations and corporations. *Conservation Biology*, 26(6): 967–977, December 2012.

Rousseau, J.-J. (2004) [original 1754]. *Discourse on Inequality.* Kessinger Publishing, Whitefish, MT.

Rudel, T., Coomes, O., Moran, E., Achard, F., Angelsen, A., Xu, J. and Lambin, E. (2005). Forest transitions: towards a global understanding of land use change. *Global Environmental Change*, 15: 23–31.

Russell, D. (2011). *Social Dimensions of REDD+ Workshop: Overview.* Open Forum. USAID Office of Natural Resource Management. Washington, DC.

Sahlins, M. (1974). *Stone Age Economics.* Routledge, London and New York.

Salafsky, N. (2010). *Measuring Effectiveness: An Overture.* PowerPoint presentation at the Measuring Conservation Success Summit. 5–6. Gordon and Betty Moore Foundation, Palo Alto, CA.

Salafsky, N., Salzer, D., Ervin, J., Boucher, T. and Ostlie, W. (2003). *Conventions for Defining, Naming, Measuring, Combining, and Mapping Threats in Conservation: An Initial Proposal for a Standard System.* Conservation Measures Partnership, Washington, DC.

Samyn, J-M, Gasana, J. and Pousse, E. (2012). *The Forest Sector in the Congo Basin countries: 20 years of AFD intervention.* Agence Française de Développement (AFD). Paris, France. May 2012.

Sandel, M. (2012). *What Money Can't Buy: The Moral Limits of Markets.* Farrar, Strauss and Giroux, New York.

Sandor, R. (2012). *Good Derivatives: A Story of Financial and Environmental Innovation.* John Wiley & Sons, Hoboken, NJ.

Sathaye, J. and S. Cobb (2012). "Papua New Guinea Draft RPP: TAP Comments & Recommendations", October 22, 2012. Powerpoint Presentation for FCPF Participants Committee 12th meeting, Brazzaville, Congo. Available at: http://www.forestcarbonpartnership.org/sites/forestcarbonpartnership.org/files/Documents/PDF/Nov2012/TAP%20PNG-RPP%20Presubmission%2021%20Oct%202012-JS.pdf.

Scherr, S. and McNeely, J. (2008). Biodiversity conservation and agricultural sustainability: towards a new paradigm of "ecoagriculture" landscapes. *Phil. Trans. R. Soc. B*, 363, 477–494.

Schieber, G., Gottret, P., Fleisher, L. and Leive, A. (2007). Financing global health: mission unaccomplished. *Health Affairs* 26(4): 921–934.

Schneider, A. (2011). *What Shall We Do Without Our Land? Land Grabs and Rural Resistance in Cambodia.* Global Land Grabbing Conference, Land Deal Politics Initiative, University of Sussex.

Schreckenberg, K., Camargo, I., Withnall, K., Corrigan, C., Franks, P., Roe, D., Scherl, L. and Richardson, V. (2010). *Social Assessment of Conservation Initiatives: A Review of Rapid Methodologies.* An output of the Social Assessment of Protected Areas (SAPA) Initiative. Institute for International Environment and Development, London.

Schumacher, E. (1973). *Small is Beautiful: A Study of Economics as if People Mattered.* Blond & Briggs, London.

Schutt, C. (2009). *Changing the world by changing ourselves: Reflections from a bunch of BINGOs.* IDS Practice Paper 3, IDS. Available at: http://www.ids.ac.uk/idspublication/changing-the-world-by-changing-ourselves-reflections-from-a-bunch-of-bingos.

Schutt, R. (2009). *Investigating the Social World.* 6th edition. Pine Forge, Thousand Oaks, CA.

Schwartzman, S., Environmental Defense et al. (2008). *Getting REDD Right: Reducing Emissions from Deforestation and Forest Degradation (REDD) in the United Nations Framework Convention on Climate Change (UNFCCC).* Environmental Defence Fund, New York.

Schweigert, B. (2012). *Lessons of the bubble.* Minnesota 20|20. March 15. Available at http://www.mn2020.org/issues-that-matter/economic-development/lessons-of-the-bubble.

Scientific Certification Systems (SCS). (2011). *Final CCBA validation report: April Salumei, East Sepik, Papua New Guinea.* Available at: https://s3.amazonaws.com/CCBA/Projects/April_Salumei_Sustainable_Forest_Management_Project/CCB_RMA_AprilSalumei_Validation_RPT_Final_Signed_062411.pdf.

Scott, J. (1977). *The Moral Economy of the Peasant.* Yale University Press, New Haven, CT.

Seifert-Granzin, J. (2011). REDD guidance: Technical project design. In Ebeling, J. and Olander, J. (eds), *Building Forest Carbon Projects.* Forest Trends, Washington, DC.

Seymour, F. (2012). REDD reckoning: A review of research on a rapidly moving target, *CAB Reviews*, 7(32), 1–13.

Seymour, F. and Angelsen, A. (2012). Summary and conclusion: REDD+ without regrets. In Angelsen, A. et al. (eds), *Analyzing REDD+: Challenges and Choices.* CIFOR, Bogor, Indonesia.

Sharma, S. (2012). *The hype versus the reality of carbon markets and land-based offsets: Lessons for the new Africa carbon exchange.* Institute for Agriculture and Trade Policy. Available at: http://www.iatp.org/documents/the-hype-versus-the-reality-of-carbon-markets-and-land-based-offsets-lessons-for-the-new-a.

Sharpe, L. (1952). Steel axes for stone age Australians. *Human Organization*, 11(2): 446–460.

Shepherd, A., Ivins, E., Geruo, A., Barletta, V., Bentley, M., Bettadpur, S., Briggs, K., Bromwich, D., Forsberg, R., Galin, N., Horwath, M., Jacobs, S., Joughin, I., King, M., Lenaerts, J., Li, J., Ligtenberg, S., Luckman, A., Luthcke, S., McMillan, M., Meister, R., Milne, G., Mouginot, J., Muir, A., Nicolas, J., Paden, J., Payne, A., Pritchard, H., Rignot, E., Rott, H., Sandberg Sørensen, L., Scambos, T., Scheuchl, B., Schrama, E., Smith, B., Sundal, A., van Angelen, J., van de Berg, W., van den Broeke, M., Vaughan, D., Velicogna, I., Wahr, J., Whitehouse, P., Wingham, D., Yi, D., Young, D. and Zwally, H. J. (2012). A reconciled estimate of ice-sheet mass balance. *Science*, November 30, 1183–1189.

Sikor, T. and Stahl, J. (eds) (2011). *Forests and People: Property, Governance, and Human Rights*. Earthscan, Abingdon, UK.

Silva-Chavez, G. (2012). *REDD+ almost at the finish line: Doha preview*. Environmental Defense Fund blog. Available at: http://blogs.edf.org/climatetalks/2012/11/21/redd-almost-at-the-finish-line-doha-preview/.

Simula, M. (2010). *Analysis of REDD+ Financing Gaps and Overlaps*. REDD+ Partnership, Washington, DC.

Smith, J. et al. (2009). Assessing dangerous climate change through an update of the Intergovernmental Panel on Climate Change (IPCC) "reasons for concern". *PNAS*, 106(11), 4133–4137. Available at: http://www.pnas.org/content/106/11/4133.full. pdf.

Sokolov, A. et al. (2009). Probabilistic forecast for twenty-first-century climate based on uncertainties in emissions (without policy) and climate parameters. *J. Climate*, 22, 5175–5204. doi: http://dx.doi.org/10.1175/2009JCLI2863.1.

Solomon, S., Qin, D., Manning, M., Alley, R. B., Berntsen, T., Bindoff, N. L., Chen, Z., Chidthaisong, A., Gregory, J. M., Hegerl, G. C., Heimann, M., Hewitson, B., Hoskins, B. J., Joos, F., Jouzel, J., Kattsov, V., Lohmann, U., Matsuno, T., Molina, M., Nicholls, N., Overpeck, J., Raga, G., Ramaswamy, V., Ren, J., Rusticucci, M., Somerville, R., Stocker, T. F., Whetton, P., Wood R. A. and Wratt, D. (2007). Technical summary. In Solomon, S., Qin, D., Manning, M., Chen, Z., Marquis, M., Averyt, K. B., Tignor M. and Miller, H. L. (eds), *Climate Change 2007: The Physical Science Basis. Contribution of Working Group I to the Fourth Assessment Report of the Intergovernmental Panel on Climate Change*. Cambridge University Press, Cambridge, UK and New York.

Sommerville, M. (2011). *Land Tenure and REDD+: Risks to Property Rights and Opportunities for Economic Growth*. Property Rights and Resource Governance Briefing Paper #11. USAID Issue Brief. Washington, DC.

Southall, A. (1976). Nuer and Dinka are people: Ecology, ethnicity and logical possibility. *Man*, 11(4), 463–491. Royal Anthropological Institute of Great Britain and Ireland.

Springer, J. (2009). *Addressing the social impacts of conservation: Lessons from experience and future directions. Conservation and Society*, 7(1). Available at: http://www.conservationandsociety. org/article.asp?issn=0972-4923;year=2009;volume=7;issue=1;spage=26;epage=29;au last=Springer.

Standing, A. (2012). *Corruption and REDD+: Identifying risks amid complexity*. Anti-corruption Resource Center, Chr. Michelsen Institute. Available at: www.U4.no.

Stern, N. (2006). *Stern Review on the Economics of Climate Change*. UK Treasury. London.

Stern, N. (2007). *The Economics of Climate Change: The Stern Review*. Cambridge University Press, Cambridge, UK.

Stern, N. (2009). *The Global Deal: Climate Change and the Creation of a New Era of Progress and Prosperity*. Public Affairs, New York.

Stockholm Resilience Center. (2011). *Governance of social-ecological systems in an increasingly uncertain world needs to be collaborative, flexible and learning-based*. Insight #3, Adaptive

Governance. Stockholm University. Available at: http://www.stockholmresilience.org/download/18.3e9bddec1373daf16fa439/Insights_adaptive_governance_120111-2.pdf.

Sukhdev, P. (2012). Quoted in Global Forest Coalition and the Global Justice Ecology Project. Available at: http://www.redd-monitor.org/2012/01/25/new-video-a-darker-shade-of-green-redd-alert-and-the-future-of-forests/?utm_source=feedburner&utm_medium=email&utm_campaign=Feed%3A+Redd-monitor+%28REDD-Monitor%29.

Sukhdev, P., Prabhu, R., Kumar, P., Bassi, A., Patwa-Shah, W., Enters, T., Labbate, G. and Greenwalt, J. (2012). *REDD+ and a Green Economy: Opportunities for a Mutually Supportive Relationship*. UN-REDD Programme Policy Briefs.

Sullivan, S. (2010). *The Environmentality of 'Earth Incorporated': On Contemporary Primitive Accumulation and the Financialisation of Environmental Conservation*. Paper presented at An Environmental History of Neoliberalism Conference, Lund University, May 6–8. Available at: http://www.worldecologyresearch.org/papers2010/Sullivan_financialisation_conservation.pdf.

Sullivan, S. (2011). *Banking Nature?* Working Papers Series #8. ISSN 2045-5763. Available at: http://openanthcoop.net/press/2011/03/11/banking-nature/#sdfootnote11sym.

Sunderland, T., Sayer, J. and Hoang, M.-H. (2012). *Evidence-based Conservation: Lessons from the Lower Mekong*. Earthscan Forest Library, Abingdon, UK.

Sunderlin, W. and Atmadja, S. (2009). Is REDD+ an idea whose time has come, or gone? In Angelsen, A. with Brockhaus, M., Kanninen, M., Sills, E., Sunderlin, W. D. and Wertz-Kanounnikoff, S. (eds), *Realising REDD+: National Strategy and Policy Options*. CIFOR, Bogor, Indonesia.

Survival International. (2009). *The most inconvenient truth of all: Climate change and indigenous people*. London. Available at: http://assets.survivalinternational.org/documents/132/survival_climate_change_report_english.pdf.

Survival International. (2012). "Indonesia denies it has any indigenous peoples". 1 October 2012. Available at: http://www.survivalinternational.org/news/8710.

Swan, S. (2012). *Commitments and Options for Safeguarding Biodiversity in REDD+. Biodiversity & REDD+ Review*. March. FCA-SNV BioREDD Brief No.1. Forest Carbon Asia, Manila, Philippines.

Swedish Society for Nature Conservation (2013). REDD Plus or REDD "Light"? – Biodiversity, carbon, and forest certification. Available at: http://www.naturskyddsforeningen.se/sites/default/files/dokument-media/REDD%20Plus%20or%20REDD%20Light.pdf.

Swingland, I. (ed.) (2003). *Capturing Carbon and Conserving Biodiversity: The Market Approach*. Earthscan, London.

Taleb, N. (2007). *The Black Swan: The Impact of the Highly Improbable*. Random House, New York.

Taleb, N. (2009). Preface to Pablo Triana, *Lecturing Birds on Flying: Can Mathematical Theories Destroy the Financial Markets?* John Wiley & Sons, Hoboken, NJ.

Tan, N., Truong, L., Van, N., K'Tip with Enters, T., Yasmi, Y. and Vickers, B. (2010). *Evaluation and Verification of the Free, Prior and Informed Consent Process under the UN-REDD Programme in Lam Dong Province, Viet Nam*. November. RECOFTC – The Center for People and Forests, Bangkok, Thailand.

Terra Global Capital LLC. (2010). *An Integrated REDD Offset Program (IREDD) – version 2.0 (draft for discussion)*. Available at: http://terraglobalcapital.com/press/Terra%20Global%20Integrated%20REDD%20Paper%20Version%202.0.pdf.

Terra Global Capital LLC. (2012). *Newsletter – Terra Bella Fund and forest and land-use carbon sector update*. Edition 5, Q1 – Q2 2012. Available at: http://www.terraglobalcapital.com/press/Terra%20Global%20Capital%20-%20Investor%20Newsletter%20Q1-Q2%202012.pdf.

Teyssier, A., Ramarojohn, L. and Ratsialonana, R. A. (2010). Des terres pour l'agro-industrie internationale ? Un dilemme pour la politique foncière malgache. *EchoGéo*, 11, February 2010: http://farmlandgrab.org/11420.

The Ecologist, UK. (2010). *Fears of corruption as REDD forest-protection schemes begin.* Available at: http://www.illegal-logging.info/item_single.php?it_id=4851&it=news.

The Economist. (2012). Compromise or deadlock?: The president's effort to balance the claims of forests and farms has satisfied few. An opportunity to promote sustainable farming may be missed. *The Economist*, 2 June.

The Rainforest Foundation UK. (2012). *Rainforest Roulette? Why Creating a Forest Carbon Offset Market is a Risky Bet for REDD.* September. London.

Thompson, M., Baruah, M. and Carr, E. (2011). Seeing REDD+ as a project of environmental governance. *Environmental Science and Policy*,14(2), 89–230 (March 2011).

Toni, F. de Castro, Aguilar-Støen, M., Ludewigs, T. and Rodrigues Filho, S. (2012). The impacts of climate politics on territorial and social reconfigurations in rural areas in Latin America, Analytical Framework Report, D.9.1. ENGOV, European Commission, Seventh Framework Programme.

Transparency International. (2011). *Global Corruption Report: Climate Change.* Earthscan, London and Washington, DC.

Transparency International. (2012). *Keeping Redd+ Clean: A Step-By-Step Guide to Preventing Corruption.* Transparency International, Berlin, Germany.

Trefon, T. (2011). *Congo Masquerade: The Political Culture of Aid Inefficiency and Reform Failure.* Zed Books, London and New York.

Trivers, R. (2011). *The Folly of Fools: The Logic of Deceit and Self-deception in Human Life.* Basic Books, New York.

Turnbull, C. (1962). *The Forest People.* American Museum of Natural History, New York.

Turnbull, C. (1965). *Wayward Servants: The Two Worlds of the African Pygmies.* American Museum of Natural History/the Natural History Press, Garden City, NY.

UNDP. (2012). *UNDP's Leadership in the MDGs.* Available at: http://www.undp.org/content/undp/en/home/mdgoverview.html.

UNDRIP. (2007). *United Nations Declaration on the Rights of Indigenous Peoples.* Adopted by General Assembly Resolution 61/295 on September 13, 2007. United Nations, New York.

UNEP. (2011). *Bridging the Emissions Gap.* United Nations Environment Programme, Nairobi, Kenya.

UNEP. (2012). *World Remains on Unsustainable Track Despite Hundreds of Internationally Agreed Goals and Objectives.* Available at: http://www.unep.org/Documents.Multilingual/Default.asp?DocumentID=2688&ArticleID=9158&l=en.

UNFCCC. (1992). *United Nations Framework Convention to Combat Climate Change.* UNFCCC, Bonn.

UNFCCC. (2008). Report of the Conference of the Parties on its thirteenth session, held in Bali from 3 to 15 December 2007. Addendum Part Two: Action taken by the Conference of the Parties at its thirteenth session. FCCC/CP/2007/6/Add.1. 14 March 2008. Available at: http://unfccc.int/resource/docs/2007/cop13/eng/06a01.pdf#page=8.

UNFCCC. (2009). Decision 2/CP.15, Copenhagen Accord, FCCC/CP/2009/11/Add.1. Available at: http://unfccc.int/resource/docs/2009/cop15/eng/11a01.pdf.

UNFCCC. (2010). *Fact sheet: Reducing emissions from deforestation in developing countries: Approaches to stimulate action.* Available at: http://unfccc.int/files/press/backgrounders/application/pdf/fact_sheet_reducing_emissions_from_deforestation.pdf.

UNFCCC. (2011a). Report of the Conference of the Parties on its sixteenth session, held in Cancun from November 29 to December 10, 2010. Addendum Part Two: Action taken by the Conference of the Parties at its sixteenth session, FCCC/CP/2010/7/Add.1.

UNFCCC. (2011b) Report of the Conference of the Parties on its seventeenth session, held in Durban from 28 November to 11 December 2011. Addendum, Part Two. FCCC/CP/2011/9/Add.1. Available at: http://unfccc.int/resource/docs/2011/cop17/eng/09a01.pdf.

UNFCCC. (2012a). *CDM project activity cycle.* Available at: http://cdm.unfccc.int/Projects/pac/howto/CDMProjectActivity/index.html.

UNFCCC. (2012b). Doha climate change conference – November 2012. Available at: http://unfccc.int/meetings/doha_nov_2012/meeting/6815.php.

Union of Concerned Scientists. (2012b). *The root of the problem – drivers of deforestation: What is driving tropical deforestation today?* Available at: http://www.ucsusa.org/global_warming/solutions/forest_solutions/drivers-of-deforestation.html.

United Nations. (2006). *United Nations Declaration on the Rights of Indigenous Peoples* (UNDRIP). New York. Available at: http://www.un.org/esa/socdev/unpfii/documents/DRIPS_en.pdf.

United Nations. (2009). *State of the World's Indigenous Peoples*, ST/ESA/328, New York.

United States Government. (2010). *Announcement of U.S. support for the United Nations Declaration on the Rights of Indigenous Peoples: Initiatives to promote the government-to-government relationship & improve the lives of indigenous peoples.* Available at: http://www.state.gov/documents/organization/153223.pdf.

United States Government. (2011). *Update of IFC's Policy and Performance Standards on Environmental and Social Sustainability, and Access to Information Policy.* 12 May. Available at: http://www.treasury.gov/resource-center/international/development-banks/Documents/IFC%20policy%20review%20-%20final%20policy%20May%2012%202011%20-%20US%20position%20to%20post.pdf.

UN-REDD. (2011a). *UN-REDD Programme – Viet Nam.* http://www.youtube.com/watch?v=lvJIwtNT9vc&feature=player_embedded.

UN-REDD. (2011b). *UN-REDD collaborates on new forest-carbon finance report with UNEP FI.* 7 April. Geneva, Switzerland. Available at: http://unredd.wordpress.com/2011/04/07/un-redd-collaborates-on-new-forest-carbon-finance-report-with-unep-fi/.

UN-REDD. (2011c). *The UN-REDD Programme Strategy 2011–2015.* Available at: http://www.unep.org/forests/Portals/142/docs/UN-REDD%20Programme%20Strategy.pdf.

UN-REDD. (2012a). *FAQs: What is REDD?, What is REDD+?* Available at: http://www.un-redd.org/AboutUNREDDProgramme/FAQs/tabid/586/Default.aspx.

UN-REDD (2012b). *2011 Year in Review.* UN-REDD, New York.

UN-REDD. (2012c). *The UN-REDD Programme and the World Bank's Forest Carbon Partnership Facility: Working together for better national and international coordination.* Available at: http://www.un-redd.org/NewsCentre/Newsletterhome/1Feature2/tabid/1588/language/en-US/Default.aspx.

UN-REDD. (2012d). *REDD+ and markets: Any lessons to be learned from voluntary carbon markets?* Issue 7, November. Available at: http://www.google.com/url?sa=t&rct=j&q=&esrc=s&source=web&cd=1&ved=0CDMQFjAA&url=http%3A%2F%2Fwww.unredd.net%2Findex.php%3Foption%3Dcom_docman%26task%3Ddoc_download%26gid%3D8665%26Itemid%3D53&ei=IyOhUIWXLObn0gGj14GIDg&usg=AFQjCNEsh1BLQ4p8uUZKEQtwKROMvhK_wA&sig2=UihMXVDoDnOpo-5S35pFhA.

UN-REDD Programme. (2009). *About REDD+.* Available at: http://www.un-redd.org/AboutREDD/tabid/582/Default.aspx.

USAID. (2011). *Evaluation learning from experience, USAID evaluation policy.* January. Washington, DC. Available at: http://pdf.usaid.gov/pdf_docs/pdacq800.pdf.

USAID. (2012). *Request for Proposal (RFP) SOL-OAA-12-0000-50, Measuring Impact (MI).* Issuance Date: 17 April. United States Agency for International Development, Washington, DC.

van den Berg, R. and Quarles van Ufford, P. (2005). Disjuncture and marginality – towards a new approach to development practice. In Mosse, D. and Lewis, D. (eds), *The Aid Effect*. Pluto Press, London and Ann Arbor, MI.

Vatn, A. and Vedeld, P. (2011). *Getting Ready! A study of National Governance Structure for REDD+*. Noragric Report No. 59. Norwegian University of Life Sciences. Department of International Environment and Development Studies (Noragric).

Verchot, L. (2012). Quoted in M. Kovacek, *Forests fare poorly in outcomes of Rio+20, say CIFOR scientists*. *Forests News*. Available at: http://blog.cifor.org/9945/forests-fare-poorly-in-outcomes-of-rio20-say-cifor-scientists/#.T-1SjJGNDWg.

Vhugen, D., Aguilar, S., Peskett, L. and Miner, J. (2012). *REDD+ and Carbon Rights: Lessons from the Field*. Publication produced for review by the United States Agency for International Development. February.

Viana, V. (2009). *Financing REDD: Meshing Markets with Government Funds*. IIED, London.

Vidal, J. (2008). *The great green land grab*. *The Guardian*, February 13. Available at: http://www.guardian.co.uk/environment/2008/feb/13/conservation.

Vieweg, M., Esch, A., Grießhaber, L., Fuller, F., Mersmann, F., Fallasch, F. and De Marez, L. (2012). *German Fast Start: Lessons learned for long-term finance*. Climate Analytics, Wuppertal Institute, Germanwatch e.V. Berlin, Germany. Available at: http://www.bmu-klimaschutzinitiative.de/files/2012-German_FSF_Study_1076.pdf.

Vira, B. (2012). Comment in Responding to Climate Change. Available at: http://www.rtcc.org/climate-change-tv/cbd-cop11-not-simply-redd/.

Vira, B. and Kontoleon, A. (2010). Dependence of the poor on biodiversity: Which poor, what biodiversity? In Roe, D. (ed.), *Linking Biodiversity Conservation and Poverty Alleviation: A State of Knowledge Review*. CBD Technical Series No: 55. Secretariat of the Convention on Biological Diversity, Montréal, Quebec, Canada.

Vrieze, P. and Naren, K. (2012). SOLD: In the race to exploit Cambodia's forests new maps reveal the rapid spread of plantations and mining across the country. *The Cambodia Daily*, 10–11 March, pp. 4–11.

Weale, A. (2004). Contractarian theory, deliberative democracy and general agreement. In Dowding, K., Goodin, R., Pateman, C. and Barry, B. (eds), *Justice and Democracy: Essays for Brian Barry*. Cambridge University Press, Cambridge, UK: pp. 79–96.

Weber, J., Sills, E., Bauch, S. and Pattanayak, S. (2011). Do ICDPs work? An empirical evaluation of forest-based microenterprises in the Brazilian Amazon land economics. *Land Economics*, November 87, 661–681.

Weizman, M. (2007). *The Stern Review of the economics of climate change. Book review for JEL*. Available at: http://www.economics.harvard.edu/faculty/Weitzman/files/JELSternReport.pdf.

Weizman, M. (2011). Fat-tailed uncertainty in the economics of catastrophic climate change. Available at: http://www.economics.harvard.edu/faculty/weitzman/files/Fat-Tailed%2BUncertainty.pdf.

Welch-Devine, M. (2009). *Navigating tradeoffs: Social sciences at IUCN's World Conservation Congress*. PhD thesis. University of Georgia. Available at: http://athenaeum.libs.uga.edu/bitstream/handle/10724/11711/welch-devine_meredith_l_200908_ms.pdf?sequence=1.

Wells, M. and Brandon, K. (1992). *People and Parks: Linking Protected Area Management with Local Communities*. World Bank, Washington, DC.

Wells, M. P., McShane, T. O., Dublin, H. T., O'Connor, S. and Redford, K. H. (2004). The future of integrated conservation projects: Building on what works. In McShane, T. O. and Wells, M. P. (eds), *Getting Biodiversity Projects to Work: Towards Better Conservation and Development*. Columbia University Press, New York: pp. 397–422.

Wertz-Kanounnikoff, S. and Angelsen, A. (2009). Global and national REDD+ architecture: Linking institutions and actions. In Angelsen, A. with Brockhaus, M., Kanninen, M.,

Sills, E., Sunderlin, W. D. and Wertz-Kanounnikoff, S. (eds), *Realising REDD+: National Strategy and Policy Options*. CIFOR, Bogor, Indonesia.

Wertz-Kanounnikoff, S. and Verchot, L. V. with Kanninen, M. and Murdiyarso, D. (2009). How can we monitor, report, and verify carbon emissions from forests? In Angelsen, A. with Brockhaus, M., Kanninen, M., Sills, E., Sunderlin, W. D. and Wertz-Kanounnikoff, S. (eds), *Realising REDD+: National Strategy and Policy Options*. CIFOR, Bogor, Indonesia.

West, P., Igoe, J. and Brockington, D. (2006). Parks and peoples: The social impact of protected areas. *Annual Review of Anthropology*, 35, 251–277.

White, A. (2007). *Is it time to rewrite the social contract?* Business for Social Responsibility. Available at: http://www.tellus.org/publications/files/BSR_AW_Social-Contract.pdf.

White, A., Khare, A. and Molnar, A. (2004). Who owns, who conserves and why it matters. *Arborivitae*, 26, 8–11. IUCN/WWF Forest Conservation newsletter, September.

Whitington, J. (2012). The prey of uncertainty: Climate change as opportunity. *The Atmosphere Business*, ephemera. Available at: http://www.ephemeraweb.org/journal/12-1/12-1ephemera-may12.pdf.

Winn, P. (2012). *Up for Grabs: Millions of hectares of customary land in PNG stolen for logging*. Greenpeace. Available at: http://www.greenpeace.org/australia/PageFiles/441577/Up_For_Grabs.pdf.

Wolstencroft, T. (2009). *Investing in markets for a low-carbon world*. Available at: http://www2.goldmansachs.com/our-thinking/environment-and-energy/low-carbon-world.pdf.

World Bank (2009). *Democratic Republic of Congo Strategic Framework for the Preparation of a Pygmy Development Program*. Report No. 51108. December 2009. Available at: http://siteresources.worldbank.org/INTRANETSOCIALDEVELOPMENT/Resources/244329-1264437980469/6732897-1265059058827/6753949-1265059145802/DRC.pdf.

World Bank Institute. (2011). *Estimating the opportunity costs of REDD+*. World Bank, Washington, DC. Available at: http://wbi.worldbank.org/wbi/Data/wbi/wbicms/files/drupal-acquia/wbi/REDDbrochure_v2pages.pdf.

World Meteorological Society (WMO). (2012). *WMO provisional statement on the state of global climate in 2012*. Available at: http://www.wmo.int/pages/mediacentre/press_releases/documents/966_WMOstatement.pdf.

World Rainforest Movement. (2011). *The Conservation International REDD pilot project in the Democratic Republic of Congo (DRC) – a very different kind of Walt Disney production*. WRM Bulletin No 169, August. Available at: http://wrm.org.uy/bulletin/169/REDD_DRC.html.

Wright, S. J. and Muller-Landau, H. C. (2006). The future of tropical forest species. *Biotropica*, 38(3): 287–301.

Wunder, S. (2007). The efficiency of payments for environmental services in tropical conservation. *Conservation Biology*, 21(1): 48–58.

Wunder, S. (2008). Payments for environmental services and the poor: concepts and preliminary evidence. *Environment and Development Economics*, 13(3): 279–297.

WWF. (2012). *Stimulating Interim REDD+ Demand – the Forest Finance Facility*. Gland, Switzerland.

WWF. (2013). 2020 *Priority Places Goal: By 2020, biodiversity is protected and well managed in the world's most outstanding natural places*. Available at: http://wwf.panda.org/what_we_do/where_we_work/.

Yale and George Mason. (2012). *Climate Change in the American Mind: The potential impact of global warming on the 2012 presidential election*. September. Yale Project on Climate Change Communication, New Haven, CT, George Mason University Center for Climate Change Communication, Fairfax, VA.

Yawanawa, Chief Tashka (2012). "Acre: Statement From Chief Tashka Yawanawa On REDD In Acre, Brazil". Available at: http://www.thenewscollective.org/display-related-news.php?itemHeadline=Acre%3A+Statement+From+Chief+Tashka+Yawanawa+On+REDD+In+Acre%2C+Brazil.

Zagema, B. (2011). *Land and Power. The Growing Scandal Surrounding the New Wave of Investments in Land*. 151 Oxfam Briefing Paper, September 22. Oxfam, UK.

Notes

Information contained in a figure is indicated with an *italic* page number. Information in a box or table is shown with a **bold** page number.

accountability: of BINGOs, 37; demand for, 48, 252–3, 273; hybrid approach, 223–4; incentives, 264; lack of, 124; REDD projects, 203; risks of, 236–7; for social feasibility, 36

accounting, 54, 60, 244–5

Achebe, Chinua, 119

Acre, Brazil, 186, 248

activism, 162–3, 173–4, 183, 240–1, 252–3

AD (avoided deforestation), 6, 8, 58, 61, 275; alternatives to, 101; carbon commodification, 243–4; financing, 212–26; framing, 41; opportunities, 255; opportunity costs, **110–11**; payment for ecosystems, 33; relationship to REDD, 38; rights-based approach, 49–51, 226; role of indigenous people, 42, 175–6; role of stakeholders, 262–3; social contract, 36, 44–5, 257–9

'additionality' principle, 128–9

Africa: capacity building, 206–7; Central Africa, 195–6, 246; deforestation, 89; environmental governance, 70; environmental refugees, 92; mobile phone use, 240–1; poverty alleviation, 93–4; sharecropping, 72; social feasibility, 195–6; Southern Africa, 67

agreements: bogus, 10–11; capacity building for, 18

agribusiness, 90–1, 172–3

agriculture, 87–90, 254–5

agroforestry, 210

AI (appreciative inquiry), 244, 247, 269

aid: *see* ODA (official development assistance)

Akwah, George, 98

alliances, corporate, 146, 165

Amazon, 61, 249; *see also* Brazilian Amazon

An Inconvenient Truth, Al Gore, 86

Ankeniheny-Zahamena Corridor (CAZ), 149–51

anthropogenic climate change (ACC), 15, 32, 77, 79; deniers, 249; mitigating arguments, 84–7; tipping point, 127–8; *see also* climate change

anti-capitalists, 263

anti-REDD advocates, 162–3, 173–4, 240–1, 252–3

approaches: *see* TMAs (tools, methods and approaches)

April Salumei, PNG, 153–6, 171

Arctic melting, 131, 220, 228

Asia: anti-REDD advocates, 240–1; deforestation, 89

Australia, 239–40, 242

awareness of REDD, 12, 148; *see also* capacity building; public relation campaigns

Bali (UNFCCC COP 13), 51, 78, 129

Bandundu province, Congo, 199–200

Bantu people, Zaire, 120–2

bargaining, 199–200; *see also* negotiation

bargaining zones, 69–70

BCI (Bonobo Conservation Initiative), **225**

benefit sharing: access to, 20; beneficiaries of carbon trading, 164; beneficiary identification, 97; equitable, 244–6; gender imbalance, 242; mechanism for, 17, 36, 68, 108, 224–5

best practice, 18, 58–9; of BINGOs, 46; failure, 43, 192–5; language of,